Nonlinear econometric modeling in time series

Nonlinear Econometric Modeling in Time Series presents some recent developments in that area of research. This volume is the eleventh in a series entitled *International Symposia in Economic Theory and Econometrics* under the general editorship of William Barnett. Many of the prior volumes in this series have included investigations of nonlinearity and complex dynamics in economic theory and in structural econometric modeling. This is the first one to focus on the more recent literature on nonlinear time series. Specific topics covered with respect to nonlinearity include cointegration tests, risk-related asymmetries, structural breaks and outliers, Bayesian analysis with a threshold, consistency and asymptotic normality, asymptotic inference, and error-correction models. This proceedings volume includes the most important papers presented at a conference held at the University of Aarhus in Aarhus, Denmark, December 14–16, 1995. This volume constitutes the proceedings volume of the Sixth Meeting of the European Conference Series in Quantitative Economics and Econometrics, $(EC)^2$.

William A. Barnett is Professor of Economics at Washington University in St. Louis. He is the series editor of the *International Symposia in Economic Theory and Econometrics* series in which this volume appears, and the founding editor of the journal *Macroeconomic Dynamics*, also published by Cambridge University Press. A volume of Professor Barnett's published papers with discussions, *The Theory of Monetary Aggregation*, coedited with Apostolos Serletis, will shortly be published in the North Holland monograph series *Contributions to Economic Analysis* edited by Dale Jorgenson, Jean-Jacques Laffont, and Torsten Persson.

David F. Hendry is a Fellow of Nuffield College and Leverhulme Personal Research Professor at the University of Oxford. A former president and current Council member of the Royal Economic Society, he is also a Fellow and former Council member of the Econometric Society. The author of *Forecasting Economic Time Series* (with Michael P. Clements) and coeditor of *Foundations of Econometric Analysis* (with Mary S. Morgan), both published by Cambridge University Press, Professor Hendry is the author or editor of four other books and over 100 chapters and papers in academic books and journals.

Svend Hylleberg is Professor of Economics at the University of Aarhus, Denmark. He has also served as a Visiting Professor at the University of California, San Diego, Cornell University, and the Australian National University. Professor Hylleberg's works include *Seasonality in Regression, New Approaches to Empirical Macroeconomics* (coedited with Martin Paldam), and *Modeling*

Seasonality, as well as numerous papers and chapters of books in refereed publications.

Timo Teräsvirta is Professor of Econometrics at the Stockholm School of Economics, Sweden. He has also taught at the Universities of Helsinki and Jyvaskyla, Finland, and been a visiting professor at institutes and universities in Sweden, Canada, the US, the UK, France, Austria, Norway, and the Netherlands. An author of *Modelling Nonlinear Economic Relationships* (with Clive W. J. Granger, Oxford University Press, 1993), Professor Teräsvirta has also written over 75 publications in refereed journals and books.

Dag Tjøstheim is Professor in the Department of Mathematics at the University of Bergen. He has also taught at the Norwegian Business School and held visiting positions at Stanford, the University of North Carolina, and the University of California, San Diego. Professor Tjøstheim has served as the main editor of the *Scandinavian Journal of Statistics* and is associate editor of the *Journal of Time Series Analysis*. His research focuses on panel and spatial data and nonparametric methods as well as nonlinear time series.

Allan Würtz is an assistant professor in the Department of Economics at the University of Aarhus. He received his Ph.D. from the University of Iowa and has also taught at the University of New South Wales, Australia. Professor Würtz has written and coauthored papers with Richard Parks, N. Eugene Savin, and Svend Hylleberg, among other leading econometricians and statisticians.

International Symposia in Economic Theory and Econometrics

Editor
William A. Barnett, *Washington University in St. Louis*

Nonlinear econometric modeling in time series

Proceedings of the Eleventh International Symposium in Economic Theory

Edited by

WILLIAM A. BARNETT
Washington University in St. Louis

DAVID F. HENDRY
Nuffield College, University of Oxford

SVEND HYLLEBERG
University of Aarhus

TIMO TERÄSVIRTA
Stockholm School of Economics

DAG TJØSTHEIM
University of Bergen

ALLAN WÜRTZ
University of Aarhus

CAMBRIDGE
UNIVERSITY PRESS

PUBLISHED BY THE PRESS SYNDICATE OF THE UNIVERSITY OF CAMBRIDGE
The Pitt Building, Trumpington Street, Cambridge, UK

CAMBRIDGE UNIVERSITY PRESS
The Edinburgh Building, Cambridge CB2 2RU, UK   http://www.cup.cam.ac.uk
40 West 20th Street, New York, NY 10011-4211, USA   http://www.cup.org
10 Stamford Road, Oakleigh, Melbourne 3166, Australia
Ruiz de Alarcón 13, 28014 Madrid, Spain

First published 2000

Printed in the United States of America

Typeface Times Roman 10/12 pt.   *System* LaTeX 2_ε [TB]

A catalog record for this book is available from the British Library.

Library of Congress Cataloging-in-Publication Data

Nonlinear econometric modeling in time series: Eleventh International
   Symposium in Economic Theory and Econometrics / edited by William A.
   Barnett . . . [et al.].
      p.   cm. – (International symposia in economic theory and
   econometrics)
      A selection of 7 papers presented at the conference held at the
   University of Aarhus, Denmark, on December 14–15, 1995.
      ISBN 0-521-59424-3 (alk. paper)
      1. Econometrics – Congresses.   2. Time-series analysis – Congresses.
   3. Nonlinear theories – Congresses.   I. Barnett, William A.
   II. International Symposium in Economic Theory and Econometrics
   (11th : 1995 : University of Aarhus)   III. Series.
   HB139.N663   2000
   330′.01′5195 – dc21   99-34095
                                                CIP

ISBN 0 521 59424 3 Hardback

Contents

Series editor's preface

This volume is the eleventh in a series, called *International Symposia in Economic Theory and Econometrics*. The proceedings series is under the general editorship of William Barnett. Individual volumes in the series generally have co-editors, who differ for each volume, since the topics of the conferences change each year. The conference that produced this volume was funded by the Commission for the European Community Human Capital and Mobility Programme, the Research Foundation of the University of Aarhus, and the Centre for Non-Linear Modeling in Economics at the University of Aarhus, which is financed by the Danish Social Science Research Council. The topic of the conference and focus of this book is nonlinear econometric modeling with an emphasis on nonlinear time series. Many of the prior volumes in this series were sponsored by the IC² Institute at the University of Texas at Austin, and some have been cosponsored by the RGK Foundation.[1]

While linear modeling has a long and productive history in economics, there has been growing interest in nonlinear structural and time series modeling. Research on nonlinear structural modeling has been growing since the 1950s, since structural models derived from economic theory almost always are nonlinear. Advances in that area have been constrained only by computing technology, since the commitment of economic theorists to nonlinear structural models has been clear for decades. This fact is reflected in many of this series' previous books, which have contained much research on nonlinear structural models, both in economic theory and in econometric applications, especially research on models exhibiting nonlinear dynamics. Interest in nonlinear time series is of more recent vintage, and is the focus of this book.

The first conference in this Cambridge series was co-organized by William Barnett and Ronald Gallant, who also co-edited the proceedings volume. That volume has appeared as the Volume 30, October/November 1985 issue of the

---

[1] IC² stands for Innovation, Creativity, and Capital.

Journal of Econometrics and has been reprinted as a volume in this Cambridge University Press monograph series. The topic was "New approaches to modeling, specification selection, and econometric inference."

Beginning with the second symposium in the series, the proceedings of the symposia appear exclusively as volumes in this Cambridge University Press monograph series. The co-organizers of the second symposium and co-editors of its proceedings volume were William Barnett and Kenneth Singleton. The topic was "New approaches to monetary economics." The co-organizers of the third symposium, which was on "Dynamic econometric modeling," were William Barnett and Ernst Berndt; and co-editors of that proceedings volume were William Barnett, Ernst Berndt, and Halbert White. The co-organizers of the fourth symposium and co-editors of its proceedings volume, which was on "Economic complexity: Chaos, sunspots, bubbles, and nonlinearity," were William Barnett, John Geweke, and Karl Shell. The co-organizers of the fifth symposium and co-editors of its proceedings volume, which was on "Nonparametric and semiparametric methods in econometrics and statistics," were William Barnett, James Powell, and George Tauchen. The co-organizers and proceedings co-editors of the sixth symposium, which was on "Equilibrium theory and applications," were William Barnett, Bernard Cornet, Claude d'Aspremont, Jean Gabszewicz, and Andreu Mas-Colell. The co-organizers of the seventh symposium, which was on "Political economy," were William Barnett, Melvin Hinich, Douglass North, Howard Rosenthal, and Norman Schofield. The co-editors of that proceedings volume were William Barnett, Melvin Hinich, and Norman Schofield.

The eighth symposium was part of a large scale conference on "Social choice, welfare, and ethics." That conference was held in Caen, France on June 9–12, 1993. The organizers of the conference were Maurice Salles and Herve Moulin. The co-editors of that proceedings volume were William Barnett, Herve Moulin, Maurice Salles, and Norman Schofield. The ninth volume in the series was on "Dynamic disequilibrium modeling: Theory and applications," and was organized by Claude Hillinger at the University of Munich, Giancarlo Gandolfo at the University of Rome "La Sapienza," A. R. Bergstrom at the University of Essex, and P. C. B. Phillips at Yale University. The co-editors of the proceedings volume were William Barnett, Claude Hillinger, and Giancarlo Gandolfo.

Much of the contents of the tenth volume in the series comprises the proceedings of the conference, "Nonlinear Dynamics and Economics," held at the European University Institute in Florence, Italy, on July 6–17, 1992. But the volume also includes the related invited papers presented at the annual meetings of the American Statistical Association held in San Francisco on August 8–12, 1993. The organizers of the Florence conference, which produced part of the tenth volume, were Mark Salmon and Alan Kirman at the European University Institute in Florence, and David Rand and Robert MacKay from the

Mathematics Department at Warwick University in England, while the organizer of the invited American Statistical Association sessions, which produced the other papers in the volume, was William Barnett, who was Program Chair in Economic and Business Statistics of the American Statistical Association during that year.

This eleventh volume is the proceedings of a conference held at the University of Aarhus, Denmark, on December 14–16, 1995. In addition to being the eleventh in this series, the volume is the proceedings of the Sixth Meeting of the European Conference Series in Quantitative Economics and Econometrics, $(EC)^2$. The organizer of the Aarhus conference was Svend Hylleberg at the University of Aarhus. The editors of this proceedings volume are William A. Barnett, David F. Hendry, Svend Hylleberg, Timo Teräsvirta, Dag Tjøstheim, and Allan Würtz.

The intention of the volumes in this proceedings series is to provide *refereed* journal-quality collections of research papers of unusual importance in areas of currently highly visible activity within the economics profession. Because of the refereeing requirements associated with the editing of the proceedings, the volumes in the series do not necessarily contain all of the papers presented at the corresponding symposia, and this volume is such a case. Only the best papers presented at the Aarhus conference are included in this book.

William A. Barnett
Washington University in St. Louis

Contributors

William A. Barnett
Department of Economics
Washington University in St. Louis

Dick van Dijk
Tinbergen Institute
Erasmus University Rotterdam

Alvaro Escribano
Departamento de Estadística y
    Econometría
Universidad Carlos III de Madrid

Philip Hans Franses
Rotterdam Institute for Business
    Economic Studies and
    Econometric Institute
Erasmus University Rotterdam

Giampiero M. Gallo
Dipartimento di Statistica
Università di Firenze

David F. Hendry
Nuffield College
University of Oxford

Svend Hylleberg
Department of Economics
University of Aarhus

Barry E. Jones
Department of Economics
State University of New York at
    Binghamton

Gary Koop
Department of Economics
University of Edinburgh

Michel Lubrano
GREQAM-CNRS
Centre de la Vieille Charité,
    Marseille

Helmut Lütkepohl
Institut für Statistik und Ökonometrie
Wirtschaftswissenschaftliche
    Fakultät
Humboldt-Universität, Berlin

Santiago Mira
Departamento de Fundamentos del
    Análisis Económico
Universidad Europea de Madrid

Travis D. Nesmith
Division of Monetary Affairs
Board of Governors of the Federal
    Reserve System
Washington, D.C.

Barbara Pacini
Dipartimento di Statistica
Università di Firenze

Simon Potter
Federal Reserve Bank
  of New York

Pentti Saikkonen
Department of Statistics
University of Helsinki

Timo Teräsvirta
Department of Economic Statistics
Stockholm School of Economics

Dag Tjøstheim
Department of Mathematics
University of Bergen

Allan H. Würtz
Department of Economics
University of Aarhus

CHAPTER 1

Introduction and overview

William A. Barnett, David F. Hendry, Svend Hylleberg,
Timo Teräsvirta, Dag Tjøstheim, & Allan Würtz

1 Introduction

The theme of the $(EC)^2$ conference held at the University of Aarhus in December 1995 was "Nonlinear modeling in economics". This theme was topical, as the University had just established a new economic and econometric research centre called the "Centre for Nonlinear Modeling in Economics", which is funded by the Danish Social Science Research Council and the Research Foundation of the University of Aarhus.

Economic theory is often nonlinear. Examples of nonlinearities in economics include economic processes with thresholds, capacity constraints restricting production, persistent disequilibria due to rationing, institutional restrictions such as tax brackets, multiple equilibria, and asymmetries of various kinds such as asymmetries in cyclical fluctuations of employment or unemployment due to asymmetric hiring and firing costs. The latter case is an example of the situation where the nonlinear theory is microeconomic theory, and it is not obvious what its implications are on the aggregated level. On the other hand, the literature contains examples of nonlinear macroeconomic models of employment based on or at least inspired by the microeconomic theory of asymmetric adjustment costs. Finally, it may be mentioned that nonlinear economic theory based on the mathematical theory of deterministic processes, called chaos, has also been a topic of discussion in the economics literature.

If the nonlinear economic theory is to be tested with data, the equations to be estimated may be expected to be nonlinear as well. However, a vast majority of econometric equations actually estimated in economics have been linear, most often because the relevant equations in question have been replaced by linear approximations – an approach which has been considered quite successful in practice. Another reason for using linear equations is the desire to avoid "incredible" (Sims 1980) theory and carry out the modeling with as few theory-based assumptions as possible. This has led to growing application of linear vector

1

autoregressive models that have become a very important tool for macroeconomic modelers having to do with stationary and nonstationary series. Besides, statistical theory for linear models is well developed, and in addition it has been possible to develop consistent modeling strategies based on linear models.

The application of the linear model has recently encountered several problems. Tests often reject parameter constancy, indicating so-called structural breaks, defined as changes in the parameters of a linear model. The solution has often been to dismiss the model or to pad with dummy variables in order to repair the deficiencies. As an economy, a market, a firm, or a household is much too complicated to be fully and adequately described by a few linear difference equations, one must expect a linear model to break down from time to time. Even so, frequent breakdowns imply a lack of credibility. However, just to apply a set of shift variables, called dummies, whenever a break is observed is an ad hoc and unsatisfactory solution. A much more satisfactory answer is to apply a specification which allows for nonlinearities.

The problems encountered for the linear models is one reason for the upsurge in popularity of nonlinear econometric models. A second reason is the advances in nonlinear time series analysis. New nonlinear models have been introduced, sometimes in parallel with econometric work, and some of these models have successfully been applied to economic series.

A third reason for the increasing interest in nonlinear models is the enormous growth in computational power available at a relatively low cost to an ordinary researcher. Nonlinear approaches such as nonparametric and semiparametric modeling have gained in popularity just because many methods have only recently become computationally feasible and because new computational possibilities have spurred the development of new statistical methods in the area. In a way, theory-free and flexible nonparametric models may be seen as a nonlinear counterpart of vector autoregressive models, except that the data requirements in nonparametric modeling are even greater than they are when vector autoregressive models are applied to economic series. Thus nonparametric models are likely to play an important role in financial econometrics while continuing to offer more limited possibilities to macroeconomic modelers. But whenever sufficient data is available, nonparametric analyses may also be a useful tool of preliminary analysis preceding the construction of parametric, possibly nonlinear models. See Yatchew (1998) and Tjøstheim (1999) for a review of nonparametric regression techniques.

From the economic theory point of view the choice between a linear or a nonlinear specification is clear: linear models should be used if the theory is linear or may be easily linearized without losing essential elements of it. From the practical point of view, the availability of data is important: it is hardly realistic to fit a nonlinear model to a data set consisting of, say, 20 annual observations. From the econometric point of view the choice may also be based on statistical considerations. In cases where there is a linear model nested in a nonlinear specification, it would be wise and in some cases necessary to test linearity first before considering the maintained nonlinear model. The choice between these two alternatives may also be made afterwards. It can be made by applying misspecification tests and other diagnostic devices and by comparing the out-of-sample forecasting performance of linear and nonlinear models. It should be noted, however, that any choice based on out-of-sample forecasting is a function of the forecasting period, which may or may not contain all the important (dynamic) characteristics of the estimation period. Nevertheless, the risk of overfitting is greater in nonlinear than in linear models, which underlines the importance of forecast comparisons.

Finally, although the extension of the econometrician's toolbox by nonlinear models certainly increases the possibilities for adequate and efficient modeling, criticism can be raised against any specific nonlinear form as well as any specific linear model. Firstly, the basis in theory is often vague, as a specific functional form of the estimating equation is the result of a choice of a specific functional form of the criterion function such as the utility function. The form of the criterion function is almost always chosen for analytical convenience and not because it can be justified by strong theoretical and/or empirical arguments. Secondly, the nonlinear model is also by nature a simplifying construction, which must be expected to break down from time to time. As is the case with linear models, a breakdown of the nonlinear model should lead to a total rethinking of the whole model, and not just to an ad hoc padding up of the observed deficiencies by adding variables and by even further complicating the nonlinear features of the model.

For a more elaborate discussion of nonlinear models in econometrics see the books by Granger and Teräsvirta (1993) and Tong (1990) for an introduction, and Gallant (1987) for a presentation of the statistical problems.

The articles of this volume relate in different ways to the above rather general, but by no means complete, outline of issues, problems, and solutions in nonlinear time series econometrics. A marriage of modern theoretical macroeconomics with a microeconomic foundation and cointegration analysis is suggested in the chapter by William A. Barnett, Barry E. Jones, and Travis D. Nesmith entitled "Time series cointegration tests and nonlinearity". It is argued that much economic theory implies that agents behave according to nonlinear decision rules, but that most cointegration analysis has not explored that avenue of possible nonlinear relations between macroeconomic variables. The choice of variables and the actual definitions of the monetary aggregates applied in the empirical analysis are based on the work on aggregation and index number theory by Barnett and others. This implies the use of the Törnqvist–Theil discrete time approximation to the continuous time Divisia index. In particular, the Törnqvist–Theil quantity variance is applied as a correction for the aggregation error.

The number of cointegrating vectors in a system of variables such as monetary services, the dual cost user index, the monetary service quantity variance, industrial production, and the consumer price index for two levels of aggregation is found by now standard methods based on the work of Johansen and Juselius (1990). The data are constructed using monthly seasonally adjusted data from Thornton and Yue (1992) for the period 1960:1 to 1992:12. The results indicate two cointegrating relations if the quantity variance is included and only one if the quantity variance is excluded from the set of variables in the VAR. To test whether the cointegrating relation is a linear process the frequency domain bispectrum test suggested by Hinich is applied; see Hinich (1982).

In their study "Risk-related asymmetries in foreign exchange markets" of the forward rate as an unbiased predictor of the future spot rate in foreign exchange markets, Giampiero M. Gallo and Barbara Pacini suggest a new procedure to deal with a time-varying risk-related premium. The risk premium is included in the analysis of foreign exchange markets together with interest rate parity. In its simplest form, an unbiasedness hypothesis says that the expected difference between the spot rate and the forward rate is zero. Also, the correlation over time and its variance are important for the efficient market hypothesis. Most studies have found, however, that this is not a good description of data. Instead, a time-varying risk premium can be included, without violating the efficiency market hypothesis.

The first problem is to find a measure of the time-varying risk-related term. There is no general increasing relationship between conditional variance and risk premium. In view of the lack of theory on a measure of risk, Gallo and Pacini adopt a nonparametric measure of risk. This measure is calculated by a latent variable approach, and as a result an instrumental variable is necessary. The conditional expectations are estimated by the Nadaraya–Watson kernel regression estimator. The input is obtained from a consistent parametric estimation of the residuals without instruments. Gallo and Pacini compare their approach with two competing estimators. They find that the unbiasedness hypothesis does not hold, despite the inclusion of a general time-varying risk-related term. The results do confirm, however, that the time-varying risk-related term is important in explaining the exchange rate movements, even more so when trading signals from technical analysis are inserted in the model.

As emphasized in most of the chapters, considerable care needs to be taken when departing from a linear model into classes of nonlinear models. This point is nicely illustrated by Gary Koop and Simon Potter, who address the problem of choosing among linear models and three different alternatives to linearity in their chapter entitled "Nonlinearity, structural breaks, or outliers in economic time series?". The three alternatives to linear models are motivated by empirical macroeconomics in that they allow for different effects of shocks on the dynamics of the model. If the dynamics change in a predictable way over the business cycle, the model is called nonlinear, while a model where the dynamics change in an unpredictable way is called an outlier model. An example of the first class is the threshold AR model, where the change of regime is a function of the lagged endogenous variable. If, instead, the dynamics of the model change in an unpredictable way, two kinds of models are considered, depending on whether a large shock has a permanent or a temporary effect on the dynamics. A temporary effect of an unpredictably large shock is characterized in an outlier model, whereas a structural break model is appropriate if the effect is permanent. In their simplest forms, the outlier model eliminates a few outliers, whereas the structural break model divides the sample period into subsample periods, each with a separate model.

Problems in classical econometrics of unidentified nuisance parameters under the null hypothesis in nonlinear models and selection among competing models are avoided by Koop and Potter by applying a Bayesian approach. To make the estimation tractable, they use prior distributions for which the posterior distribution can be derived analytically. As an example, Koop and Potter compare linear, nonlinear, structural breaks, and outlier models for the growth in US GDP and for the growth in the British industrial production. For each model, they also estimate a version which allows for heteroskedasticity. For the GDP series they find the structural break models to fit considerably better than the other three types of models, whereas there is evidence that the industrial production series is best described by a nonlinear threshold AR model.

Estimation of threshold models in a Bayesian context usually involves calculating high dimensional integrals using, for instance, the Gibbs sampler. In the paper by Michel Lubrano entitled "Bayesian analysis of nonlinear time series models with a threshold", he shows how to specify a threshold model such that only a low dimensional integral needs to be calculated by deterministic integration. The class of threshold models considered by Lubrano includes switching regime models where each regime is characterized by a linear index. The switching function is either a step or a smooth function of time, exogenous variables, or lagged endogenous variables. The same variables are included in all the regimes, and none of these variables enter in the switching function. The key assumption is the choice of prior on the parameters in the linear index. They could be noninformative or natural conjugate prior densities. Then the marginal posterior densities of the parameters of the linear index is calculated as a two dimensional integral, or a three dimensional integral in the case where heteroskedasticity is allowed.

The problem of unidentified parameters under the null hypothesis of a linear model is solved by assuming a particular prior on the parameters in the linear index. Lubrano proves that a noninformative prior leads to a posterior density which is infinite when there is no switching. Instead, this problem is avoided if a partially informative normal prior is imposed. The important property of the

partially informative normal prior is that it depends on the parameter characterizing the degree of smoothness in the switching function, namely, the parameter
which causes the nonidentification when it equals zero. In addition, to get an
integrable posterior density for this smoothness parameter, a convenient prior
is the truncated Cauchy density. Using these assumptions and results, Lubrano
investigates the French consumption function and US real GNP and industrial
production index by applying different types of threshold models.

The consistency and asymptotic normality of the nonlinear least squares
estimator of a nonlinear dynamic model is derived by Santiago Mira and
Alvaro Escribano, in the chapter "Nonlinear time series models: Consistency
and asymptotic normality of NLS under new conditions". The conditions are
new and easier to check than the conventional ones. Their results cover parametric nonlinear models which are used in practice, for example, state-dependent
and smooth transition autoregressive models.

The time series are allowed to be nonstationary, though excluding unit root
processes. The main assumption is that the series are strongly mixing. This
assumption replaces the common assumption of stationary ergodicity or geometric ergodicity. The strong mixing assumption allows for some degree of
heterogeneity. Both assumptions are hard to test in practice, but geometrically
ergodic series are contained in the class of strongly mixing series.

The chapter reviews the basic assumptions as given in for instance Gallant
and White (1988), before providing new assumptions which are easier to verify.
Together with the assumption on strong mixing, the new conditions on the
series are mainly moment conditions. In addition, the regression function must
be differentiable and bounded by a linear function. As an application, the new
conditions are verified for a smooth transition autoregressive model.

To interpret a cointegrated VAR model, often nonlinear functions of the
parameters are of interest, for example, impulse responses. The asymptotic
distribution of nonlinear functions of the parameters of a cointegrated VAR
model is derived by Pentti Saikkonen and Helmut Lütkepohl in their chapter
"Asymptotic inference on nonlinear functions of coefficients of infinite order
cointegrated VAR processes". With suitable normalizations, the distribution of
the nonlinear functions of the parameters is standard normal. They assume that
the cointegrated VAR system is of infinite order, but only a finite order VAR is
estimated. To derive the asymptotic distribution of the nonlinear functions of the
parameters, the finite lag length of the estimated VAR model increases as a function of the sample size. Hence, the approach can be considered nonparametric.

Using the asymptotic distribution of the nonlinear functions, tests can be
derived on hypotheses concerning the nonlinear functions. This is done for a
Wald type test. Since the approach involves infinitely many lags, it is possible to
test hypotheses with infinitely many restrictions. This is particularly interesting
when an impulse in one variable has an impact on other variables for any lead

time. A special case of that test is for a finite number of restrictions, for instance, the total effect of an impulse. In both cases, the Wald test statistic has a chi-square distribution. Although the emphasis is on the asymptotic distribution theory, the article also contains applications with different types of impulse responses.

Market frictions, transactions cost, and heterogeneity among traders can lead to nonlinear adjustments toward a long-run equilibrium path. In the standard model with cointegrated variables, the error-correction representation shows that adjustments to the equilibrium path are linear. In the chapter entitled "Nonlinear error-correction models for interest rates in the Netherlands" Dick van Dijk and Philip Hans Franses investigate different models where the adjustment to the equilibrium path is nonlinear. In particular, they consider smooth transition autoregressive adjustments. The strategy for identifying an appropriate specification is firstly to estimate a linear error-correcting model and the number of cointegration relationships. Since the linear error-correction model is misspecified in case of nonlinear adjustments, the consequences for estimating the cointegration relationships are investigated in finite samples using a Monte Carlo study. The Monte Carlo study mainly confirms the asymptotic results that the tests for cointegration and the estimates of the cointegration relationship are not affected by nonlinearity of the error-correcting term.

After finding the cointegrating relationships, the time series representing the error or equilibrium correcting relationship is tested for nonlinearity. While the cointegrating relationship is always assumed linear, the strength of the adjustment is assumed nonlinear under the alternative. The authors use an LM-type linearity test that has power against smooth transition regression. When the test is applied to a pair of Dutch interest rates, the results support the idea of a nonlinear equilibrium correction. One problem often encountered when estimating nonlinear models is that the nonlinearity mainly captures potential outliers. Therefore, the LM tests for nonlinearity are modified by estimating the equations using a robust method. This test suggests that the nonlinearity detected by the original LM test could be caused by a few outliers. Disaggregating the data, however, again provides more evidence for nonlinearity – a finding which accords with previous findings in the literature.

The editors believe that this selection of articles is a useful addition to the econometric literature on nonlinearities and nonlinear models and will lead to further investigations in this rapidly growing field.

REFERENCES

Gallant, A. R. (1987), *Nonlinear Statistical Models*. New York: Wiley.
Gallant, A. R., and H. White (1988), *A Unified Theory of Estimation and Inference for Nonlinear Dynamic Models*. New York: Blackwell.

Granger, C. W. J., and T. Teräsvirta (1993), *Modelling Nonlinear Economic Relationships*. Oxford: Oxford University Press.

Hinich, M. J. (1982), "Testing for Gaussianity and Linearity of a Stationary Time Series", *Journal of Time Series Analysis* 3: 169–76.

Johansen, S., and K. Juselius (1990), "Maximum Likelihood Estimation and Inference on Cointegration – with an Application to the Demand for Money", *Oxford Bulletin of Economics and Statistics* 52: 169–210.

Sims, C. A. (1980), "Macroeconomics and Reality", *Econometrica* 48: 1–49.

Thornton, D. L., and P. Yue (1992), "An Extended Series of Divisia Monetary Aggregates", *Federal Reserve Bank of St. Louis Review*, November/December, 35–52.

Tjøstheim, D. (1999), "Nonparametric Specification Procedures for Time Series", in S. Ghosh (ed.), *Asymptotics, Nonparametrics and Time Series*. New York: Marcel Dekker.

Tong, H. (1990), *Non-linear Time Series. A Dynamical System Approach*. Oxford: Clarendon Press.

Yatchew, A. (1998), "Nonparametric Regression Techniques in Economics", *Journal of Economic Literature* 36: 669–721.

CHAPTER 2

Time series cointegration tests
and non-linearity

William A. Barnett, Barry E. Jones, & Travis D. Nesmith

1   Introduction

Modern macroeconomic theory emphasizes the interactions among representative agents (households and firms) who are, in general, assumed to behave
according non-linear decision rules that are obtained as optimal solutions to
dynamic optimization problems. Consequently, it is reasonable to posit the
existence of non-linear relationships among macroeconomic variables.

During the last decade, as theoretical macroeconomics has been concerned
with microeconomic foundations, cointegration has become one of the most
important characterizations of macroeconomic time series. For example, real
business cycle research now commonly assumes balanced growth between output, consumption, and investment, and stable long run money demand with unit
income elasticity; see for example King and Watson (1996). These assumptions
imply the existence of cointegration relations among the key business cycle variables. Cointegration studies have, however, rarely explored the possibility of
non-linear relationships among macroeconomic variables. $I(1)$ cointegration
analysis focuses on non-stationary economic variables that are integrated of
order one, meaning that their first differences are stationary. The existence of

William Barnett is a professor of economics at Washington University in St. Louis. Barry E.
Jones and Travis D. Nesmith were Ph.D. candidates at Washington University in St. Louis and
visiting scholars at the Research Department of the Federal Reserve Bank of St. Louis when this
paper was written. Barry E. Jones and Travis Nesmith wish to thank the Federal Reserve Bank of
St. Louis for partially supporting this research. Any views expressed in this article are those of the
authors and not necessarily those of the Federal Reserve Bank of St. Louis or the Federal Reserve
System. William Barnett acknowledges partial support from National Science Foundation grant
SES9223557. The authors wish to thank Melvin Hinich for providing them with the code for his
bispectrum test and for his helpful technical advice. They wish to thank Richard Anderson, John
Keating, Lisbeth la Cour, Yi Liu, Houston Stokes, and Haiyang Xu for their helpful comments and
suggestions. Any remaining errors are the authors'.

cointegration implies that some linear combination of these integrated variables is stationary. The integrated variables may be non-linear stochastic processes – a hypothesis that is seldom entertained either as a feature of the data-generating process or as a convenient statistical description of the data in the cointegration literature. This paper begins to address this gap between the implications of modern dynamic macroeconomic theory and the cointegration literature by applying tests for non-linearity to the stationary linear combinations produced from cointegration.

A number of studies have tested for the existence of non-linearity in macroeconomic data (Hinich and Patterson 1985, Barnett and Chen 1988, Barnett and Hinich 1992, Brock and Sayers 1988); however, most of the existing non-linearity tests are univariate, and some of the available non-linearity tests are not invariant to prior linear filtering of the data.[1] In this chapter, we investigate the application of univariate non-linearity tests to stationary linear combinations of non-stationary (and possibly non-linear) macroeconomic time series, which have been identified through cointegration analysis. Thus, rather than testing the first differences of individual economic time series for non-linearity, we test the long run relationships between those series.

The remainder of this paper is organized as follows: in Section 2, we review the relevant aggregation theory and indexes number theory; in Section 3, we present the results of the cointegration analysis; in Section 4, we present the results of the non-linearity tests; and Section 5 concludes.

2   Aggregation and indexes number theory

In this section, we briefly review the monetary aggregation theory motivating the choice of variables in our empirical analysis; for more extensive reviews, see Barnett (1987, 1990), Barnett, Fisher, and Serletis (1992), and Anderson, Jones, and Nesmith (1997a), in which the most general conditions under which monetary aggregates exist are discussed.

Arrow and Hahn (1971) showed that if monetary assets are valued in general equilibrium, there exists a derived utility function containing monetary assets. If we assume that a representative agent exists and that current period monetary assets are blockwise weakly separable in that agent's utility function, a conditional second stage monetary services allocation decision exists. In that second stage, the representative agent can be viewed as solving a current period

[1] Barnett, Gallant, Hinich, Jungeilges, Kaplan, and Jensen (1994, 1995) study the power of several competing univariate non-linearity tests with artificial and monetary data respectively. Hinich and Wilson (1992) analyzes the cross bispectrum, which can detect multivariate non-linear relationships.

decision problem of the following form:

$$\text{Max } V(m_t) \qquad \text{subject to} \quad m_t^T \pi_t = M_t,$$

where V is the category sub-utility function of the (weakly separable) block of real monetary assets, m_t is the vector of monetary assets, π_t is the vector of user costs for monetary assets, and M_t is total expenditure on monetary assets allocated in the prior first stage decision. If V is linearly homogeneous, then it is the economic monetary aggregate, and the economic agent views the economic aggregate as an elementary good, which we call monetary services.[2]

In aggregation theory, monetary assets are viewed as durable goods, and thus the opportunity cost for each monetary asset is its user cost. Monetary assets provide a flow of services to the representative agent over the decision period, and the equivalent rental price of this flow is the user cost. Barnett (1978, 1980) shows that under perfect certainty the user cost of monetary assets in a household intertemporal optimization model is

$$\pi_{it} = p^* \frac{R_t - r_{it}}{1 + R_t},$$

where R_t is the risk-free rate of return on a completely illiquid asset, called the benchmark rate, r_{it} is the own rate of return on monetary asset i, and p^* is a true cost-of-living indexes.

For the remainder of this paper, we assume that monetary assets are weakly separable from the other decision variables in the representative agent's utility function, that the category sub-utility function is linearly homogeneous, and that the representative agent is an optimizing pricetaker. Under these assumptions, the economic monetary aggregate can be tracked by the Törnqvist–Theil discrete time approximation to the continuous time Divisia indexes. Diewert (1976) showed that this indexes is superlative, in the sense that it can provide a second-order approximation to any arbitrary economic aggregate in discrete time.

The Törnqvist–Theil monetary services indexes is defined as follows:

$$Q_t = Q_{t-1} \prod_i \left(\frac{m_{it}}{m_{i,t-1}} \right)^{\frac{1}{2}(s_{it} + s_{i,t-1})},$$

where $s_{it} = \pi_{it} m_{it}/M_t$ is the ith expenditure share and

$$M_t = \sum_{j=1}^{n} \pi_{jt} m_{jt}$$

is the total expenditure on monetary services (assets).

---

[2] The extension to non-homogeneous V requires the use of the distance function. See Barnett (1987) and Anderson, Jones, and Nesmith (1997a).

Under these aggregation assumptions, there exists a demand function for monetary services that is a function of the opportunity cost of monetary services, called the dual user cost. This dual user cost can be tracked by a user cost index, Π_t, that satisfies Fisher's weak factor reversal formulas:

$$\Pi_t = \Pi_{t-1} \left(\frac{M_t / M_t}{Q_t / Q_{t-1}} \right).$$

Under these aggregation assumptions, the economic aggregates contain all information about the component monetary assets and user costs that is relevant to other variables. Consequently, the dispersion of the component growth rates contains no additional macroeconomic information. See Barnett and Serletis (1990) and Anderson, Jones, and Nesmith (1997a) for discussions of these issues.

Although the aggregation conditions are typically maintained in macroeconomics, the assumptions for exact aggregation are strong. Barnett and Serletis (1990) have argued that a test for the existence of additional information in measures of component dispersion would be a diagnostic test for the existence of aggregation error. They also suggest adding such measures to an economic model as a correction for aggregation error.

The log change of the monetary services indexes is

$$DQ_t = \sum_{i=1}^{n} \bar{s}_{it} Dm_{it},$$

where $\bar{s}_{it} = \frac{1}{2}(s_{it} + s_{i,t-1})$, and D is the log change operator defined by $Dx_t = \log x_t - \log x_{t-1}$. The log change (growth rate) of the monetary services indexes is a weighted average of the log changes (growth rates) of the component asset stocks. Because the weights are average expenditure shares, Theil (1967) observed that the growth rate of the Törnqvist–Theil indexes has the interpretation of a share weighted mean of component growth rates. This is the intuition behind stochastic indexes number theory; the average shares induce a valid probability measure, and the growth rate of the quantity indexes are means of component quantity growth rates.[3] Theil (1967) defined the Törnqvist–Theil second moments, which have a direct interpretation as measures of component dispersion. In particular, the monetary services quantity variance, K, is defined as

$$K_t = \sum_{i=1}^{n} \bar{s}_{it} (Dm_{it} - DQ_t)^2.$$

[3] The Törnqvist–Theil user cost and expenditure share indexes have a similar interpretation. See Theil (1967). Diewert (1995) presents measures of "functional form error", of which the second moments are a specific example.

Törnqvist–Theil second moments were first used empirically with monetary data in Barnett, Offenbacher, and Spindt (1984). Törnqvist–Theil second moments have been found to contain information relevant to macroeconomic variables in two empirical studies: Barnett and Serletis (1990) and Barnett, Jones, and Nesmith (1996). These papers suggest that aggregation error is present in monetary models and that inclusion of the monetary services second moments in monetary models can correct for that aggregation error. We include the Törnqvist–Theil quantity variance in our cointegration analysis as a correction for aggregation error.

3   Cointegration analysis

Cointegration has been widely used to study monetary variables, for example by Johansen (1992b, 1995b). Only a few such studies have used theoretically consistent monetary aggregates. For example, using data for the United States, Barnett and Xu (1996) and Chrystal and MacDonald (1994) found that monetary services indexes are cointegrated with income, prices, and interest rates. Jones (1998) conducted an extensive $I(2)$ analysis of theoretically consistent monetary systems. Additional studies that used monetary services indexes for other countries include Serletis and King (1993) for Canada, and la Cour (1995) for Denmark. The cointegration analysis in this paper extends these previous studies in two ways. Our system includes the dual user cost indexes and the monetary services quantity variance, which can provide a correction for certain types of aggregation error.

3.1   *Theory*

Granger (1981, 1983) first defined the concept of cointegration. Let $X(t)$ be a vector stochastic process, $X(t) = (x_1(t), \ldots, x_n(t))^T$, and let $d = (d_1, \ldots, d_n)^T$ be a vector of integer values, where each of the system variables, $x_i(t)$, is integrated of order d_i, denoted $x_i(t) \sim I(d_i)$. By this we mean that the d_ith difference of $x_i(t)$ is stationary, denoted $\Delta^{d_i} x_i(t) \sim I(0)$, for all i. In this paper, we assume that d is a vector of zeros and ones, and therefore the system is said to be $I(1)$. There may exist some linear combinations of elements of $X(t)$ that are $I(0)$, which is the definition of linear $I(1)$ cointegration. Specifically, an $I(1)$ system $X(t)$ is said to be cointegrated if there exists a non-zero vector β_i such that $\beta_i^T X(t)$ is $I(0)$, and β_i is called a linear cointegration vector. Granger (1991) has suggested a number of generalizations of cointegration, including non-linear cointegration, in which there is a non-linear function f such that $f(x_1(t), \ldots, x_n(t))$ is stationary. If any of the variables in the system are $I(0)$, there will be a trivial cointegrating vector consisting of zeros in all entries except the entry corresponding to the stationary variable.

Non-trivial linear cointegration is an interesting property, because linear combinations of non-stationary processes are generally non-stationary. Although cointegration produces stationary stochastic processes $\beta_i^T X(t)$, these processes may be non-linear, because, like non-stationarity, non-linearity is a dominant property.

Johansen (1988, 1991) provides a maximum likelihood estimation procedure for determining the number of significant (linear) cointegration vectors, under the assumption that $X(t)$ is a vector of processes that are multivariate Gaussian as well as $I(1)$. We use the Johansen procedure in the following analysis.

The Johansen procedure starts from the vector error correction model (VECM):

$$\Delta X(t) = \mu + \sum_{i=1}^{q-1} \Gamma_i \, \Delta X(t-i) - \alpha \beta^T X(t-1) + \varepsilon_t,$$

where $X(t)$ is the p dimensional $I(1)$ vector stochastic process, q is the lag length of the underlying vector autoregression (VAR), μ is a p by one vector of constants, the Γ_i's are p by p coefficient matrices of short run effects, α and β are p by r matrices, and ε_t is the serially independent Gaussian residual of the underlying VAR, for $t = 1, \dots, T$ and $X(1-q), \dots, X(0)$ fixed.[4]

The hypothesis H_{r_0} that there are at most r_0 linear cointegrating vectors can be stated as the reduced rank condition on the matrices α and β, that $r = r_0$ is strictly less than p. The r estimated cointegration vectors constitute the rows of β^T. This reduced rank hypothesis can be tested and the α and β matrices can be estimated using the maximum likelihood procedure described in Johansen and Juselius (1990). The number of significant cointegrating vectors can be determined using either the trace test or the maximum eigenvalue test statistics described in Johansen and Juselius (1990).

3.2   *Empirical results*

The variables estimated are: monetary service indexes (at both the M2 and L levels of aggregation, denoted MSIM2 and MSIL), the dual user cost indexes (at M2 and L levels of aggregation, denoted DUALM2 and DUALL), the monetary services quantity variance (KM2 and KL), industrial production (IP), and the consumer price indexes (CPI). The monetary services indexes, their dual user cost indexes, and their second moments are calculated using the monthly seasonally adjusted quantity, own rate, and benchmark rate data from Thornton

---

[4] Similar tests and representation theorems for $I(2)$ systems have recently been worked out in Johansen (1992a, 1992b, 1995b), Parulo (1996), and Jorgensen, Kongsted, and Rahbek (1996).

and Yue (1992) for the periods 1960:1–1992:12.[5] IP proxies for monthly output as the income variable, and the CPI proxies for the appropriate monthly output deflator.[6] We estimate the model in log form (except for the quantity variances, which are not logged), and the system is estimated both with and without K. All variables have been seasonally adjusted, based on the X-11 procedure, and deterministic seasonal dummies are not included in the system. The lag length q in the VECM is set by a sequence of Sims' corrected likelihood ratio tests. The likelihood ratio tests chose a lag length of three for all four estimations.

In order to implement the VECM analysis we test the hypothesis that the system is $I(1)$. One method of ensuring that the system is not stationary is to test the hypothesis that there are trivial cointegration vectors, β_i, satisfying $\beta_i^T X(t) = x_i(t)$, i.e., that $x_i(t)$ is itself stationary. This test rejects stationarity for all of the variables tested, ensuring that the system is at least $I(1)$. Univariate augmented Dickey–Fuller (ADF) tests confirm that the variables are non-stationary, with the possible exception of the quantity variances. The results of univariate testing are sensitive to the lag lengths in the regression from which the ADF tests are conducted. Most variables appear to be $I(1)$, but at lag lengths significantly larger than three there is some evidence that the CPI is integrated of higher order.

In this paper, we maintain that the system is $I(1)$. This maintained hypothesis is consistent with most previous research on cointegration; see Miyao (1996) for a survey. In contrast, King, Plosser, Stock, and Watson (1991) and Friedman and Kuttner (1992) have assumed that US monetary aggregates and the price level are $I(2)$, but that real monetary aggregates are $I(1)$. Recently, Johansen (1995a,b), Jorgenson, Kongsted, and Rahbek (1996), and Parulo (1996) have developed maximum likelihood estimation procedures that can be used to test for the existence of $I(2)$ roots within the VECM framework. These $I(2)$ procedures generalize the $I(1)$ procedures used in this paper. Jones (1998) conducted a systematic $I(2)$ VECM analysis of US monetary systems, using both conventional and aggregation theoretic monetary aggregates, for the period 1954–1997. Jones found that previous empirical characterizations from $I(1)$ analyses are

---

[5] The Federal Reserve Bank of St. Louis has extensively revised the monetary services quantity and user cost indexes, for the period 1960–present. These revisions are detailed in Anderson, Jones, and Nesmith (1997a,b). All underlying non-confidential source data, as well as the quantity and user cost indexes and the Divisia second moments, are publicly available from the Federal Reserve Economic Database (FRED) at http://www.stls.frb.org/research/msi/index.html. The monetary services indexes, at the M2 and L level of aggregation, is reported in the Federal Reserve Bank of St. Louis' monthly publication *Monetary Trends*.

[6] If we strictly followed Barnett and Xu (1995), we would use GDP and the GDP deflator, but the larger number of observations at the monthly frequency is desirable for the Hinich bispectrum test. The use of industrial production as a proxy for GDP follows Christiano (1986), Fisher (1989), and Serletis (1987).

Table 1. *Trace and maximum eigenvalue tests:*
log(MSIM2), log(*user cost of* MSIM2), log(IP), log(CPI)

r^a	Test statistics		90% critical values	
	L_{max}	Trace	L_{max} (90%)	Trace (90%)
0	52.84	76.44	17.14	43.84
1	13.4	23.6	13.39	23.7
2	8.77	10.2	10.6	13.31
3	1.43	1.43	2.71	2.71

[a] The null $H_0 : r = n$ means that there are n cointegration vectors in the system. The null is rejected if the test statistic exceeds the 90% critical value. The alternative hypotheses are slightly different for the two test statistics.

not robust in an $I(2)$ VECM analysis, and that nominal monetary aggregates and prices may be $I(2)$.

Most published cointegration studies report the results of various whiteness tests. We implemented the following univariate residual tests: a corrected Jarque–Bera (normality) test, LM tests for ARCH, and LM tests for serial correlation, as discussed in L. G. Godfrey (1988). We also implemented the multivariate LM serial correlation tests, as described in Johansen (1995b). We eliminated models at the M1 and M3 levels of aggregation, based on evidence of serially correlated residuals. The M2 and L models performed very well on all serial correlation tests, but the univariate residuals showed evidence of both non-normality and ARCH. As noted in Johansen (1995b), the rejection of normality is not serious, because the asymptotic properties of the maximum likelihood estimator only depend on the i.i.d. assumptions. But for the same reason, the finding ARCH may be more damaging.[7]

Based on the above findings, we assume that $X(t) = $ (MSIM2 or MSIL, DUALM2 or DUALL, CPI, IP, KM2 or KL) is an $I(1)$ vector stochastic process, and it is therefore appropriate to implement the Johansen estimation procedure. In Tables 1–4, we report the trace test statistic and the maximum eigenvalue test statistic for the number of significant cointegration vectors in the various models. Both tests indicate the existence of a unique cointegration vector at both levels of aggregation for the models without the monetary services quantity variances. When the quantity variances are added to the models, two cointegration vectors are significant at each level of aggregation. We report the significant vectors for all the models in Table 5.

[7] Lee and Tse (1996) find that generalized autoregressive conditional heteroskedasticity (GARCH) causes over-rejection of the null of no cointegration, but they state that the problem is generally not very serious.

Table 2. *Trace and maximum eigenvalue tests:* log(MSIL), log(*user cost of* MSIL), log(IP), log(CPI)

r[a]	Test statistics		90% critical values	
	L_{max}	Trace	L_{max} (90%)	Trace (90%)
0	39.33	61.18	17.14	43.84
1	13.22	21.85	13.39	23.7
2	6.79	8.63	10.6	13.31
3	1.84	1.84	2.71	2.71

[a] The null $H_0 : r = n$ means that there are n cointegration vectors in the system. The null is rejected if the test statistic exceeds the 90% critical value. The alternative hypotheses are slightly different for the two test statistics.

Table 3. *Trace and maximum eigenvalue tests:* log(MSIM2), log(*user cost of* MSIM2), log(IP), log(CPI), K(M2)

r[a]	Test statistics		90% critical values	
	L_{max}	Trace	L_{max} (90%)	Trace (90%)
0	109.12	186.82	20.90	64.74
1	54.11	77.7	17.14	43.84
2	13.25	23.59	13.39	26.70
3	8.85	10.34	10.6	13.31
4	1.49	1.49	2.71	2.71

[a] The null $H_0 : r = n$ means that there are n cointegration vectors in the system. The null is rejected if the test statistic exceeds the 90% critical value. The alternative hypotheses are slightly different for the two test statistics.

The cointegration analysis produces stationary linear combinations of variables, from the $I(1)$ system. However, the stationary linear combinations are not necessarily linear stochastic processes. In testing for non-linearity, data is typically differenced until it appears stationary and the transformed data is then tested for non-linearity, whereas cointegration analysis provides a model based method of producing stationary series. In the remainder of this chapter, we will apply Hinich's bispectrum test to the stationary linear combinations produced by this cointegration analysis.

The interpretation of cointegration relationships can be tenuous, although it is possible to state and hypothesis test identifying restrictions in the model. In the system we are examining, the logical approach would be to attempt an identification of the cointegration relations as linearized money demand functions

Table 4. *Trace and maximum eigenvalue tests:* log(MSIL), log(*user cost of* MSIL), log(IP), log(CPI), K(L)

r^{a}	Test statistics		90% critical values	
	L_{max}	Trace	L_{max} (90%)	Trace (90%)
0	106.85	168.79	20.90	64.74
1	39.98	61.93	17.14	43.84
2	13.15	21.96	13.39	26.70
3	6.94	8.81	10.6	13.31
4	1.87	1.87	2.71	2.71

[a] The null $H_0 : r = n$ means that there are n cointegration vectors in the system. The null is rejected if the test statistic exceeds the 90% critical value. The alternative hypotheses are slightly different for the two test statistics.

Table 5. *Estimated cointegration vectors with and without the MSI variance added* [a]

Level	CPI	IP	MSI	Dual user cost	K
	Without K				
M2	−13.6	−5.118	13.597	2.527	NA
L	−12.907	−2.18	11.333	2.351	NA
	With K				
M2	−3.218	1.479	2.632	−0.839	1865.3
	13.603	5.434	−13.743	−2.570	80.483
L	3.216	−0.805	−3.002	0.923	−2349
	−13.091	−2.417	11.624	2.361	−27.97

[a] At each level of aggregation the most significant vector appears first.

in implicit form. The variables in our system are consistent with the aggregation theory discussed in Section 2, and would therefore be more amenable to such an identification than an analysis using theoretically inconsistent monetary aggregates. Such an analysis, although interesting, is beyond the scope of this chapter.

4   Testing for non-linearity

In this section, we review bispectrum theory and estimation. Basic discussions on higher-order polyspectra can be found in Nikias and Raghuveer (1987), Mendel (1991), Brillinger and Rosenblatt (1967a,b), and Brillinger (1965). The first use of the bispectrum in economics was by M. D. Godfrey (1965). The following sections have drawn upon Hinich and Patterson (1989), Stokes (1991), and Barnett and Hinich (1992).

4.1   Theory

We assume that $X = \{x(t)\}$ is a real, mean zero, third-order stationary stochastic process.[8] Third-order stationarity implies that the mean function, $c_x(t) = E[x(t)]$, is zero for all t, the covariance function, $c_{xx}(t_1, t_2) = E[x(t_1)x(t_2)]$, is a function only of $\tau_1 = t_1 - t_2$, and the general third-order moments, $c_{xxx}(t_1, t_2, t_3) = E[x(t_1)x(t_2)x(t_3)]$, are a function of only two variables, $\tau_1 = t_1 - t_2$ and $\tau_2 = t_2 - t_3$.

If $c_{xx}(\tau_1) = 0$ for all $\tau_1 \neq 0$, then the process is said to be white noise. If the distribution of $\{x(n_1), \ldots, x(n_T)\}$ is multivariate normal for all $n_1, \ldots, n_T$, then X is said to be Gaussian. X is said to be a pure white noise process if $\{x(n_1), \ldots, x(n_T)\}$ are independent random variables for all values of $\{n_1, \ldots, n_T\}$. Priestly (1981) and Hinich and Patterson (1985) have emphasized that stochastic independence and whiteness are not the same. All pure white noise processes are white, but white noise processes are not, in general, pure white noise processes, although for Gaussian processes the conditions are equivalent. If the process is not Gaussian, then, in general, the third-order moments are not zero. Testing for pure white noise using only the covariance function implicitly assumes that the process is Gaussian. Such a procedure ignores the potential existence of non-linear serial dependence, which can only be detected using higher-order moments. It is thus necessary to test residuals for both Gaussianity and non-linearity in addition to whiteness, since structural disturbances often are assumed to be pure white noise.[9]

The following definitions of the power spectrum and bispectrum are for discrete stochastic processes, although these definitions can be generalized; see for example Hinich and Messer (1995). The power spectrum (second-order cumulant polyspectrum), $P(\omega)$, is defined as the Fourier transform of the covariance function, i.e.

$$P(\omega) = \sum_{\tau=-\infty}^{\infty} c_{xx}(\tau) \exp[-i(\omega\tau)] \qquad \text{for frequencies (in radians) } |\omega| < \pi.$$

The bispectrum (third-order cumulant polyspectrum) is defined as the second-order Fourier transform of the third-order moment function, i.e.,

$$B_{xxx}(\omega_1, \omega_2) = \sum_{r=-\infty}^{\infty} \sum_{s=-\infty}^{\infty} c_{xxx}(r, s) \exp[-i(\omega_1 r + \omega_2 s)]$$

---

[8] The assumptions of real and mean zero can be relaxed; see Hinich and Messer (1995).

[9] The following comment by Johansen (1995b) is particularly relevant, "The methods derived [for cointegration] are based on the Gaussian likelihood but the asymptotic properties of the model depend only on the i.i.d. assumption of the errors" (p. 29).

for frequencies in the principal domain, Ω, which is defined as[10]

$$\Omega = \{(\omega_1, \omega_2) : 0 \leq \omega_1 \leq \pi, \quad \omega_2 \leq \omega_1, \quad 2\omega_1 + \omega_2 \leq 2\pi\}.$$

For extensive discussion of the principal domain and the symmetries of the bispectrum see Hinich and Messer (1995).[11]

The bispectrum can be interpreted by considering the Cramér spectral representation of X,

$$x(t) = \frac{1}{2\pi} \int_{-\infty}^{\infty} e^{i\omega t} \, dZ(\omega),$$

where

$$E[dZ(\omega)] = 0, \qquad E[dZ(\omega_1) \, dZ^*(\omega_2)] = \begin{cases} 0, & \omega_1 \neq \omega_2, \\ 2\pi \, P(\omega) \, d\omega, & \omega_1 = \omega_2 = \omega, \end{cases}$$

and[12]

$$E[dZ(\omega_1) \, dZ(\omega_2) \, dZ^*(\omega_3)] = \begin{cases} 0, & \omega_1 + \omega_2 \neq \omega_3, \\ B(\omega_1, \omega_2) \, d\omega_1 \, d\omega_2, & \omega_1 + \omega_2 = \omega_3. \end{cases}$$

As noted by Nikias and Raghuveer (1987), the power spectrum describes the contribution to the expectation of the product of two Fourier components whose frequencies are the same, whereas the bispectrum describes the contribution to the expectation of the product of three Fourier components where one frequency is equal to the sum of the other two.[13]

It is now easy to motivate the use of the bispectrum in our analysis. If $W = \{w(t)\}$ is a mean zero, third-order stationary, pure white noise process, it will have both a flat power spectrum and a flat bispectrum. If X is a linear process (i.e., it is the output of a linear time invariant filter applied to a pure white noise process) such that $x(t) = \sum_{n=0}^{\infty} a(n) w(t-n)$, then the normalized squared

---

[10] It is assumed that c_{xx} and c_{xxx} are absolutely summable.

[11] In general, the kth-order polyspectrum is the Fourier transform of the kth-order cumulant function. The power spectrum is thus the second-order polyspectrum, while the bispectrum is the third-order polyspectrum. In the case of the third-order polyspectrum, the third-order cumulants equal the third-order moments, although this is not true for fourth and higher orders. One reason to prefer cumulants over moments is that Gaussian processes have all higher-order ($N > 2$) cumulants equal to zero.

[12] Here, and throughout this chapter, $*$ denotes the complex conjugate operation.

[13] As is commonly noted, the second-order moment sequence can be recovered from the power spectrum by inverting the Fourier transform. The integral of the power spectrum is equal to $c_{xx}(0)$, the variance of the process. Thus, the power spectrum can be viewed as a decomposition of the variance by frequency. Similarly, the bispectrum can be viewed as a decomposition of $c_{xxx}(0, 0)$, the skewness of the process, by frequency pairs.

skewness function, $\Gamma(\omega_1, \omega_2)$, will be equal to a constant for all frequency pairs in the principal domain, where[14]

$$\Gamma^2(\omega_1, \omega_2) = \frac{|B_{xxx}(\omega_1, \omega_2)|^2}{P_{xx}(\omega_1)P_{xx}(\omega_2)P_{xx}(\omega_1 + \omega_2)}.$$

If, in addition, $X(W)$ is Gaussian, then $\Gamma(\omega_1, \omega_2) = 0$ for all frequency pairs in the principal domain.[15]

4.2   *Estimation*

The mathematical theory relating the normalized squared skewness function to linearity and Gaussianity has been used to derive testing procedures by Hinich (1982) and Rao and Gabr (1980). The procedure used in this paper is the one derived in Hinich (1982). Details of the Hinich test are also discussed in Hinich and Patterson (1985, 1989) and Ashley, Patterson, and Hinich (1986).

The conventional methods of bispectrum estimation are reviewed in Nikias and Raghuveer (1987). The bispectrum can be estimated consistently from a finite sample $\{x(1), \ldots, x(N)\}$ by the following procedure. Segment the record of N observations into K (non-overlapped) blocks of L observations each; L is called the block length.[16] The parameter $K/N = 1/L$ is the resolution bandwidth.[17] Define, for $k = 1, \ldots, K$, the bi-periodogram as

$$G_k(f_i, f_j) = \frac{1}{L}X_k(f_i)X_k(f_j)X_k^*(f_i + f_j),$$

where $X_k(f) = \sum_{n=(k-1)L+1}^{kL} x(n)\exp[-i2\pi f n/N]$. A consistent and asymptotically normal estimator for the bispectrum is

$$\hat{B}_{xxx}(f_i, f_j) = \frac{1}{K}\sum_{k=1}^{k} G_k(f_i, f_j), \qquad \text{where} \quad 2f_i + f_j < N$$

$$\text{and} \quad 0 < f_j < f_i < N, \quad \text{and} \quad f_i = i/L \quad (i = 1, 2, \ldots, L);$$

---

[14] The operation | | denotes the complex modulus, because the bispectrum is in general complex. This result is due to Brillinger (1965) and is based on the fact that under the stated assumptions $B_{xx}(\omega_1, \omega_2) = c_{www}(0, 0)A(\omega_1)A(\omega_2)A^*(\omega_1 + \omega_2)$, where $A(\ )$ denotes the filter transfer function.

[15] A second important property of polyspectra is that under a hypothesis of time reversibility the imaginary part of all polyspectra is zero; see Brillinger and Rosenblatt (1967a) and Hinich and Rothman (1997). As noted in Brillinger and Rosenblatt (1967a), for stationary Markov processes this property implies that the backward transition probability operator of the Markov process must be the same as the forward transition probability operator.

[16] Melvin Hinich, in personal correspondence, has suggested that the block length be set to ensure that $(\ln L)/(\ln N) \approx 0.4$. Consistency of the estimators requires that the parameter $e = (\ln L)/(\ln N) < 0.5$.

[17] If the last frame is incomplete, it is dropped from the calculation of the estimator.

see Hinich and Messer (1995) for details on the estimator.[18] This type of estimator is analogous to the direct estimator of the power spectrum described in Welch (1967) and Groves and Hannan (1968), in which the data record is segmented into frames, and periodograms are computed frame by frame and then averaged at each frequency. Hence, the power spectrum estimator is

$$\hat{P}_{xx}(f_i) = \frac{1}{K} \sum_{k=1}^{k} I_k(f_i),$$

where the periodogram is defined as $I_k(f_i) = (1/2\pi L)X_k(f_i)X_k^*(f_i)$, $k = 1, 2, \ldots, K$.[19] In the bispectrum case, bi-periodograms are computed frame by frame and then averaged at each frequency pair. It is the final averaging step which leads to consistency of the estimator in both cases. The variance is reduced by averaging over more frames, but a cost of reduced resolution.[20]

As will be detailed below, we estimate the bispectrum over a range of values for the block length L, in accordance with a suggestion of Stokes (1991). The suggested range of block lengths is $(N/3)^{1/2}$ to $N^{1/2}$, which for our sample size ($N = 396$) corresponds to block lengths between 12 and 19. See Stokes (1991) for an example using a well-known gas data model. $L = 12$ corresponds to $N^{0.42}$, and is the closest to Hinich's suggestion of $N^{0.4}$.

The Hinich test for non-linearity produces a test statistic Z, which is distributed as the standard normal under the null hypothesis of constant skewness (linearity). The Hinich Gaussianity test also produces a test statistic G, which is a standard normal under the null of zero skewness (Gaussianity). Both tests are one sided, and the null is rejected if the test statistics are large.

The Hinich test is extremely conservative. If the stochastic processes X is linear, then all of its polyspectra of order greater than two are constant. The Hinich test is based only on the bispectrum. A rejection of its null would be a strong result, because the null includes all linear processes and some non-linear processes. Consequently, the Hinich test cannot confirm linearity, it can only fail to reject it. In principle, we could test for non-linearities using polyspectra of higher order than the bispectrum, but estimating even the trispectrum would be not be feasible for the sample sizes of most economic data sets.

The conservatism of the Hinich test has been reflected in empirical studies. For example, Barnett, Gallant, Hinich, Jungeilges, Kaplan, and Jensen (1995a,b) find that the Hinich test was much less likely to reject its null than

---

[18] For highly kurtotic stochastic processes, Hinich and Messer (1995) state that the use of the asymptotic distribution may not be warranted.

[19] We employ a trapezoidal taper in order to reduce sidelobe distortion. Some modification of these formulas is therefore required.

[20] Koopmans (1975) called this tradeoff the Grenander uncertainty principle. For a discussion of power spectral estimation, see Kay and Marple (1981).

other competing tests, such as the BDS test (Brock, Dechert, Scheinkman, and LeBaron 1996), which is based on the correlation integral from chaos theory, and the Kaplan (1993) test. In particular, Hong (1996) notes that the third-order cumulants of an autoregressive conditional heteroskedastic (ARCH) process can be identically zero, in which case the bispectrum test would fail to reject linearity. Barnett, Gallant, Hinich, Jungeilges, Kaplan, and Jensen (1995a,b) demonstrate that empirically the Hinich test has low power against ARCH. Nevertheless, Ashley, Patterson, and Hinich (1986) show that the Hinich non-linearity test does have substantial power (at reasonable sample sizes) against many commonly considered forms of non-linear serial dependence.[21]

The Hinich test has been applied previously in economic analysis. Barnett and Hinich (1992) find that Divisia monetary aggregate data exhibit deep non-linearity at the M1 level of aggregation. Hinich and Patterson (1989) examine trade by trade stock market data for evidence of non-linearity.

4.3   Cointegration and non-linearity

Cointegration in an $I(1)$ economic system implies the existence of stationary linear combinations of the $I(1)$ system variables. If $X(t) \sim I(1)$ and β is a cointegration vector, then the derived stochastic process $\zeta_t = \beta^T X(t)$ is stationary. According to the Wold decomposition, a stationary stochastic process with no deterministic components can be represented as a moving average process, MA(∞); see Brockwell and Davis (1991, p. 187) and Engle and Granger (1987). Consequently, the derived stochastic process can be represented as the output of a linear time invariant filter applied to a white noise input process:

$$\zeta_t = \beta^T X(t) = \sum_{n=0}^{\infty} a(n) u(t - n),$$

where $U = \{u(t)\}$ is a white noise process. The derived stationary stochastic process is not a linear stochastic process, unless U is pure white noise.

The Hinich bispectrum test for non-linearity was designed to test stationary stochastic processes for non-linearity, and hence can be applied to the stationary linear combinations we found in Section 3.[22] We implement the Hinich bispectrum test to investigate the possibility that the components of $\beta^T X(t)$ are stationary non-linear stochastic processes. The results are reported in the next section.

---

[21] As noted in Nikias and Raghuveer (1987), the bispectrum can be particularly useful in identifying quadratic phase coupling, resulting from interaction between two harmonic components at their sum and/or difference frequencies.

[22] Ashley, Patterson, and Hinich (1986) also argue that the bispectrum based test can be applied without modification to the residuals of a linear model.

Table 6. *Bispectrum results: Z-statistics for cointegrated linear combinations without the MSI quantity variance included*

	Z	
Block size	Level of aggregation : M2	L
12	−0.44	−0.09
13	−0.11	−0.25
14	0.44	0.30
15	−0.23	−0.39
16	0.21	−0.07
17	−0.15	0.39
18	−0.37	0.58
19	0.33	0.29
Average	−0.04	0.095

If the null hypothesis of linearity is rejected, then there is neglected structural non-linearity in the economic system. Linear combinations of non-linear stochastic processes are, in general, non-linear stochastic processes. Thus, if any of the components of $X(t)$ are non-linear, we would expect the components of the derived process, ζ_t, to also be non-linear. Rejection of the null could also occur if the maintained hypothesis of the test is invalid. The maintained hypotheses of the Hinich test is third-order stationarity of the process, and absolute summability of the second- and third-order moments. If the conclusions of the maximum likelihood analysis were in error, then the maintained hypothesis of the Hinich test might be invalid. The maximum likelihood analysis is based on the Gaussian likelihood function, although the asymptotic properties of the analysis only depend on the assumption that the errors are independent and identically distributed. The existence of non-linear serial dependence in the error structure of the VECM could invalidate the estimation and rank identification of the model. Although most published cointegration studies report the results of tests for serial correlation in the residuals of the VECM, as well as the results of ARCH tests, most studies do not report the results of tests for general non-linear serial dependence in the residuals of the VECM.[23]

4.4   Empirical results

The results of the Hinich bispectrum test for our models are reported in Tables 6 and 7. We find that in the linear combinations which do not include the quantity variance, there is no evidence of non-linearity. In the linear combinations which

---

[23] The Johansen maximum likelihood test has also been criticized on the basis of poor finite sample performance; see Watson (1994, p. 2893).

Table 7. *Bispectrum results: Z-statistics for cointegrated linear combinations with the divisia quantity variance included* [a]

	Z			
	Level of aggregation : M2		L	
Block size	Vector 1	Vector 2	Vector 1	Vector 2
12	1.50	−0.45	1.50	1.50
13	1.50	0.14	1.50	1.50
14	1.58	0.83	1.64	1.73
15	1.56	−0.63	1.58	1.73
16	1.67	−0.25	1.82	1.73
17	1.57	−0.45	1.82	0.86
18	1.63	−0.38	1.68	2.23
19	1.30	0.35	1.29	2.23
Average	1.535	−0.104	1.61	1.69

[a] Vector 1 is the most significant cointegration relationship. Vector 2 is the second most significant cointegration relationship.

Table 8. *Bispectrum results: Z-statistics for the* MSI *quantity variance*

	Z	
Block size	Level of Aggregation : M2	L
12	1.50	1.50
13	1.49	1.49
14	1.59	1.56
15	1.57	1.58
16	1.73	1.74
17	1.75	1.82
18	1.47	1.51
19	1.39	1.39
Average	1.56	1.57

do include the quantity variance, we find evidence of non-linearity. Specifically, for the more significant linear combination, which heavily weights the quantity variance, we find some evidence of non-linearity at every level of aggregation. This is not surprising, because the univariate tests detect non-linearity in the quantity variances (see Table 8). We find the strongest evidence of non-linearity in the second linear combination at the L level of aggregation. This result does not hold at the M2 level of aggregation; the second linear combination at the M2 level of aggregation shows no evidence of non-linearity.

5   Conclusion

In this chapter, we have tested for the existence of non-linearity in the cointegration relations of a system containing money demand variables, by applying the Hinich bispectrum test. We find some evidence of non-linearity in spite of that test's inherent conservatism, and therefore demonstrate that the issue is empirically relevant. Although the testing of identification restrictions is beyond the scope of the current paper, the most probable source of the cointegration we investigate would be the existence of long run money demand relations. This interpretation would be more reasonable in our study than in many other studies, due to our use of aggregation theory in the selection of the variables in our system. There are many avenues for further research on this topic. The testing of non-linearity in cointegration relations could be extended to other contexts, notably systems including consumption and investment. In addition, greater attention could be paid to formal identification of these relations. In future applications of cointegration analysis, we suggest that non-linear cointegration models be investigated, as suggested by Granger (1991). Approaches that have the ability to remove non-linear structure by modeling the source of the non-linearity within the equations should also be investigated. We conclude that stationarity of the linear combination of cointegrated variables should not be viewed as sufficient for linearity of the derived process without testing that null.

REFERENCES

Anderson, R. G., B. E. Jones, and T. D. Nesmith (1997a), "Monetary Aggregation Theory and Statistical Index Numbers", *Federal Reserve Bank of St. Louis Review*, Jan/Feb.
  (1997b), "Building New Monetary Services Indexes: Data and Methods", *Federal Reserve Bank of St. Louis Review*, Jan/Feb.
Arrow, K. F., and F. H. Hahn (1971), *General Competitive Analysis*. San Francisco: Holden Day.
Ashley, R., D. Patterson, and M. Hinich (1986), "A Diagnostic Test for Non-linear Serial Dependence in Time Series Fitting Errors", *Journal of Time Series Analysis* 7(3): 165–78.
Barnett, W. A. (1978), "The User Cost of Money", *Economic Letters* 1(2): 145–9.
  (1980), "Economic Monetary Aggregates: An Application of Index Number and Aggregation Theory", *Journal of Econometrics* 14: 11–48.
  (1987), "The Microeconomic Theory of Monetary Aggregation", in W. Barnett and K. Singleton (eds.), *New Approaches to Monetary Economics*. Cambridge University Press.

(1990), "Developments in Monetary Aggregation Theory", *Journal of Policy Modeling* 12(2): 205–57.

(1991), "A Reply to Julio Rotemberg", in *Monetary Policy on the 75th Anniversary of the Federal Reserve System,* Michael T. Belongia (ed.). Kluwer Academic Publishers, pp. 232–45.

Barnett, W. A., and P. Chen (1988), "The Aggregation Theoretic Monetary Aggregates Are Chaotic and Have Strange Attractors: An Econometric Application of Mathematical Chaos", in W. A. Barnett, E. Berndt, and H. White (eds.), *Dynamic Econometric Modeling, Proc. 3rd Int. Symp. on Economic Theory and Econometrics.* Cambridge: Cambridge University Press, pp. 199–246.

Barnett, W. A., D. Fisher, and A. Serletis (1992), "Consumer Theory and the Demand for Money", *Journal of Economic Literature,* Dec., pp. 2086–2119.

Barnett, W. A., R. Gallant, M. Hinich, J. Jungeilges, D. Kaplan, and M. Jensen (1995), "Robustness of Non-linearity and Chaos Tests to Measurement Error, Inference Method, and Sample Size", *Journal of Economic Behavior and Organization* 27: 301–20.

(1997), "A Single Blind Controlled Competition among Tests For Non-linearity and Chaos", *Journal of Econometrics* 77: 297–302.

Barnett, W. A., and M. J. Hinich (1992), "Empirical Chaotic Dynamics in Economics", *Annals of Operations Research* 37: 1–15.

Barnett, W. A., B. E. Jones, and T. D. Nesmith (1996), "Divisia Second Moments: An Application of Stochastic Index Number Theory", *International Review of Comparative Public Policy* 8: 115–38.

Barnett, W. A., E. K. Offenbacher, and P. A. Spindt (1984), "The New Divisia Monetary Aggregates", *Journal of Political Economy* 92: 1049–85.

Barnett, W. A., and A. Serletis (1990), "A Dispersion Dependency Diagnostic Test for Aggregation Error: With Applications to Monetary Economics and Income Distribution", *Journal of Econometrics* 43: 5–43.

Barnett, W., and H. Xu (1996), "An Investigation of Recent Empirical Paradoxes in Monetary Economics", *International Review of Comparative Public Policy* 8: 139–55.

Brillinger, D. R. (1965), "An Introduction to Polyspectrum", *Annals of Mathematical Statistics* 36: 1351–74.

Brillinger, D. R., and M. Rosenblatt (1967a), "Asymptotic Theory of k-th Order Spectra", in *Spectral Analysis of Time Series,* B. Harris (ed.), New York: Wiley, pp. 153–88.

(1967b), "Computation and Interpretation of k-th Order Spectra", in *Spectral Analysis of Time Series,* B. Harris (ed.), New York: Wiley, pp. 189–232.

Brock, W. A., W. D. Dechert, J. Scheinkman, and B. LeBaron (1996), "A Test for Independence Based on the Correlation Dimension", *Econometric Reviews* 15(3): 197–235.

Brock, W. A., and C. L. Sayers (1988), "Is the Business Cycle Characterized by Deterministic Chaos?", *Journal of Monetary Economics* 22(1): 71–90.

Brockwell, P., and R. Davis (1991), *Time Series: Theory and Methods.* New York: Springer.

Christiano, L. J. (1986), "Money and the U.S. Economy in the 1980's: A Break from the Past", *Federal Reserve Bank of Minneapolis Quarterly Review,* Summer, pp. 2–13.

Chrystal, K. A., and R. MacDonald (1994), "Empirical Evidence on the Recent Behavior and Usefulness of the Simple Sum and Weighted Measures of the Money Stock", *Federal Reserve Bank of St. Louis Review,* Mar./Apr., pp. 73–109.

Diewert, E. (1976), "Exact and Superlative Index Numbers", *Journal of Econometrics* 4: 115–45.

(1995), "On the Stochastic Approach to Index Numbers", Discussion Paper DP95-31, University of British Columbia.

Engle, R., and C. W. J. Granger (1987), "Co-integration and Error Correction: Representation, Estimation, and Testing", *Econometrica* 55(2): 251–76.

Fisher, D. (1989), "Velocity and the Growth of Money in the United States, 1970–1985", *Journal of Macroeconomics* 11(3): 323–32.

Friedman, B. M., and K. N. Kuttner (1992), "Money, Income, Prices, and Interest Rates", *American Economic Review* 82: 472–92.

Godfrey, L. G. (1988), *Misspecification Tests in Econometrics*. Cambridge: Cambridge University Press.

Godfrey, M. D. (1965), "An Exploratory Study of the Bispectrum of Economic Time Series", *Applied Statistics* 14: 48–69.

Granger, C. W. J. (1981), "Some Properties of Time Series Data and their Use in Econometric Model Specification", *Journal of Econometrics* 16: 121–30.

(1983), "Cointegrated Variables and Error Correction Models", Discussion Paper 83-13a, University of California, San Diego.

(1991), "Some Recent Generalizations of Cointegration and the Analysis of Long-Run Relationships", in *Long-Run Economic Relationships*, R. F. Engle and C. W. J. Granger (eds.), Oxford University Press, pp. 277–87.

Groves, G. W., and E. J. Hannan (1968), "Time Series Regression of Sea Level on Weather", *Reviews of Geophysics* 6(2): 129–74.

Hinich, M. J. (1982), "Testing for Gaussianity and Linearity of a Stationary Time Series", *Journal of Time Series Analysis* 3(3): 169–76.

Hinich, M. J., and G. R. Messer (1995), "On the Principle Domain of the Discrete Bispectrum of a Stationary Signal", *IEEE Transactions on Signal Processing* 43(9): 2130–4.

Hinich, M. J., and D. Patterson (1985), "Identification of the Coefficients in a Non-linear Time Series of the Quadratic Type", *Journal of Econometrics* 30: 269–88.

(1989), "Evidence of Non-linearity in the Trade by Trade Stock Market Return Generating Process", in W. Barnett, J. Geweke, and K. Shell (eds.), *Economic Complexity: Chaos, Sunspots, Bubbles and Non-linearity, Proc. 4th Int. Symp. on Economic Theory and Econometrics*. Cambridge: Cambridge University Press.

Hinich, M. J., and P. Rothman (1997), "A Frequency Domain Test of Time Reversibility", Working Paper, University of Texas at Austin.

Hinich, M. J., and G. R. Wilson (1992), "Time Delay Estimation Using the Cross Bispectrum", *IEEE Transactions on Signal Processing* 40(1): 106–13.

Hong, Y. (1996), "Consistent Testing for Serial Correlation of Unknown Form", *Econometrica* 64: 837–64.

Johansen, S. (1988), "Statistical Analysis of Cointegrating Vectors", *Journal of Economic Dynamics and Control* 12: 231–54.

(1991), "Estimation and Hypothesis Testing of Cointegration Vectors in Gaussian Vector Autoregressive Models", *Econometrica* 59(6): 1551–80.

(1992a), "A Representation of Vector Autoregressive Processes Integrated of Order 2", *Econometric Theory* 8: 188–202.

(1992b), "Testing for Weak Exogeneity and the Order of Cointegration in UK Money Demand Data", *Journal of Policy Modeling* 14(3): 313–34.

(1995a), "A Statistical Analysis of Cointegration for $I(2)$ Variables", *Econometric Theory* 11: 25–59.

(1995b), *Likelihood-Based Inference in Cointegrated Vector Auto-regressive Models.* Oxford: Oxford University Press.

Johansen, S., and K. Juselius (1990), "Maximum Likelihood Estimation and Inference on Cointegration – with an Application to the Demand for Money", *Oxford Bulletin of Economics and Statistics* 52: 169–210.

Jones, B. E. (1998), "An $I(2)$ Cointegration Analysis of US Monetary Aggregates", Working Paper, Washington University in St. Louis.

Jorgensen, C., H. C. Kongsted, and A. Rahbek (1996), "Trend Stationarity in the $I(2)$ Cointegration Model", Working Paper, Washington University in St. Louis.

Kaplan, D. T. (1993), "Exceptional Events as Evidence for Determinism", *Physica D,* to appear.

Kay, S. M., and S. L. Marple, Jr. (1981), "Spectrum Analysis – A Modern Perspective", *Proceedings of the IEEE* 69(11), 1380–419.

King, R. G., C. I. Plosser, J. H. Stock, and M.W. Watson (1991), "Stochastic Trends and Economic Fluctuations", *American Economic Review* 81(4): 819–40.

King, R. G., and M. W. Watson (1996), "Money, Prices, Interest Rates, and the Business Cycle", *The Review of Economics and Statistics* 78(1): 35–53.

Koopmans, L. H. (1975), *The Spectral Analysis of Time Series.* New York: Academic Press.

la Cour, L. F. (1995), "On the Measurement Problem of 'Money': Results from the Experience with Divisia Monetary Aggregates for Denmark and some Methodological Considerations of the Comparison of Money Demand Relations Based on Alternative Monetary Aggregates", Working Paper, Copenhagen Business School.

Lee, T.-Hwy, and Y. Tse (1996), "Cointegration Tests with Conditional Heteroskedasticity", *Journal of Econometrics* 73: 401–10.

Mendel, J. M. (1991), "Tutorial on Higher-Order Statistics (Spectra) in Signal Processing and System Theory", *Proceedings of the IEEE* 79(3): 278–305.

Miyao, R. (1996), "Does a Cointegrating M2 Demand Relation Really Exist in the United States?", *Journal of Money, Credit, and Banking* 28(3): 365–80.

Nikias, C. L., and M. R. Raghuveer (1987), "Bispectrum Estimation: A Digital Signal Processing Framework", *Proceedings of the IEEE* 75(7): 869–91.

Parulo, P. (1996), "On the Determination of Integration Indices in $I(2)$ Systems", *Journal of Econometrics* 72(1–2): 313–56.

Priestly, M. (1981), *Spectral Analysis and Time Series,* Vol. 2. New York: Academic Press.

Rao, S. T., and M. Gabr (1980), "A Test for Linearity of Stationary Time Series", *Journal of Time Series Analysis* 1: 145–58.

Serletis, A. (1987), "Monetary Asset Separability Tests", in W. Barnett and K. Singleton (eds.), *New Approaches to Monetary Economics, Proceedings of the Second International Symposium in Economic Theory and Econometrics.* Cambridge University Press.

Serletis, A., and M. King (1993), "The Role of Money in Canada", *Journal of Macroeconomics* 15(1), 91–107.

Stokes, H. (1991), *Specifying and Diagnostically Testing Econometric Models.* London: Quorum Books.

Theil, H. (1967), *Economics and Information Theory.* Amsterdam: North Holland.

Thornton, D. L., and P. Yue (1992), "An Extended Series of Divisia Monetary Aggregates", *Federal Reserve Bank of St. Louis Review*, Nov./Dec. pp. 35–52.

Watson, M. W. (1994), "Vector Autoregressions and Cointegration", in *Handbook of Econometrics*, Vol. 4., Chap. 47, Robert F. Engle and Daniel L. McFadden (eds.), Elsevier.

Welch, P. D. (1967), "The Use of Fast Fourier Transform for the Estimation of Power Spectra: A Method Based on Time Averaging over Short Modified Periodograms", *IEEE Transactions on Audio and Electroacoustics* AU-15(2): 70–3.

CHAPTER 3

Risk-related asymmetries in foreign exchange markets

Giampiero M. Gallo & Barbara Pacini

1   Introduction

In the literature on foreign exchange markets, the reader is customarily briefed on the largely documented untenability of the hypothesis which asserts that the forward rate is an unbiased predictor of the future spot rate. In this paper we seek to investigate the nature of the time-varying risk-related term often inserted in the spot–forward relationship, in its possible links to market inefficiencies, bounded rationality, or nonlinearities as reasons for the breakdown of the unbiasedness hypothesis. We will concentrate in particular on the suggestions to evaluate this risk-related term by linking it to the conditional variance, exploiting some of the recent developments in nonparametric and semiparametric estimation.[1]

One suggestion advanced in evaluating the presence of this risk-related term in the spot–forward relationship has been to consider an ARCH-M framework where the autoregressive conditional variance term enters as its proxy in the mean equation. Such a parameterization of the conditional variance may turn out to be restrictive in that it imposes specific assumptions about how the information available can be processed to provide a measure of risk. In fact, the

A preliminary version of this paper was presented as *Semiparametric Evaluation of Foreign Exchange Risk* at the ESEM 94, Maastricht. Thanks are due to Renzo G. Avesani, Lucia Buzzigoli, Giorgio Calzolari, Gabriele Fiorentini, Alan Kirman, Grayham Mizon, and Mark Salmon for useful comments and suggestions. The comments of an anonymous referee substantially improved the presentation. Special thanks to Miguel Delgado, who kindly provided his nonparametric estimation routines modified by us for this context. Financial support from the Italian MURST and CNR is gratefully acknowledged.

[1] Although the presence of stochastic heteroskedasticity is largely documented in the exchange-rate analysis, it is not clear whether conditional heteroskedasticity is a structural characteristic of the data-generating process or, rather, is an effect of an incorrect linear specification of the conditional-mean function. A different stream of research suggests seeking nonlinearities in the mean equation with nonlinear and chaotic models. Nonparametric estimates of the conditional-mean equation are proposed in Diebold and Nason (1990), Hsieh (1993), and Mizrach (1993).

empirical evidence provided by Domowitz and Hakkio (1985) with monthly data and by Baillie and Bollerslev (1990) with weekly data has shown the weakness of a parametric approach to the evaluation of risk-premium effects in a spot–forward rates relationship and signals the need to explore other routes. As noted by Froot and Thaler (1990), this approach belongs to the class of statistical models of risk which are not derived from an asset pricing theory where time-varying risk is related to intertemporal optimization and attitude toward risk. Nevertheless, its adoption is usually justified on the grounds that risk is related to uncertainty, and the latter to volatility. In our case, a measure of volatility in the market, conditional on an information set and derived from weekly data without imposing a specific parameterization, is used to reflect the prevailing level of uncertainty, given the recent experience on the markets. The outcome is a flexible nonlinear moving average where previous surprises relative to the spot–forward relationship are processed nonlinearly.

There is no consensus as to whether the presence of a conditional volatility term in the mean equation should be referred to as a risk-premium or more generically as a risk-related term: the theoretical foundations of a model with a risk-premium fade away when its testable implications need to be derived (cf. Hansen and Hodrick 1983); moreover, as shown by Backus and Gregory (1993), a monotonic, increasing relationship between conditional variance and risk-premium cannot be derived in general. Apart from terminology, the essence of the matter is that the trends, prevailing in the short run on the markets, are surrounded by uncertainty. The presence of a large number of heterogeneous agents is reflected at times by disparate (but evolving) beliefs about these trends, which may show up in clustering of volatility which affects the mean equation. With this in mind, for ease of reference we will use the two terms as synonyms.

Moreover, the pure risk-premium argument would not explain market behavior such as periods of strong appreciations and sharp depreciations of the US dollar in the 1980s, as noted also by Froot and Thaler (1990). Also, since the behavior of the exchange rates mirrors reputation as regards stability and credibility of monetary and fiscal policies, a working hypothesis is that agents hold different attitudes toward observed periods of relative strength and weakness of a currency. For currencies such as the Italian lira, for example, for which progressive devaluations vis-à-vis the major currencies in the 1970s and the 1980s have been the rule rather than the exception, some agents might maintain expectations of a future depreciation even when the signals coming from interest-rate differentials would suggest otherwise. An interest-rate differential of about 2% between Eurodeposit rates on the Italian lira and the Deutsche mark was the norm when the behavior of the lira in the ERM of the European Monetary System was fairly stable and was not accompanied by any expectation of specific lira movements. At other times higher differentials were seen as a sign of distress for the lira, in the presence of expectations of a depreciation. High differentials do not imply necessarily an impending depreciation: in fact, such a situation was observed between the Deutsche mark and the US dollar in the early 1990s, but the former long maintained a position of strength relative to the latter.

The reputation of each currency is related to many economic and political elements under consideration by the markets, among which are the anti-inflationary stances taken by monetary authorities. These elements vary across countries and time, so that it is of interest to investigate the effect of the risk-related term on the exchange-rate movements in an attempt to isolate asymmetry of behavior in some specific market situations. Here we will adopt three common trading rules (cf. Le Baron 1993b) which provide signals of action, in an effort to sift through the different attitudes held by market operators toward a given currency as reflected in buy or sell actions on the market. To avoid confusion, the trading rules are not examined here for their profitability or to show possible market inefficiencies, but as indicators of situations which the market may not interpret univocally. The movements in the conditional variance (risk-related term) will then have different effects on exchange-rate movements if asymmetry is present. In fact, if symmetry in the reactions were to hold, we would observe an outright reversal of sign in the coefficient of the risk-related term. Otherwise, we should interpret the evidence as a sign of the presence of a "reputation" or "prejudice" effect attached to the currency which filters the signals coming from the market. Note that in our analysis the only economic fundamentals taken into consideration are the interest-rate differentials, so that there is no explicit reference to learning about changes in monetary policy as in Lewis (1989). The evolution of beliefs concerning the credibility of the monetary authority's actions is taken to be reflected in the behavior of conditional volatility.

Our approach differs from previous parametric studies not only in the particular interpretation given to the conditional variance term, and the interpretation given in this context, but also in that, unlike Baillie and Bollerslev (1990), we allow the MA coefficients to be freely varying, and we estimate the risk term nonparametrically on a larger sample size of 1077 weeks (from 1973 to 1994).[2] We differ from previous semiparametric estimations of the spot–forward relationship (Pagan and Ullah 1988, Pagan and Hong 1991) in the choice of estimator and in the provision of a comparison of the three methods.

The paper is organized as follows: after recalling some theoretical issues surrounding the relationship between forward spot markets, the econometric treatment of risk evaluation in the parametric case is presented in Section 3. Section 4 discusses the nonparametric estimation of the risk-premium, suggesting an original instrumental variable framework given the error-in-variable

---

[2] We avoid altogether the sample selection problem, which led Hansen and Hodrick (1983) to exclude the years up to 1976 on the grounds that up to that point the free-float system was still being perfected.

problem affecting the use of generated variables for the risk term. A comparison with other methods of choosing the instruments for the estimator at hand (Pagan and Ullah 1988, Pagan and Hong 1991) is discussed in Section 5, which leads to the empirical application of the procedures to the bilateral exchange rate of five currencies vis-à-vis the US dollar.[3]

In Section 6 we address the explicit question of asymmetric expectations when periods are formed relative to three trading rules (LeBaron 1993b): the first is based on the interest-rate differentials, the second on short-term and long-term moving averages, and the third on short-term and long-term variances.

2   The spot–forward relationship

The cornerstone of the analysis of foreign-exchange market efficiency is the theory of interest-rate parity, which, in its covered form, conveniently provides a link between spot and forward rates and interest-rate differentials,

$$f_{t,k} - s_t = i_{t,k}^*, \tag{1}$$

which is widely accepted and forms the pricing basis for the forward rates. In its uncovered version, it establishes a relationship between spot rates and expectations about future spot rates and the interest-rate differential,

$$E_t(s_{t+k}) - s_t = i_{t,k}^*, \tag{2}$$

where $f_{t,k}$ is the (logarithm of the) forward exchange rate at time t for delivery at time $t + k$; s_t is the (logarithm of the) spot rate at time t expressed as units of foreign currency per unit of domestic currency; E_t is the expected value conditional on the relevant information set at time t;

$$i_{t,k}^* = \log\left(1 + i_{t,k}^d\right) - \log\left(1 + i_{t,k}^f\right) \approx i_{t,k}^d - i_{t,k}^f;$$

$i_{t,k}^d$ is the interest rate on the domestic currency between t and $t + k$; and $i_{t,k}^f$ is the interest rate on the foreign currency on the same horizon and on foreign assets perfectly substitutable with domestic ones.

Thus, according to the theory, in the absence of market frictions, transaction costs, capital controls, and so on, when faced with the need of availability of foreign currency k periods into the future one would be indifferent (in *ex ante* expected terms) between holding domestic currency (earning domestic interest rates) and purchasing a forward contract or purchasing foreign currency (earning foreign interest rates) right away.

The issues of whether the error term $\epsilon_{t,k} = s_{t+k} - f_{t,k}$ has a zero mean (un-biasedness hypothesis), is uncorrelated, or has a constant variance have often

---

[3] Several studies have highlighted the rejection of a "dollar phenomenon" so that the results would not be dependent on the choice of the numeraire.

surfaced in the literature of the past fifteen years, with a wide array of results according to which currency was under consideration and for what period. In sampling the data at a frequency higher than the interest-rate maturity (for example, 30-day contracts with weekly or daily data), an additional complication arises from the operation of matching data on the forward rates with the corresponding future spot rates. This is not only a problem of determining the appropriate timing of the contract (Fama 1984, for example, incorrectly takes Friday data for both spot and forward rates four weeks apart). Depending on the actual terms of the problem, $\epsilon_{t,k}$ follows either a MA(k) process or a MA($k-1$) (cf. Baillie and Bollerslev 1990), because of the sampling at a higher frequency than the maturity of the forward contract.

In empirical applications, the relationship is tested as[4]

$$s_{t+k} - s_t = \alpha = \beta i^*_{t,k}$$

or, which is the same,

$$s_{t+k} - s_t = \alpha = \beta(f_{t,k} - s_t).$$

The simple unbiasedness hypothesis ($\alpha = 0$, $\beta = 1$) is seldom accepted in empirical applications, despite the fact that it is widely recognized by now that its rejection does not imply market inefficiency. In fact, using the Lucas (1982) model of intertemporal asset pricing in a two-country world, it has often been shown (e.g. Hodrick and Srivastava 1984) that uncertainty about the future purchasing power of domestic and foreign monies and about future marginal utility of the domestic good translates into uncertainty about the intertemporal rate of substitution of domestic currency between t and $t+k$. The presence of a conditional covariance term between this rate of substitution (multiplied by the risk-free return) and the future spot rate is used to support the argument for the existence of a time-varying risk-premium. Stockman (1978) was probably the first to stress the sign changes in the influence of the risk-related factor, with a division of his sample into subperiods.

Various assumptions are needed to translate Lucas's economic model into an empirically testable model (Hansen and Hodrick 1983, Domowitz and Hakkio 1985). Various statistical models have a less clearly interpretable economic foundation, but aim at reproducing some stylized facts and at extracting an economically interpretable signal from $\epsilon_{t,k}$. The latent nature of this term calls for appropriate econometrics (Pagan and Ullah 1988, Pagan and Hong 1991). The definition and measurement of risk is the object of the present investigation, where nonparametric measures of risk are taken into account. The empirical

---

[4] The covered-interest-rate parity relationship is generally considered as valid, hence the equivalence between the two formulations.

interest in the present paper is to compare these alternative measures and subsequently to investigate the importance of the risk term in the relationship of the spot exchange rate to forward rates. As noted by Cumby (1988), for example, the error term contains both the uncertainty in *ex ante* profits relative to an information set and the error-in-variable problem between the unobservable *ex ante* profits and their realized counterparts. We start from the expression

$$E_t s_{t+k} = \mathrm{RP}_{t,k} + f_{t,k},$$

where forward rate unbiasedness implies $\mathrm{RP}_{t,k} = 0$. By subtracting s_t from both sides we get a different expression

$$s_{t+k} - s_t = \mathrm{RP}_{t,k} + (f_{t,k} - s_t) + \epsilon_{t+k},$$

where $\mathrm{RP}_{t,k}$ is taken to represent the risk-premium of the theory, assumed to be linked to the conditional variance in the ϵ_{t+k}, an $\mathrm{MA}(k-1)$ pure forecast error. As mentioned previously, a monotonic and increasing relationship between conditional variance and risk-premium has been recently challenged by Backus and Gregory (1993), who show that the convenient insertion of the conditional variance in the mean equation has little theoretical foundation from existing dynamic asset-pricing models, and that the use of the conditional variance as a proxy for the risk-premium can be justified on the basis of a specific structure of the economy, but is by no means general. With this caution, we will continue to refer to the risk-related term as $\mathrm{RP}_{t,k}$. Nevertheless, the interpretation given in the present context maintains a relationship between the evolution of conditional volatility, of uncertainty on the markets, and the perception of risk.

This issue is quite separate from the motivation for a nonparametric treatment of the risk-related term, which mainly stems from the limitations of a linear specification for the mean equation in the ARCH-M model. A nonlinear mapping between the conditional variance and the information set is more likely to be captured in a flexible context (cf. Pagan and Hong 1991). Moreover, the performance of the ARCH-M model by Domowitz and Hakkio (1985) is somewhat unsatisfactory, failing to assess the importance of the risk-related term, although the endogenous dynamics introduced does present some appealing elements. An explicit parameterization of the risk term introduces uncertainty about the interpretability of the results because of the possible misspecification of the model.

The change in the effects of the risk-premium and the frequent changes in sign discovered by Stockman (1978) were interpreted as being related to the nature of the stochastic processes ruling the state variables. Adding to that the highly nonlinear nature of the transformations these processes undergo in the intertemporal asset-pricing models, the adoption of a nonparametric measure of risk seems to buy a lot of flexibility relative to a parametric specification which is not derived from the theory anyway.

Traditionally, a number of suggestions have been advanced in the literature (mostly without success) in this and other fields where risk plays a role. Moving variances have been proposed by French, Schwert, and Stambaugh (1987) to model inflation risk; Pagan, Hall, and Trivedi (1983) derive measures of risk which highlight the relationship between individual and aggregate prices; also, survey data on business expectation have been used (Levi and Makin 1979) to infer a measure of risk. Alternative nonparametric measures of risk-related volatility are suggested by Pagan and Schwert (1990).

From an econometric point of view, the use of proxy variables determines an error-in-variable problem, since the proxy variable is correlated with the disturbance. A way to avoid this problem is to parameterize the second moment according to a model of risk determination. Some authors model the conditional variance as a function of some variable z_t, as $\sigma_t^2 = \sigma^2 + z_t'\alpha$, although the choice is not clear-cut either for the set of variables or for the linearity of the functional form. Since, as noted, economic theory is not clear as to what relationship the predictable component of market volatility – and thus of risk – has to the relevant information set Ψ_t, referring to all publicly available information does not help to determine which variables should be used in the empirical analysis. Below we will adopt an information set limited to the spot and forward rates at one maturity (one month), relying on the results by Hakkio (1981) in assuming that further maturities would not add to the analysis in terms of surprises from other forward premia.

3   The nature of the risk-premium

As noted before, the existence of a time-varying risk-premium makes it possible that the efficiency hypothesis of the speculative markets still holds. Domowitz and Hakkio (1985) were the first to model this time-varying term within an ARCH-M framework (Engle, Lilien, and Robins 1987), obtaining results which point to the time variability of the influence of the risk-premium, but fail to isolate clearly its contribution to the magnitude of the forecast error. Their model is

$$\frac{S_{t+k} - S_t}{S_t} = RP_{t,k} + \beta_1 \frac{F_{t,k} - S_t}{S_t} + \epsilon_{t+k}, \tag{3}$$

$$RP_{t,k} = \beta_0 + \theta h_{t+k},$$

$$\epsilon_{t+k} \mid \Psi_t \sim N(0, h_{t+k}),$$

$$h_{t+k} = \alpha_0 + \sum_{i=1}^{p} \alpha_i \epsilon_{t+k-i}^2,$$

where Ψ_t is the information set available at time t.[5]

---

[5] In the expression for the conditional variance one needs to impose constraints to preclude negative variances. A sufficient condition is $\alpha_0 > 0$ and $\alpha_i \geq 0$; alternatively the specification should be changed to read $\log h_{t+k} = \alpha_0 + \sum_{i=1}^{p} \alpha_i \log \epsilon_{t+k-i}^2$.

In such a model the conditional variance is evolving as a function of its own past and enters the equation for the mean as well. By its own nature, this term is time-varying and lends itself to interpretation as a risk term once the signs of β_0 and θ are determined. In fact, given the structure of the model, and considering lowercase letters as indicating the logs of the variables in equation (3), we have that

$$
\begin{aligned}
h_{t+k} &= \text{var}(\epsilon_{t+k}|\Psi_{t+k-1}) \\
&= \text{var}(s_{t+k} - s_t|\Psi_{t+k-1}) + \theta^2 \text{var}(h_{t+k}|\Psi_{t+k-1}) \\
&\quad + \beta_1^2 \text{var}(f_{t,k} - s_t|\Psi_{t+k-1}) - 2\theta \text{cov}(s_{t+k} - s_t, h_{t+k}|\Psi_{t+k-1}) \\
&\quad + 2\beta_1 \text{cov}(s_{t+k} - s_t, f_{t,k} - s_t|\Psi_{t+k-1}) \\
&\quad + 2\theta\beta_1 \text{cov}(h_{t+k}, f_{t,k} - s_t|\Psi_{t+k-1}),
\end{aligned}
$$

thus stressing the dependence of the risk term on higher-order conditional moments of the forecast error, and on conditional variances and covariances of the exchange rate and the forward premium.

The ARCH-M model, in the formulation by Domowitz and Hakkio, allows for testing some hypotheses about the behavior of the risk-premium: in particular, to test the hypothesis $\theta = 0$ means to verify the role played by the conditional variance in determining the difference between forward and expected spot rates. Assuming that $\beta_1 = 1$ and ϵ_{t+k} is MA($k - 1$), then $\beta_0 = 0$ and $\theta = 0$ implies the absence of a risk-premium, $\beta_0 \neq 0$ and $\theta = 0$ implies the presence of a constant risk-premium, and $\beta_0 \neq 0$ and $\theta \neq 0$ implies the presence of a time-varying risk term. Note that model (3) implies that the movements in the risk-premium can only be introduced through changes in the conditional variance. Moreover, $\text{RP}_{t,k}$ can be either positive or negative according to the values of β and θ. The disappointing results of the analysis by Domowitz and Hakkio, which fail to lend support to the relationship between conditional variance and risk-premium, have been attributed to the use of monthly data; other authors think that the univariate framework is too restrictive, while in a multivariate framework one could take into consideration not only the conditional variances but also the covariances among the various currencies in the market. Bollerslev (1990) and Baillie and Bollerslev (1990), for example, use a multivariate GARCH model on weekly data, but do not achieve strong results.

Pagan and Hong (1991) have proposed to estimate flexible forms for the ARCH-M model, estimated in a nonparametric fashion on monthly data. In what follows we will discuss the instrumental variable procedure and suggest an alternative way to select the instrument for the risk-related term. Our suggestion and the estimators proposed by Pagan and Ullah (1988) and by Pagan and Hong (1991) are then compared using weekly data.

4   Semiparametric IV estimation

4.1   *A suggestion*

Formulated in more general terms, the problem at hand can be seen as

$$s_{t+k} - s_t = \beta_0 + \beta_1(f_t - s_t) + \sigma_{t+k}^2\delta + \epsilon_{t+k}, \qquad t = 1, \ldots, T. \qquad (4)$$

We are interested in the estimation of β_0, β_1, and δ; σ_{t+k}^2 is unobservable and must be estimated on the basis of the information at time t. As previously stressed, in recent years the general tendency has been that of specifying a parametric model for σ_{t+k}^2, the most popular choice being an ARCH process ($\sigma_{t+k}^2 = h_{t+k}$). This model can be efficiently estimated by maximum likelihood under correct specification, but the estimator is inconsistent if the functional form of σ_{t+k}^2 is misspecified, since σ_{t+k}^2 (or a function of it) is a constituent part of the conditional mean of $s_{t+k} - s_t$.

Pagan and Ullah (1988) considered issues related to the estimation of a linear model containing a risk term as a regressor. They questioned the form of the mapping between risk terms and information set available to agents and suggested the use of an instrumental variable estimator. As σ_{t+k}^2 is unobservable, assume another variable ϕ_{t+k} exists, such that $E(\phi_{t+k} \mid \Psi_t) = \sigma_{t+k}^2$. In the ARCH-M model such a variable is given by the series of squared innovations ϵ_{t+k}^2. Instead of ϵ_{t+k}^2, some residuals $\hat{\epsilon}_{t+k}^2$ may be used without affecting the asymptotic properties of the estimator (Pagan and Ullah 1988). By substitution we arrive at

$$
\begin{aligned}
s_{t+k} - s_t &= \beta_0 + \beta_1(f_t - s_t) + \phi_{t+k}\delta + \left(\sigma_{t+k}^2 - \phi_{t+k}\right)\delta + \epsilon_{t+k} \\
&\equiv \mathbf{x}_t'\beta + \phi_{t+k}\delta + \left(\sigma_{t+k}^2 - \phi_{t+k}\right)\delta + \epsilon_{t+k} \qquad (5) \\
&\equiv \mathbf{z}_{t+k}'\theta + u_{t+k}, \qquad (6)
\end{aligned}
$$

where $\mathbf{x}_t = (1, (f_t - s_t))$, $\beta = (\beta_0, \beta_1)$, $\mathbf{z}_{t+k} = (\mathbf{x}_t, \phi_{t+k})$, $\theta = (\beta, \delta)$, and $u_{t+k} = (\sigma_{t+k}^2 - \phi_{t+k})\delta + \epsilon_{t+k}$, and where $E(\mathbf{x}_t\epsilon_{t+k}) = 0$, $E(\phi_{t+k}\epsilon_{t+k}) = 0$, $E(\phi_{t+k}u_{t+k}) \neq 0$, and $E(\mathbf{z}_{t+k}u_{t+k}) \neq 0$, so that the OLS estimation of model (6) is inconsistent. Pagan and Ullah show that, in the case of stationary time series, consistent and asymptotically normal estimates can be obtained via nonparametric estimation of the instruments (Proposition 5, p. 94).

Efficiency improvements are related to the choice of the instruments. The definition of the instruments plays an important role in the instrumental-variable estimation of a parametric model. BNL2SLS (best nonlinear two-stage least squares) and BNL3SLS (best nonlinear three-stage least squares), proposed by Amemiya (1974, 1977), rely on the choice of optimal instruments minimizing the asymptotic covariance matrix of the estimates. Computational problems, due to the presence of nonlinear functions or unknown conditional distributions of the endogenous variables, led to the development of semiparametric instrumental variable estimation methods.

Recent results suggest the potential use of nonparametric regression techniques to estimate the conditional expectation of the endogenous variables, which appears in the Amemiya optimal-instruments formulation. In Newey (1990) two different kinds of nonparametric regression estimators are proposed. The first one is based on a local approximation of the conditional expectation, using the nearest neighbor method; the second relies on global approximation criteria using series expansion techniques.

Robinson (1991) proved that the optimal instruments can be estimated using a (not necessarily random) sampling without replacement from the empirical distribution of the residuals of a preliminary consistent estimation. Therefore it seems worthwhile to rely on a semiparametric specification of models containing risk terms (the ARCH-M model in this particular case) to allow for a flexible form in modeling the risk-premium. Since

$$E(\mathbf{u}(\mathbf{z}_{t+k}, \boldsymbol{\theta}) \mid \Psi_t) = \mathbf{0},$$

where **u** is the residual vector from (6), there will be a function of the information set $\mathbf{g}(\Psi_t)$ such that

$$E(\mathbf{u}(\mathbf{z}_{t+k}, \boldsymbol{\theta}) \mathbf{g}(\Psi_t)) = \mathbf{0}.$$

An optimal choice of instruments is given by

$$\mathbf{g}(\Psi_t) = (\hat{\boldsymbol{\Omega}}_t)^{-1} \hat{\mathbf{Q}}_t,$$

where

$$\mathbf{Q}_t = E\left(\frac{\partial \mathbf{u}(\mathbf{z}_{t+k}, \boldsymbol{\theta})}{\partial \boldsymbol{\theta}} \,\middle|\, \Psi_t \right),$$

$$\boldsymbol{\Omega}_t = E(\mathbf{u}(\mathbf{z}_{t+k}, \boldsymbol{\theta}) \mathbf{u}(\mathbf{z}_{t+k}, \boldsymbol{\theta})' \mid \Psi_t).$$

Both these conditional expectations[6] can be estimated in a nonparametric way. In particular, instead of the nearest-neighbor method suggested by Newey, we prefer the kernel method, using the Nadaraya–Watson kernel regression estimator in the leave-one-out version. This kind of regression estimator is based on a linear combination of the response variable with coefficients depending on a differentiable kernel function and a bandwidth parameter. Robinson (1983) proved consistency and asymptotic normality of such an estimator in a time-series context, under weak conditions, which are likely to be satisfied

---

[6] Note that they have subscript t for the information set on the basis of which they are calculated.

in our case.[7] Comparable results are not yet available for the nearest-neighbor method.

Let us consider now the specific heteroskedastic regression model, which contains a risk term among the regressors:

$$s_{t+k} - s_t = \text{RP}_{t,k} + \beta_1(f_{t,k} - s_t) + \epsilon_{t+k} + \sum_{j=1}^{k-1} \gamma_j \epsilon_{t+k-j},$$

$$\text{RP}_{t,k} = \beta_0 + \theta \sigma_{t+k}^2, \tag{7}$$

$$\sigma_{t+k}^2 = \text{Var}[\epsilon_{t+k} \mid \Psi_t] = g(\epsilon_t, \epsilon_{t-1}, \ldots, \epsilon_{t-p}).$$

The conditional variance σ_t^2 is a measure of the predictable component of market volatility, and is used to approximate the time-varying risk term.

Note that, upon substitution of the various terms in (7), the right-hand side contains, besides the forward premium, the time-$(t + k)$ disturbance, a linear function of the past disturbances, and a nonlinear function of the past disturbances, with only the last entering the risk term. Independently of the overlapping observations problem (where the linear MA term appears explicitly), the interpretation of the conditional variance as reflecting the level of uncertainty on the markets shows that in this model the past forecast errors exert their effects in a nonlinear way. Thus the model could be interpreted as a nonlinear MA model. Its specific form is left unspecified in our case to allow for the consideration of the various ways in which the relevant information is processed to forecast market volatility.

Following the semiparametric approach suggested in Pagan and Ullah (1988), an observed (or consistently estimated) series can be used instead of σ_t^2. It turns out that

$$s_{t+k} - s_t = \beta_0 + \theta \sigma_{t+k}^2 + \beta_1(f_{t,k} - s_t) + \epsilon_{t+k} + \sum_{j=1}^{k-1} \gamma_j \epsilon_{t+k-j} \tag{8}$$

---

[7] With the help of higher-order kernels (Bartlett 1963) and the device to trim out small density estimates, under suitable regularity conditions a reasonable conjecture is that

$$T^{-1/2}(\tilde{\theta} - \theta) \overset{a}{\sim} N(\mathbf{0}, \Phi^{-1}),$$

$$\tilde{\theta} = \bar{\theta} - \left(\sum_t \dot{\mathbf{z}}_t \mathbf{z}_t' \right)^{-1} \sum_t \dot{\mathbf{z}}_t \dot{\mathbf{u}}_t,$$

where $\bar{\theta}$ is a preliminary consistent estimate of θ, $\dot{\mathbf{z}}_t$ is the nonparametric regression of $\mathbf{z}_t$ on the information set, the $\dot{\mathbf{u}}_t$ are the residuals of that regression, and Φ^{-1} achieves the semiparametric efficiency bound (Chamberlain 1987).

becomes

$$s_{t+k} - s_t = \beta_0 + \theta \epsilon_{t+k}^2 + \beta_1 (f_{t,k} - s_t) + u_{t+k} + \sum_{j=1}^{k-1} \gamma_j \epsilon_{t+k-j},$$

$$u_{t+k} = \theta \left(\sigma_{t+k}^2 - \epsilon_{t+k}^2 \right) + \epsilon_{t+k},$$

where the error term and the new regressor are correlated. The instrumental variable estimation procedure is implemented here using three types of nonparametric instruments:

1. The estimated conditional variance is obtained as a nonparametric regression function of $\hat{\epsilon}_{t+k}^2$, given $\hat{\epsilon}_t, \hat{\epsilon}_{t-1}, \ldots, \hat{\epsilon}_{t-k+1}$, where $\hat{\epsilon}_t$ are the consistent residuals of a preliminary OLS parametric regression and $\hat{\epsilon}_t^2$ are used as ϕ_t.
2. The conditional expectation of $y_t = (s_{t+k} - s_t) - (f_{t,k} - s_t)$ given $\Psi_{t-1} = \{y_{t-1}, y_{t-2}, \ldots, y_{t-k+1}\}$ is computed; the squared nonparametric residuals are then used as ϕ_t, and the conditional variance is obtained as $E(y_t^2 \mid \Psi_{t-1}) - (E(y_t \mid \Psi_{t-1}))^2$. This way of proceeding corresponds to imposing $\beta_1 = 1$ in the model of risk determination, as in Pagan and Ullah (1988).
3. Without imposing $\beta_1 = 1$, the same procedure as in type 2 is carried out, calculating the nonparametric conditional expectation of $s_{t+k} - s_t$ given its lagged values and $f_{t,k} - s_t$ (Pagan and Hong 1991).

4.2   A comparison among estimators

The purpose of this subsection is to evaluate the importance that the risk term has in the spot–forward relationship, and to outline the differences in the various nonparametric estimation methods, deferring to the following section the issues related to the possible asymmetries of risk-related effects.

We have considered weekly spot and 30-day forward rates from June 1973 to February 1994 relative to five currencies: the French franc (FF), the Italian lira (ItL), the Japanese yen (JY), the British pound (BP), and the Deutsche mark (DM), all against the US dollar. The data are 12 noon bid (spot) and ask (forward) prices[8] from the New York Foreign Exchange Market for a total of 1077 observations. We have decided to use the definition of the exchange-rate as foreign currency per one US dollar for all currencies (there including the BP for comparison's sake).

---

[8] Alan Kirman has pointed out to us that these quotes, although widely used, are not necessarily equilibrium prices and might yield a higher measured volatility of returns than the actual ones on the markets.

Table 1. *Estimation with unrestricted MA(4)*

Exch. rate	Constant $\times 10^3$	$f_{t,k} - s_t$	AC (12)	HS	ARCH (4)
FF/\$	0.6981	0.3466	106.9	0.6	37.5
	(5.4220)	(0.0293)	[0.000]	[0.438]	[0.000]
ItL/\$	3.2135	0.2970	136.2	9.5	71.6
	(0.6145)	(0.0551)	[0.000]	[0.002]	[0.000]
JY/\$	−4.2055	0.0894	56.1	1.7	43.3
	(0.4130)	(0.0470)	[0.000]	[0.192]	[0.000]
BP/\$	2.4703	0.0234	19.5	14.8	27.9
	(0.4992)	(0.0502)	[0.077]	[0.000]	[0.000]
DM/\$	−1.0471	0.4234	151.4	1.4	50.8
	(0.5520)	(0.0606)	[0.000]	[0.236]	[0.000]

Standard errors of coefficients in parentheses; p-values for tests in square brackets.
AC(12): Ljung–Box test for autocorrelation (χ^2_{12}).
HS: White test for heteroskedasticity, χ^2_l, $l = p(p + 1)/2$ (p = number of regressors).
ARCH(4): Test for ARCH(4) effect (χ^2_4).

As customary, the forward rates taken on Tuesday refer to the spot rates four weeks and two days later on Thursday (referred to as f_t and s_{t+k}). An MA(4) term is inserted in the conditional-mean equation, but, contrary to Baillie and Bollerslev (1990), we do not impose the values found under the hypothesis that the (continuous-time) data-generating process for the spot rate is a Brownian motion.[9] In what follows, the order of the ARCH process is assumed equal to four throughout.

In Table 1 the results of a preliminary maximum-likelihood estimation in the absence of a risk term are reported to be used as a benchmark to show the departure from the efficiency hypothesis. Note that some of the coefficients on the forward term are not significant, and that the joint hypothesis $\beta_0 = 0$ and $\beta_1 = 1$ can be rejected. The diagnostics on autocorrelation, heteroskedasticity, and ARCH confirm earlier findings on the model's inadequacy.

The subsequent Tables (2–4) are devoted to the presentation of the results of instrumental-variable estimation using the different methods of estimating σ_t^2, previously outlined. The coefficient of the risk term can be interpreted as reflecting the marginal impact of volatility on the currency, or the degree of the prevailing marginal risk aversion toward the currency: if positive, an increase in volatility would push toward an appreciation of the US dollar; if negative, the opposite would apply.

---

[9] In fact, the empirical evidence (not shown but available upon request) suggests that the values for the MA coefficients are quite far from the values implied by Baillie and Bollerslev's model.

Table 2. *IV MA(4): instruments chosen according to our method*

Exch. rate	Constant $\times 10^3$	$f_{t,k} - s_t$	$\hat{\epsilon}_t^2$	AC (12)	HS	ARCH (4)
FF/$	3.8603	0.3228	−2.5532	28.0	9.7	12.4
	(0.8308)	(0.0566)	(0.4888)	[0.005]	[0.021]	[0.014]
ItL/$	−1.7363	0.4118	3.6192	24.2	8.5	10.7
	(0.7604)	(0.0521)	(0.3588)	[0.019]	[0.036]	[0.039]
JY/$	1.7736	0.4163	−4.8256	24.7	9.9	12.3
	(0.7256)	(0.0568)	(0.4004)	[0.016]	[0.019]	[0.015]
BP/$	−4.2734	0.3710	5.2676	83.1	10.8	30.8
	(1.6649)	(0.0761)	(1.1232)	[0.000]	[0.012]	[0.000]
DM/$	0.5945	0.2945	−1.2584	22.8	11.0	13.1
	(0.7720)	(0.0541)	(0.4524)	[0.029]	[0.011]	[0.010]

Standard errors of coefficients in parentheses; p-values for tests in square brackets.
AC(12): Ljung–Box test for autocorrelation (χ_{12}^2).
HS: White test for heteroskedasticity, χ_l^2, $l = p(p + 1)/2$ (p = number of regressors).
ARCH(4): Test for ARCH(4) effect (χ_4^2).

Although our suggestion relies on a principle of optimal choice of the instruments,[10] we can comment on the economic interpretation of the results for the various currencies. In fact, method 1 detects the presence of a significant effect of the risk term for all currencies, whereas the other two methods produce more mixed results, with a change reversal for the French franc and the Deutsche mark and nonsignificant effects for the French franc and the Italian lira for method 3.

The analysis was repeated by adding a term $(f_{t,k} - s_t)^2$ in order to check for possible nonlinear effects captured by the squared forward premium. The only currencies for which the addition was relevant were the French franc, for Pagan and Ullah's method, and the British pound, for our method. The changes in the other coefficients were not such as to change the sign of the risk term or the significance of the estimates.

The impact of volatility as captured by the measured risk term shows that there is an alternation of positive and negative values, as discussed by Stockman (1978) and Domowitz and Hakkio (1985). In fact, the French franc, the yen, and the mark have a positive constant and a negative slope coefficient, while the lira and the pound show reversed signs. The range within which the risk-related term varies shows different values for the various currencies, stressing how differently the various currencies are affected by time-varying

[10] As suggested by a referee, a Monte Carlo experiment would provide better guidelines as to which estimator performs better. However, note that this is not a straightforward exercise, due to the difficulty in devising an "impartial" data-generating process, and therefore it is not performed here.

Table 3. *IV MA(4): instruments chosen according to Pagan and Ullah (1988)*

Exch. rate	Constant $\times 10^3$	$f_{t,k} - s_t$	$\hat{\epsilon}_t^2$	AC (12)	HS	ARCH (4)
FF/$	2.8940	0.4750	−5.5120	35.1	10.8	13.0
	(1.5299)	(0.0672)	(2.7040)	[0.000]	[0.012]	[0.011]
ItL/$	−1.8841	0.4642	8.6116	29.7	12.4	10.2
	(1.1298)	(0.0628)	(1.8824)	[0.003]	[0.006]	[0.040]
JY/$	0.1277	0.4740	−11.2840	31.0	11.9	9.9
	(1.9886)	(0.0692)	(4.7892)	[0.001]	[0.007]	[0.042]
BP/$	−2.2013	0.3618	7.0252	89.7	14.3	22.0
	(0.9758)	(0.0643)	(1.4040)	[0.000]	[0.002]	[0.000]
DM/$	−6.8065	0.4632	12.8700	34.6	14.0	17.1
	(3.2817)	(0.0765)	(6.5624)	[0.000]	[0.002]	[0.001]

Standard errors of coefficients in parentheses; p-values for tests in square brackets.
AC(12): Ljung–Box test for autocorrelation (χ_{12}^2).
HS: White test for heteroskedasticity, χ_l^2, $l = p(p + 1)/2$ ($p =$ number of regressors).
ARCH(4): Test for ARCH(4) effect (χ_4^2).

volatility. In fact, on the basis of our estimates, the same three currencies that have a negative slope coefficient on the risk term exhibit a moderate positive impact (0.17% for the yen, 0.38% for the franc, and 0.59% for the DM), and a wider range on the negative side (−1.74% for the DM, −4.30% for the franc, and −8.47% for the yen). The two "weaker" currencies are the lira (ranging from −0.17% to 13.26%) and the pound (ranging from −0.42% to 16.86%), for which the impact of volatility on the positive side reached quite strong levels in the direction of their depreciation vis-à-vis the US dollar.

The overall residual diagnostics show that there is still some information left in the residuals; normality tests (not reported) reject the null in all cases. Therefore, we need to look for alternative specifications which might capture this extra structure contained in the residuals.

The question we turn to now is whether it is possible to discern more recognizable patterns in the behavior of the risk term using a classification of regimes under which the agents' reactions can be expected to be different.

5   Asymmetries in risk effects

As noted in Section 2, if interest parity theory were to hold, market operators should be indifferent between purchasing a forward contract and purchasing foreign currency right away. However, it is often remarked that a common rule when facing a need for currency at a future date (see, for example, Froot and

Table 4. *IV MA(4): instruments chosen according to Pagan and Hong (1991)*

Exch. rate	Constant $\times 10^3$	$f_{t,k} - s_t$	$\hat{\epsilon}_t^2$	AC (12)	HS	ARCH (4)
FF/$	1.4540	0.1084	−0.1508	37.0	11.2	13.4
	(1.0011)	(0.0497)	(1.1232)	[0.000]	[0.010]	[0.011]
ItL/$	3.5500	0.1405	0.8372	28.9	15.6	10.8
	(0.7981)	(0.0483)	(0.8060)	[0.004]	[0.001]	[0.001]
JY/$	−0.2120	0.4863	−6.6404	30.7	13.2	18.1
	(1.6049)	(0.0630)	(2.4336)	[0.004]	[0.004]	[0.004]
BP/$	−1.6873	0.3045	4.3940	79.4	12.0	20.2
	(1.4067)	(0.0633)	(1.4612)	[0.000]	[0.007]	[0.000]
DM/$	−5.3237	0.3926	5.8396	25.5	11.8	12.6
	(1.9147)	(0.0684)	(2.3764)	[0.012]	[0.008]	[0.013]

Standard errors of coefficients in parentheses; p-values for tests in square brackets.
AC(12): Ljung–Box test for autocorrelation (χ_{12}^2).
HS: White test for heteroskedasticity, χ_l^2, $l = p(p + 1)/2$ (p = number of regressors).
ARCH(4): Test for ARCH(4) effect (χ_4^2).

Thaler 1990) is to invest the money where the interest rate is higher. This, like other trading rules examined below, seems to provide a profitable outcome (cf. LeBaron 1993a, 1993b), in apparent contradiction to the theory.

Technical analysis is receiving considerable attention in the academic literature, as the focus shifts from a representative individual to heterogeneity of beliefs in the markets. The results reported by Taylor and Allen (1992), for example, show that there is a high proportion of traders relying on technical analysis to determine their position on the markets. The main focus of recent research is the expectation formation process and the possibility of expectational errors, or of fads as the results of mutual influence by participants in the markets (Lehmann 1990, Kirman 1993). In particular, the signals hitting the markets require interpretation and translation into actions which, in turn, will affect the exchange rate movements.

As noted before, the risk-premium argument for reconciling the untenability of the unbiasedness hypothesis relies on inflation expectations, as the term derived from the Lucas model involves expectations on relative real returns as contributing to the asset price. Higher interest rates may mean expectations of higher inflation and hence of a loss of purchasing power, but may also reflect the result of a strong antiinflationary stance by the monetary authority. The strength and weakness of the US dollar during the early 1980s occurred in a situation of constantly higher US dollar interest rates, and thus a switch in expectation formation must not have surfaced in the forward premium. In this respect, the uncertainty is actually about future monetary policy and the way the monetary authorities will react to nominal or real shocks.

In our model the only fundamentals considered are the interest-rate differentials, and so what enters the information set is the filtered outcome of the combined effect of interventions (which build reputation) and of expectations (which reflect that reputation). However, given our heterogeneous worldview, we assume that, in forming expectations, traders influence one another through their interactions. The clustering of volatility observed in the exchange-rate returns can be interpreted as the outcome of contrasting beliefs, since the $RP_{t,k}$ term is measured as a constant plus a time-varying portion which measures the effect of an increase of one percentage point in volatility on exchange-rate movements. If disparate beliefs are present among agents, their effects may be exerted differently according to the specific situation on the market.

Clearly, the question is more complex than the mere assessment of the presence or absence of a risk-premium term in the spot–forward relationship, and this is not only due to the difficulties of interpreting the time-varying term as the risk-premium in theoretical models. The empirical interest in the present analysis is, rather, focused on the presence of uncertainty and on the perception of risk, and hence on the (possibly nonlinear) effects that risk has on exchange-rate determination. In order to extract these effects, we characterize various market situations on the basis of signals referred to by technical analysts as *buy*, *sell*, and *hold the position* and which give rise to actions when meshed with the agents' perceptions of the market trends. The various trading rules need not provide the same signal, as we will also see from the empirical results: in fact, these (at times contradictory) signals received by the agents have to be accompanied by a process of further information gathering where reputation about the strength and weakness of a given currency also plays a role.

Another way of justifying the importance of a currency's reputation derives from analysis of the impact that the conditional volatility has in correspondence with a buy or sell signal. If this perception were irrelevant, the risk-related term would have the same impact irrespective of the nature of the signal. To perform this evaluation we apply the previous analysis defining various regimes in accordance to three trading rules[11]:

1. The first rule selects regime 1 when the foreign interest rate is higher than the numeraire (in our case the US dollar). It provides a buy signal for the higher-interest-rate currency. Neglecting, for simplicity's sake, the rare instances when the interest-rate differential is exactly zero,

---

[11] For all rules, a band of neutrality can be built, implying stronger signals before they are considered as impetus to action. Also, other existing trading rules can be applied, or others may be devised, based on whether the exchange rate is above or below the PPP level or on current account levels.

expression (8) is modified as

$$
\begin{aligned}
s_{t+k} - s_t = {} & \beta_0 + \beta_0^1 D_{1t} + \theta \sigma_{t+k}^2 + \theta^1 \sigma_{t+k}^2 D_{1t} \\
& + \beta_1 (f_{t,k} - s_t) + \beta_1^1 (f_{t,k} - s_t) D_{1t}
\end{aligned}
$$

plus the MA error term. $D_{1t} = 1$ characterizes the periods when $i_t^* > 0$, i.e., domestic rates are higher than the US dollar. Under the null hypothesis of symmetry, of course, $\beta_0^1 = \theta^1 = beta_1^1 = 0$. Also, if there is a mere switch of sign, but the effect stays the same, we should have that $\theta + \theta^1 = -\theta$.

2. The second rule is based on the comparison between short- and long-term moving averages of exchange rates s_t, and selects as belonging to regime 1 the periods characterized by a short-term moving average that is higher than that found in the long term. Usually the short term contains just the observation itself, and the long term is chosen here to contain 10 observations. Its occurrence is interpreted as an unusual depreciation of the currency relative to the US dollar and hence as a buy signal for the US dollar. Also in this case the regimes are considered as mutually exclusive. The same interpretation for the coefficients follows.

3. The third rule is based on short- and long-term moving variances of the exchange rate returns, defined respectively in our case as

$$
\mathrm{MVS}_t = \frac{1}{10} \sum_{j=0}^{9} (s_{t-j} - f_{t-j-1})^2
$$

and

$$
\mathrm{MVL}_t = \frac{1}{100} \sum_{j=0}^{99} (s_{t-j} - f_{t-j-1})^2.
$$

Action is called for when $\mathrm{MVS}_t < \mathrm{MVL}_t$, i.e., the short-term volatility is lower than the long-term one. Regime 1 is characterized by periods when the previous return was positive (hence appreciation of the US dollar), and regime 2 by periods when the previous return was negative. In this case we will have a base period assumed as neutral, and the two regimes:

$$
\begin{aligned}
s_{t+k} - s_t = {} & \beta_0 + \beta_0^1 D_{1t} + \beta_0^2 D_{2t} + \theta \sigma_{t+k}^2 \\
& + \theta^1 \sigma_{t+k}^2 D_{1t} + \theta^2 \sigma_{t+k}^2 D_{2t} + \beta_1 (f_{t,k} - s_t) \\
& + \beta_1^1 (f_{t,k} - s_t) D_{1t} + \beta_1^2 (f_{t,k} - s_t) D_{2t}
\end{aligned}
$$

plus the MA error term. $D_{1t} = 1$ characterizes regime 1, and $D_{2t} = 1$ characterizes regime 2.

Table 5. *French franc/US dollar: analysis by regimes, model with constrained* $f_{t,k} - s_t$

Quantity	i_t^*	MA	MV
Constant $\times 10^3$	4.8140	-0.7203	3.6193
	(1.5250)	(1.1384)	(1.6753)
Dummy (1) $\times 10^3$	-2.9969^a	5.8383^a	2.2104
	(1.9594)	(1.5255)	(3.0209)
Dummy (2) $\times 10^3$			2.5710
			(2.8750)
$f_{t,k} - s_t$	0.1942	0.1759	0.4423
	(0.0761)	(0.0481)	(0.0752)
Risk term	-0.7067	-13.0732	-0.6893
	(0.8561)	(0.6903)	(0.7277)
Risk term (1)	3.5971^a	22.1599^a	0.3892^a
	(1.0129)	(0.8422)	(3.1237)
Risk term (2)			-15.9962^b
			(2.5787)
Fraction spent in (1)	0.6729	0.5216	0.2833
Fraction spent in (2)	0.3215	0.4784	0.2782
Switching	411	159	133
AC(12)	21.5 [0.043]	20.1 [0.065]	19.0 [0.088]
HS	25.7 [0.011]	23.1 [0.026]	42.9 [0.004]
ARCH(4)	9.1 [0.058]	8.9 [0.063]	8.4 [0.077]

Standard errors of coefficients in parentheses; p-values for tests in square brackets.
[a] Sum of two regime coefficients significantly different from zero at 5% significance level.
[b] Two regime coefficients significantly different from each other at 5% significance level.

6   The empirical evidence

The main empirical results derive from the estimation of the spot–forward relationship, including in the analysis appropriate dummies corresponding to the regimes. Tables 5 to 9 contain the estimated coefficients with the appropriate standard errors for each currency. We omit the results for the MA coefficients, as they are of no interest in this context, despite their being highly significant. Since a differentiation of the coefficient on the forward premium across regimes turned out to be not significant, the reported estimates are obtained restricting the coefficient on the forward premium to be the same.

The results for overall significance of the discrimination across regimes on the risk term show that the interest-rate differential rule is uninteresting. In fact,

Giampiero M. Gallo & Barbara Pacini

Table 6. *Italian lira/US dollar: analysis by regimes, model with constrained $f_{t,k} - s_t$*

Quantity	i_t^*		MA		MV	
Constant $\times 10^3$	−2.8174		−2.9156		1.0012	
	(1.4612)		(1.2354)		(1.5071)	
Dummy (1) $\times 10^3$	2.3620[a]		8.7736		1.1478[a]	
	(1.9025)		(1.6130)		(2.5759)	
Dummy (2) $\times 10^3$					5.3625	
					(2.2358)	
$f_{t,k} - s_t$	0.3824		0.3407		0.3750	
	(0.0690)		(0.0536)		(0.0628)	
Risk term	3.8312		−11.4842		1.7138	
	(0.6431)		(0.8369)		(0.5205)	
Risk term (1)	−0.9276[a]		19.4250[a]		5.8428[a]	
	(0.7707)		(0.9378)		(2.1581)	
Risk term (2)					−18.0806[b]	
					(2.1968)	
Fraction spent in (1)	0.7596		0.5461		0.2669	
Fraction spent in (2)	0.1976		0.4539		0.3069	
Switching	322		157		127	
AC(12)	17.9	[0.118]	18.0	[0.115]	17.5	[0.131]
HS	15.8	[0.200]	19.9	[0.069]	30.0	[0.118]
ARCH(4)	8.7	[0.069]	7.5	[0.111]	7.6	[0.107]

Standard errors of coefficients in parentheses; p-values for tests in square brackets.
[a] Sum of two regime coefficients significantly different from zero at 5% significance level.
[b] Two regime coefficients significantly different from each other at 5% significance level.

only the French franc exhibits a slope different from zero in regime 1, while all the other currencies would see the two regimes as not distinguishable from one another. This is not surprising, given the large number of switches from one regime to another (which would imply changing position and incurring transaction costs) and the fact that, as previously noted, an interest-rate differential of a certain sign can be consistent with both an appreciation and a depreciation of the foreign currency. The other two regimes are significant on the basis of a joint test on the coefficients.

According to the results (examined by currency), the sign reversal in the measure of the impact of the conditional volatility on the exchange rate returns is by no means preserved. For example, for the French franc in regime 1 it has a purely positive impact for the MA and MV rules, and in regime 2 a purely

Table 7. *Japanese yen/US dollar: analysis by regimes, model with constrained $f_{t,k} - s_t$*

Quantity	i_t^*		MA		MV	
Constant $\times 10^3$	1.1057		−1.9504		2.3773	
	(1.0855)		(0.9756)		(1.4568)	
Dummy (1) $\times 10^3$	1.9905[a]		1.3849		−0.9250	
	(1.7819)		(1.5635)		(2.2816)	
Dummy (2) $\times 10^3$					−1.8855	
					(2.3542)	
$f_{t,k} - s_t$	0.3987		−0.0615		0.3255	
	(0.0789)		(0.0501)		(0.0731)	
Risk term	−4.4426		−11.5259		−6.7304	
	(0.5038)		(0.3949)		(0.5585)	
Risk term (1)	−1.2349[a]		26.0993[a]		13.8318[a]	
	(0.8406)		(0.9721)		(2.1276)	
Risk term (2)					−4.5727[b]	
					(1.4772)	
Fraction spent in (1)	0.4287		0.4934		0.3336	
Fraction spent in (2)	0.5555		0.5066		0.2710	
Switching	438		166		137	
AC(12)	18.7	[0.096]	14.8	[0.252]	14.2	[0.288]
HS	22.1	[0.036]	21.5	[0.043]	33.7	[0.052]
ARCH(4)	7.6	[0.107]	4.5	[0.342]	3.4	[0.493]

Standard errors of coefficients in parentheses; p-values for tests in square brackets.
[a] Sum of two regime coefficients significantly different from zero at 5% significance level.
[b] Two regime coefficients significantly different from each other at 5% significance level.

negative one for the MA. Also, the size of the impact changes remarkably. For the Italian lira the maximum impact of the volatility in the so-called neutral period of the MV rule increases to 15.41% (from 13.26% derived from Table 2), but the minimum goes from −0.17% to −31.3% in regime 2. Similar occurrences appear for the other currencies as well and will be pointed out below when the regimes are analyzed in greater detail. In general we can rule out symmetry when we observe a significance of the regime coefficients, but also, most importantly, when the sum of the two regime coefficients is significantly different from zero, implying that the same conditional volatility would have a different impact on the foreign-exchange returns.

The coefficient on the forward premium changes considerably relative to the constrained counterpart under the absence of regimes (Table 2). In fact, it

Table 8. *British pound/US dollar: analysis by regimes,*
model with constrained $f_{t,k} - s_t$

Quantity	i_t^*		MA		MV	
Constant $\times 10^3$	−3.8009		−7.7332		−5.4812	
	(2.4331)		(2.3716)		(3.8470)	
Dummy (1) $\times 10^3$	−1.0064[a]		6.9898[a]		6.3644	
	(3.5002)		(2.6424)		(5.2258)	
Dummy (2) $\times 10^3$					3.0099	
					(4.5187)	
$f_{t,k} - s_t$	0.2901		0.3353		0.4859	
	(0.1026)		(0.0636)		(0.0812)	
Risk term	4.0510		0.4342		4.5824	
	(1.2921)		(1.8605)		(1.6714)	
Risk term (1)	2.5244[a]		7.4856[a]		2.4530	
	(2.2382)		(1.9600)		(3.9958)	
Risk term (2)					−11.5195[b]	
					(3.4154)	
Fraction spent in (1)	0.6636		0.5160		0.2792	
Fraction spent in (2)	0.3346		0.4840		0.3244	
Switching	431		173		129	
AC(12)	20.9	[0.065]	19.4	[0.087]	17.8	[0.121]
HS	24.5	[0.017]	22.3	[0.034]	36.4	[0.027]
ARCH(4)	9.5	[0.049]	8.8	[0.066]	8.2	[0.084]

Standard errors of coefficients in parentheses; p-values for tests in square brackets.
[a] Sum of two regime coefficients significantly different from zero at 5% significance level.
[b] Two regime coefficients significantly different from each other at 5% significance level.

decreases (to the point of being negative, although insignificant for the yen) under rule MA, with more mixed behavior under rule MV.

Note that the residual diagnostics signal an improvement of the specification, with occasional problems occurring for some currencies. Normality of the residuals is still rejected for all currencies.

The comparative analysis by rule is best illustrated by depicting graphically the joint range of impact of the risk-related term on the foreign-exchange returns, where each axis corresponds to a regime. This analysis ignores the neutral period for the MV rule, and clearly there is no temporal correspondence between the minima and the maxima in each regime. Also, the usual word of caution applies regarding the unequal scales, between axes in some pictures.

Each box corresponds to a different currency, with the sides representing the range between the minimum and the maximum impact of the conditional

Table 9. *Deutsche mark/US dollar: analysis by regimes, model with constrained* $f_{t,k} - s_t$

Quantity	i_t^*		MA		MV	
Constant $\times 10^3$	1.2900		−4.2945		7.0817	
	(1.1535)		(0.9914)		(1.5231)	
Dummy (1) $\times 10^3$	−1.2173		6.9466		−9.9169[a]	
	(1.9060)		(1.4533)		(2.6262)	
Dummy (2) $\times 10^3$					−8.5539[b]	
					(2.4080)	
$f_{t,k} - s_t$	0.3363		0.3348		0.3408	
	(0.0781)		(0.0492)		(0.0652)	
Risk term	−1.4016		−9.0723		−3.2556	
	(0.5789)		(0.5740)		(0.6508)	
Risk term (1)	0.1101[a]		19.0692		12.0268[a]	
	(0.9323)		(0.7760)		(2.7309)	
Risk term (2)					−4.3846[b]	
					(1.7185)	
Fraction spent in (1)	0.4315		0.4859		0.2720	
Fraction spent in (2)	0.5610		0.5141		0.2823	
Switching	479		149		124	
AC(12)	20.7	[0.065]	17.2	[0.148]	13.0	[0.369]
HS	24.1	[0.019]	19.6	[0.075]	26.9	[0.215]
ARCH(4)	9.1	[0.058]	2.4	[0.662]	5.1	[0.277]

Standard errors of coefficients in parentheses; p-values for tests in square brackets.
[a] Sum of two regime coefficients significantly different from zero at 5% significance level.
[b] Two regime coefficients significantly different from each other at 5% significance level.

volatility. Ruling out symmetry on the basis of the analysis of the coefficients implies that the constants (minimum impact) and the slopes (marginal impact) are different across regimes. In practice, the analysis must be complemented by the inspection of the various situations, once the actual values of the conditional volatility are considered. In fact, it is possible that the higher coefficients (in absolute value) are associated with lower values of conditional volatility, thus reequilibrating the overall effect.

We start by commenting on Figure 1, where the interest-rate differential rule is reported. Recall that for this rule regime coefficients do not achieve statistical significance. In fact, the picture shows a clustering of the various currencies around the origin, with different shapes and sizes, the smallest corresponding to the French franc and the largest to the pound.

More interestingly, Figure 2, corresponding to the MA rule, shows that the currencies all belong to the second quadrant, i.e., there is a positive impact of

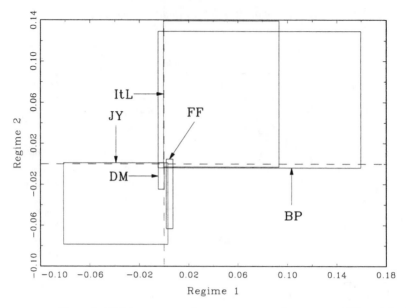

Figure 1. Risk impact on exchange-rate returns: interest-rate differential rule.

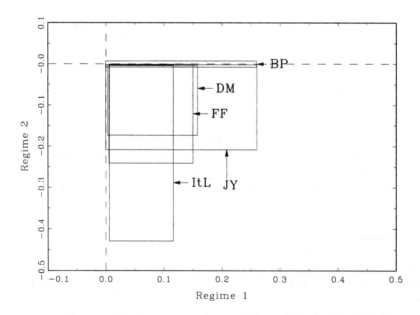

Figure 2. Risk impact on exchange-rate returns: moving-average rule.

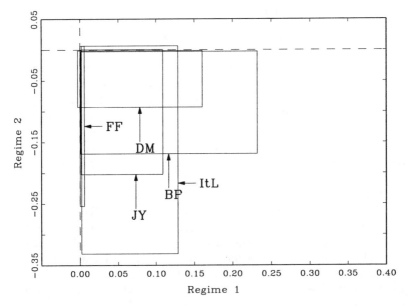

Figure 3. Risk impact on exchange-rate returns: moving-variance rule.

volatility (i.e., toward depreciation of the foreign currency) when the signals relate to selling the currency and buying dollars (short-term MA above long-term
MA), and a negative impact of volatility for the other regime. Note, however,
that the impact of the volatility is not symmetric, the clearest case in point being
the pound, for which the volatility impact in regime 2 is just a fraction of the
impact in regime 1. Note how similar the impacts are for the yen and the pound
in regime 1, and how different they are in regime 2. However, the ranges of
impacts for the franc and the mark are fairly similar in the two regimes, while
for the Italian lira the appreciating impact of the volatility in regime 2 is much
higher (in absolute value) than the depreciating one in regime 1.

The MV rule exhibits (Figure 3) a similar range of outcomes, in that also
in this case all currencies belong to the second quadrant, with minuscule sign
reversals for the franc and the mark (regime 1) and the lira (regime 2). The
striking feature of the results is the wide response range across currencies. The
lira and the pound have the largest effects in the two regimes, although it occurs
for different regimes. The impact for the franc is small in regime 1 due to an
insignificant slope coefficient, but it is quite high in regime 2. We interpret this
as suggesting that in periods of low volatility (characterizing the MV rule) the
signals leading to regime 1 (previous forward excess return greater than 0, that
is, a surprise depreciation) are received less clearly (hence with smaller impact)
than those leading to regime 2.

7   Concluding remarks

In line with several studies on the subject, the research question addressed in this chapter is to explain the failure of the unbiasedness hypothesis in the foreign-exchange market even when conditional volatility is inserted among the regressors. In the present analysis we suggest an instrumental-variable estimator which accounts for the unobservability of the risk-related term. The comparison of its performance with that of two other semiparametric estimators shows that our method appears to be more stable across countries and provides a better economic interpretation of the outcomes, pointing to the importance of the conditional volatility term. A simulation exercise would be required to fully evaluate the different properties of each method.

A more substantial question was whether the information set that is customarily used for testing the relevance of a risk-related term should be supplemented by elements which could highlight the different attitudes of agents with respect to market situations, or their different signal processing. We think it is natural to consider that asymmetries might exist in relation to economic and political factors, such as the reputation attached to the antiinflationary stances of the monetary authorities. While we do not propose a model in which reputation emerges in its effects on foreign-exchange determination, we select a number of market situations likely to determine different attitudes on the markets. These situations are characterized as signals based on the observed behavior of the exchange rates in their relation to interest-rate differentials, moving averages of levels, or moving variances of excess returns. Buy or sell signals from each rule may be contradictory and may not be received uniformly on the markets, due to a disparity of expectations. The reputation effect which we have in mind would then filter the signal in a nonhomogeneous way, giving rise to clusters of volatility. The analysis of the effect of the risk-related term on foreign-exchange returns then mirrors this disparity in translating the signals into action.

The results show that in the case of the interest-rate differential rule there is no significant change in the impact of the risk term with respect to regimes. This is not surprising given the coexistence of consistently positive (or negative) differentials and expectations of appreciation or depreciation of a currency. Although there is no difference between the regimes, the overall effect shows a sign reversal, implying the existence of a threshold beyond which a higher volatility produces an inversion of tendency. This inversion of tendency is absent for the moving-average rule, where there is a certain homogeneity of behavior in accordance with the definition of the regimes, but quite a difference in the measured impact across currencies. A more differentiated behavior of the risk-related term arises in the moving-variance rule, with the impact varying across regimes and across currencies.

Even if the unbiasedness hypothesis is still rejected, the evidence is in favor of the relevance of a risk-related term which behaves asymmetrically according to various situations on the markets. The analysis performed here used the US dollar as numeraire. Although previous studies have excluded the dependence of the results on such a choice, further insights into the nature of the risk-related term can be gained by applying the same methodology to the exchange-rate mechanism in the EMS using the DM as a reference, to investigate the evolution and the impact of the risk term relative to the position of the exchange rate in the band.

REFERENCES

Amemiya, T. (1974), "The Nonlinear Two-Stage Least-Squares Estimator", *Journal of Econometrics* 2: 105–10.

(1977), "The Maximum Likelihood and the Nonlinear Three-Stage Least-Squares Estimator in the General Nonlinear Simultaneous Equation Model", *Econometrica* 45: 955–68.

Backus, D. K., and A. W. Gregory (1993), "Theoretical Relations between Risk-Premiums and Conditional Variances", *Journal of Business and Economic Statistics* 11: 177–85.

Baillie, R. T., and T. Bollerslev (1990), "A Multivariate Generalized ARCH Approach to Modelling Risk Premia in Forward Foreign Exchange Rate Markets", *Journal of International Money and Finance* 9: 307–27.

Baillie, R. T., and P. McMahon (1989), *The Foreign Exchange Market. Theory and Econometric Evidence*. Cambridge: Cambridge University Press.

Bartlett, M. S. (1963), "Statistical Estimation of Density Functions", *Sankhya Series A* 25: 145–54.

Bollerslev, T. (1986), "Generalized Autoregressive Conditional Heteroskedasticity", *Journal of Econometrics* 31: 307–27.

(1990), "Modelling the Coherence in Short-Run Nominal Exchange Rates: A Multivariate Generalized ARCH Model", *Review of Economics and Statistics* 72: 498–505.

Bollerslev, T., R. Y. Chou, and K. F. Kroner (1992), "ARCH Modeling in Finance", *Journal of Econometrics* 52: 5–59.

Bollerslev, T., and R. Hodrick (1992), "Financial Market Efficiency Tests", Working Paper 4108, National Bureau of Economic Research.

Chamberlain, G. (1987), "Asymptotic Efficiency in Estimation with Conditional Moment Restrictions", *Journal of Econometrics* 34: 305–34.

Clark, P. K. (1973), "A Subordinated Stochastic Process Model with Finite Variance for Speculative Prices", *Econometrica* 41: 135–55.

Cumby, R. E. (1988), "Is It Risk?", *Journal of Monetary Economics* 22: 279–99.

Cumby, R. F., and M. Obstfeld (1981), "Exchange Rate Expectations and Nominal Interest Differentials: A Test of the Fisher Hypothesis", *Journal of Finance* 36: 697–704.

Delgado, M. (1992), "Computing Nonparametric Functional Estimates in Semiparametric Problems", Working Paper 92-17, Universidad Carlos III de Madrid, Madrid.

Diebold, F. X. (1986), "Modeling Persistence in Conditional Variances: A Comment", *Econometric Reviews* 5: 51–6.

Diebold, F. X., and J. A. Nason (1990), "Nonparametric Exchange Rate Prediction", *Journal of International Economics* 28: 333–47.

Domowitz, I., and C. S. Hakkio (1985), "Conditional Variance and the Risk-Premium in the Foreign Exchange Market", *Journal of International Economics* 19: 47–66.

Engle, R. F. (1982), "Autoregressive Conditional Heteroskedasticity with Estimates of the Variance of U.K. Inflation", *Econometrica* 50: 987–1008.

Engle, R. F., D. M. Lilien, and R. P. Robins (1987), "Estimating Time-Varying Risk Premia in the Term Structure: the ARCH-M Model", *Econometrica* 55: 391–407.

Fama, E. F. (1984), "Forward and Spot Exchange Rates," *Journal of Monetary Economics* 14: 319–338.

French, K. R., G. V. Schwert, and R. F. Stambaugh (1987), "Expected Stock Return and Volatility", *Journal of Financial Economics* 19: 3–29.

Froot, K. A., and R. H. Thaler (1990), "Anomalies: Foreign Exchange", *Journal of Economic Perspectives* 4: 179–92.

Gallant, R., D. Hsieh, and G. Tauchen (1991), "On Fitting a Recalcitrant Series: The Pound/Dollar Exchange Rate 1974–1983", in W. Barnett, J. Powell, and G. Tauchen (eds.), *Semiparametric and Nonparametric Methods in Econometrics and Statistics*, Cambridge: Cambridge University Press.

Grauer, F. L. A., R. H. Litzenberger, and R. E. Stehle (1976), "Sharing Rules and Equilibrium in an International Trade Capital Market under Uncertainty", *Journal of Financial Economics* 4: 233–56.

Hakkio, C. S. (1981), "The Term Structure of the Forward Premium," *Journal of Monetary Economics* 8: 41–58.

Hansen, L. P., and R. J. Hodrick (1983), "Risk Averse Speculation in the Forward Foreign Exchange Markets: An Econometric Analysis of Linear Models", in J. A. Frenkel (ed.), *Exchange Rates and International Macroeconomics*, Chicago: University of Chicago Press.

Hodrick, R., and S. Srivastava (1984), "An Investigation of Risk and Return in Forward Foreign Exchange", *Journal of International Money and Finance* 3: 5–30.

Hsieh, D. A. (1993), "Using Nonlinear Methods to Search for Risk Premia in Currency Futures", *Journal of International Economics* 35: 113–32.

Kirman, A. (1993), "Ants, Rationality and Recruitment", *Quarterly Journal of Economics* 108: 137–56.

LeBaron, B. (1993a), "Forecast Improvements Using a Volatility Index", in M. H. Pesaran and S. M. Potter (eds.), *Nonlinear Dynamics, Chaos and Econometrics*, New York: Wiley.

   (1993b), "Practical Comparisons of Foreign Exchange Forecasts", *Neural Network World* 6: 770–90.

Lehmann, B. (1990), "Fads, Martingales and Market Efficiency", *Quarterly Journal of Economics* 105: 1–28.

Levi, M., and J. Makin (1979), "Fisher, Phillips, Friedman and the Measured Impact of Inflation on Interest", *Journal of Finance* 34: 35–52.

Lewis, K. K. (1989), "Changing Beliefs and Systematic Rational Forecast Errors with Evidence from Foreign Exchange", *American Economic Review* 79: 621–36.

Lucas, R. (1982), "Interest Rates and Currency Prices in a Two-Country World", *Journal of Monetary Economics* 10: 335–60.

Mizrach, B. (1993), "Multivariate Nearest-Neighbor Forecast of EMS Exchange Rates", in M. Pesaran and S. Potter (eds.), *Nonlinear Dynamics, Chaos and Econometrics*, New York: Wiley.

Newey, W. K. (1990), "Efficient Instrumental Variables Estimation of Nonlinear Models", *Econometrica*, 58(4): 809–37.

Pagan, A. R. (1984), "Econometric Issues in the Analysis of Regressions with Generated Regressors", *International Economic Review* 25(1): 221–47.

Pagan, A. R., A. Hall, and P. Trivedi (1983), "Assessing the Variability of Inflation", *Review of Economic Studies* 50: 585–96.

Pagan, A. R., and Y. S. Hong (1991), "Nonparametric Estimation and the Risk-Premium", in W. Barnett, J. Powell, and G. Tauchen (eds.), *Semiparametric and Nonparametric Methods in Econometrics and Statistics*, Cambridge: Cambridge University Press.

Pagan, A. R., and G. W. Schwert (1990), "Alternative Models for Conditional Stock Volatility", *Journal of Econometrics* 45: 267–90.

Pagan, A. R., and A. Ullah (1988), "The Econometric Analysis of Models with Risk Terms", *Journal of Applied Econometrics* 3: 87–105.

Robinson, P. M. (1983), "Nonparametric Estimators for Time Series", *Journal of Time Series Analysis* 4(3): 185–207.

(1991), "Best Nonlinear Three-Stage Least Squares Estimation of Certain Econometric Models", *Econometrica* 59(3): 755–86.

Stockman, A. C. (1978), "Risk, Information and Forward Exchange Rates", in J. A. Frenkel and H. G. Johnson (eds.), *The Economics of Exchange Rates*, Reading, MA: Addison-Wesley, pp. 159–78.

Taylor, M. P., and H. Allen (1992), "The Use of Technical Analysis in the Foreign Exchange Market", *Journal of International Money and Finance*, 11: 304–14.

CHAPTER 4

Nonlinearity, structural breaks, or outliers in economic time series?

Gary Koop & Simon Potter

1 Introduction

Time series models are often used to predict or improve understanding of the dynamic properties of macroeconomic data. Parametric models are commonly used which assume a constant linear dynamic structure over time. For instance, much of modern macroeconomics, including the unit root literature, have been based on autoregressive or autoregressive moving average models. However, this assumption may be inappropriate. For example, US post-World War II macroeconomic series cover a time span which includes wars, rapid technical change, the breakdown of the Bretton Woods agreement, the change in the Fed's operating behavior under Paul Volcker, and numerous other shifts in monetary and fiscal policy.

All of these factors may imply that it is inappropriate to assume the same dynamic model exists for the 1950s, say, as for the 1980s. We refer to models where the dynamics change permanently in a way that cannot be predicted by the history of the series as *structural break* models. Alternatively, it is possible that dynamic properties can vary over the business cycle. For instance, it is likely that shocks to real output have different effects if the economy is operating below capacity (i.e., is in a recession), from those if the economy is in normal times or is overheating (i.e., is expanding rapidly). With some abuse of terminology, we refer to models which allow for dynamics which vary over the business

This paper has its motivation from discussions at the (EC)[2] conference in 1995. The views expressed in the paper are those of the authors and are not necessarily reflective of views at the Federal Reserve Bank of New York or the Federal Reserve System. Financial support for G. K. from the Social Sciences and Humanities Research Council of Canada is gratefully acknowledged, as are helpful discussions with Sid Chib and Les Oxley. Financial support for S. P. from the Academic Senate UCLA and Center for Computable Economics, UCLA, is gratefully acknowledged.

cycle in a predictable way as *nonlinear* models. Still another possibility is that apparent departures from linearity are due to unpredictable large shocks which have only temporary effects. We refer to models which have this property as *outlier* models.

Nonlinear models provide very different understanding of the effects of shocks over the business cycle and lead to forecasts depending on the state of economic activity. For example, as shown by Beaudry and Koop (1993), a simple nonlinear model implies that positive shocks to US output growth are more persistent than negative ones. In contrast, structural break models have no predictable changes in regime over the business cycle. Hence, a consequence of adopting them is that the current forecasting model is estimated only using data observed since the most recent break. Outlier models are useful for removing the influence of rare events from the estimates used for forecasting and analyzing dynamics.

Since they have very different consequences for forecasting and understanding business cycles, it is important to test whether nonlinearities, structural breaks, or outliers are present in economic time series, and if they are, to estimate models which incorporate them. A significant literature now exists which tests for structural breaks or nonlinearity in macroeconomic time series. However, this literature, with a few exceptions, takes a classical econometric approach and concentrates on only one of the three possible classes of model considered in this paper. For example, an important recent paper, Stock and Watson (1996), uses a battery of classical tests on 76 US postwar quarterly time series and finds significant evidence of structural instability in many cases. However, it is possible that this apparent structural instability is, in reality, a reflection of some form of nonlinearity. Hence, it is important not only to compare linear with structural break models, but also to compare structural break with nonlinear models and outlier models.

Such a comparison of many possibly non-nested models is theoretically challenging using classical techniques but theoretically straightforward using Bayesian approaches. Bayesian methods for choosing between models are based on the marginal likelihood associated with each model. The marginal likelihood is the average of the likelihood function across parameter values with respect to the prior distribution of the parameters. The comparison of multiple, possibly non-nested models can be accomplished by constructing posterior model probabilities directly from these marginal likelihoods.[1] The number or nature of the models under consideration does not affect the logic of the calculation. The two main drawbacks to such a Bayesian approach are the need to

---

[1] That is, under the assumption of equal prior probabilities for each model, one adds the marginal likelihood across all models under consideration and then divides the individual marginal likelihoods by the value of this sum.

specify informative priors and the difficulty in calculating the marginal likelihood. Before discussing these drawbacks it is useful to outline other advantages of the Bayesian approach for nonlinear, structural break, and outlier models.

1.1 Advantages of a Bayesian approach

In previous work (Koop and Potter 1999) we note that, with many nonlinear models, likelihood functions are nonsmooth and multimodal. Similar observations apply to structural break and outlier models. Bayesian methods, by using information from the entire parameter space, capture this finite sample uncertainty about the true parameter values. In contrast, standard classical maximum likelihood methods choose one point (i.e., the MLE in sample) in the parameter space and use a normal asymptotic approximation to capture the local uncertainty around this point. Uncertainty produced by multiple peaks in the likelihood function is ignored.

In addition, posterior model probabilities can be used to combine dynamic features of different models. For example, instead of "accepting" or "rejecting" the linear model one can include its dynamics weighted by the posterior probability for linearity. This means that forecasts and impulse response functions will be more reflective of the underlying model and parameter uncertainty when based on Bayesian methods of analysis.

Another issue is the choice of lag length. It is standard to choose a particular lag length at which to conduct the analysis by using information criteria or by testing the significance of additional lags. This leads to issues of data mining and reduces the credibility of the results. Rather than picking a particular lag length to work with, Bayesian methods allow one to combine information from a range of lag lengths for each model.

In constructing classical statistical tests of the null hypothesis of a linear model versus the three alternatives, one runs into Davies' problem: nuisance parameters which are not identified under the null. This leads to the difficult classical statistical issues addressed in Andrews (1993) and Andrews and Ploberger (1994). Optimal classical solutions to Davies' problem involve integrating out nuisance parameters with respect to an *a priori* weighting function, and calculation of critical values for classical test statistics can be computationally demanding – involving complicated simulation methods (see Hansen 1996). Koop and Potter (1999) show how the presence of nuisance parameters that are unidentified under the null poses no problems for a Bayesian analysis, and the unidentified nuisance parameters can be integrated out using an *a posteriori* weighting function.

In the case of structural break and/or outlier models, Bayesian methods, unlike the classical approach, extend directly to making inferences regarding the possibility of multiple breaks or outliers.

1.2 *Drawbacks of the Bayesian approach*

A major drawback of using Bayesian methods is that they can be computation-ally very demanding. For instance, in the case of structural break models, they require specification of a parametric likelihood function before and after the breakpoints (see, e.g., Barry and Hartigan 1993, Carlin, Gelfand, and Smith 1992, and Stephens 1994). For many common specifications (e.g., ARMA models, regression models with AR errors, and models with non-normal errors), analytic posterior properties do not exist and Bayesian analysis requires the use of simulation-based methods such as the Gibbs sampler. This implies, if the breakpoint is unknown, that likelihood evaluations at every possible breakpoint are often required at every pass through the Gibbs sampler. If the breakpoint can occur at any time, the sample size is T, and the Gibbs sampler is run for S passes, then approximately ST likelihood evaluations must be made. For typical values of S and T (e.g., 10,000 and 200), the computational burden be-comes quite large. If two breakpoints are allowed for, the number of likelihood function evaluations becomes approximately ST^2. In addition, researchers are often interested in investigating many different models, further increasing the computational burden.

In the present paper, for one of our data sets, we work with six different lag lengths and allow for 0, 1, and 2 breakpoints. Furthermore, we consider nonlinear and outlier models with homoscedastic and heteroscedastic versions of most models. In total we estimate 66 models. In cases of this sort, any Bayesian procedure requiring extensive simulation is virtually impossible given the present level of computer technology. Similar arguments hold for many common nonlinear specifications (e.g., the Markov switching model; see Albert and Chib 1993). In practice, then, applied economists interested in a thorough data analysis involving a wide variety of nonlinear and structural break models will be interested in Bayesian methods which can be used analytically or at most require numerical integration over a low dimensional subset of the parameter space. This, of course, places restrictions on the types of models analyzed and the priors used.

This leads us to the second possible drawback of the Bayesian approach: the use of proper informative priors. It is well known in the context of nested hypothesis testing that improper noninformative priors over the parameters of interest lead to Bayes factors which always prefer the restricted model (see Poirier 1995, Section 9.10). Since the Bayes factor is the ratio of the marginal likelihood of the linear model to the marginal likelihood of the alternative model in our examples, the use of the standard uninformative priors will lead to all the posterior probability being in favor of the linear model. By continuity, relatively flat priors over the parameter space will thus tend to favor the restricted (i.e., linear) model. Koop and Potter (1999) argue that this is an attractive feature of the Bayesian analysis in this context, since it incorporates a strong reward for the relative parsimony of linear models. Although it is probably unrealistic to expect researchers to universally embrace the use of informative priors, we are confident that those used in this paper will be thought reasonable by a wide variety of readers. Furthermore, optimal classical tests also use a subjective weighting function (see Andrews and Ploberger 1994).

1.3 Two examples

We apply our methods to two commonly analyzed data sets. The first is a postwar quarterly US real GDP growth series. We use CITIBASE series GDP from 1954:1 to 1995:1 in 1987 prices. A number of papers have found evidence of nonlinearity in this data set (see Pesaran and Potter 1997 for a review), and there is considerable debate about the possibility of a structural break in the time series in the early 1970s (see Perron 1989). The second is a long annual UK industrial production growth series. This latter series runs from 1700 to 1992 and has been extensively investigated by economic historians (see Greasley and Oxley 1994 and Mills and Crafts 1996) who examine whether or not the industrial revolution was a distinct epoch (i.e., whether a structural break occurred in this series).

We work with a similar, relatively noninformative prior for both data sets. We give examples of how the priors could be altered in our discussion of the models and techniques required to calculate the marginal likelihood.

2 Tools for analyzing switching regime models

2.1 A simple class of switching regime models

Most of the nonlinear models commonly used by macroeconomists are based on autoregressive specifications (i.e., there are no moving average components) which vary across regimes or states. Prominent examples include Markov switching (Hamilton 1989) and threshold autoregressive (or TAR; see Potter 1995) models. In this paper we use the latter class of models, since computationally they are much simpler to deal with. Structural break models can also be interpreted as switching regime models where the regimes are defined by a fixed but unknown breakpoint in time. Similarly, in our specification outlier models have temporary regimes that occur for one period only.

These considerations give the following simple regime switching specification:

$$Y_t = \begin{cases} \phi_{10} + \phi_{1p}(L)Y_{t-1} + \sigma_1 V_t & \text{if} \quad I_t = 1, \\ \phi_{00} + \phi_{0p}(L)Y_{t-1} + \sigma_0 V_t & \text{if} \quad I_t = 0, \\ \phi_{20} + \phi_{2p}(L)Y_{t-1} + \sigma_2 V_t & \text{if} \quad I_t = 2, \end{cases}$$

where I_t is an indicator variable for the regimes and $\phi_{ip}(L)$ is a polynomial of order p in the lag operator.[2] V_t is assumed to be standard normal and independent over time. The Gaussianity assumption is strong, but it will allow us to obtain analytical results.

We consider four ways of defining I_t:

1. The linear Gaussian AR model is obtained if we set $I_t = 0$ for all t.
2. A three regime TAR model is obtained if we set $I_t = 1$ if $Y_{t-d} > r_1$, $I_t = 2$ if $Y_{t-d} < r_2$, and $I_t = 0$ if $r_2 \leq Y_{t-d} \leq r_1$. If $\phi_{20} = \phi_{00}$, $\phi_{0p}(L) = \phi_{2p}(L)$, and $\sigma_0 = \sigma_2$, then a two regime model is obtained.
3. The structural break model is obtained if we set $I_t = 0$ if $t < \tau_1$, $I_t = 1$ if $\tau_1 \leq t < \tau_2$, and $I_t = 2$ if $t \geq \tau_2$. Again, under suitable parameter restrictions it reduces to a single break model.
4. The outlier model is obtained if $I_t \neq 0$ for only two values of t and $\phi_{1p}(L) = \phi_{2p}(L) = \phi_{0p}(L)$, $\sigma_1 = \sigma_2 = \sigma_0$, but $\phi_{10} \neq \phi_{20} \neq \phi_{00}$.

Of course, there are many other ways of defining I_t, and hence the tools developed in this paper can be used in a much wider class of models. Note that these models are potentially heteroscedastic, since the error variance may differ across regimes. In the empirical section of this chapter, we also consider homoscedastic versions of the TAR and structural break models. That is, our homoscedastic models assume $\sigma_1 = \sigma_2 = \sigma_0$. For future reference, let $\phi_i = (\phi_{i0}, \phi_{ip}')'(i = 0, 1, 2)$, $\tau = [\tau_1, \tau_2]'$, and $\gamma = [r_1, r_2, d]'$.

In constructing a classical test for the presence of a structural break or outlier, the null hypothesis is the AR model and the parameter vector τ is unidentified. In the TAR, the parameter vector γ is unidentified under the AR null. These are the sources of the classical statistical problems discussed previously. Similarly, in testing the two regime models versus three regime models or single break versus two breaks or single outliers versus two outliers, r_2 or τ_2 are unidentified under the null hypothesis.

2.2   Analytical expressions for marginal likelihoods

The major advantage of these simple classes of regime switching models is that analytical expressions for the posterior distribution are available conditional on the parameter vectors γ or τ, if we use independent (across i) natural conjugate priors for ϕ_i, σ_i. Hence, it is feasible to carry out a thorough analysis of model choice (i.e., choice of I_t or even choice of heteroscedasticity vs. homoscedasticity), lag length, and prior sensitivity within the natural conjugate class. It is worth stressing that, with typical sample sizes and lag lengths, our approach

---

[2] For the sake of simplicity we assume the lag length is the same in each regime – an assumption that can be easily relaxed.

allows for such an analysis to be done in minutes of computer time on a Pentium PC, whereas other approaches (e.g., classical approaches using the optimal tests of Andrews and Ploberger 1994 and the simulation approach of Hansen 1996, or Bayesian analysis of Markov switching models as in Albert and Chib 1993) would take days or weeks of computer time.

Intuitively, the existence of analytical results is due to the fact that, conditional on knowing I_t, the regime switching model breaks down into three AR specifications. If we treat initial conditions as fixed, it is well known that the AR model can be analyzed in a similar manner to the standard linear regression model. Hence, analytical posterior results exist for parameters of each regime, ϕ_i, σ_i, if we work with natural conjugate priors for each regime. To obtain posterior results which are not conditional on I_t, we need to know the marginal posterior for the parameters which define I_t (i.e., τ for the structural break model and γ for the TAR). For example, in the case of the single break model τ_1 the possible values are the integers from 1 to T. Since these parameters are discrete, it is simple to calculate their marginal probability for each possible value. In the case of the TAR, r_1 and r_2 are continuous parameters, but their effect on the likelihood function is the same as if they were discrete. This is because one can set the possible values of r_1, r_2 to the observed data points. For example, suppose r_2 is chosen to be the 25th smallest observation. Then as we vary r_2 up to the value of the 26th smallest observation, the regime split between regime 0 and 2 remains the same. Thus the likelihood function is flat between these values and the posterior for (ϕ_2, σ_2) is the same across all these values for r_2.

Included in the class of priors for which analytical results are available are the improper priors: $p(\phi_i) \propto c_i, p(\sigma_i) \propto 1/\sigma_i$, where $c_i > 0$ is an arbitrary constant. As discussed above, these priors will lead to all the posterior probability being in favor of the linear model. Thus, we assume an informative prior of the form

$$p(\phi_0, \phi_1, \phi_2, \sigma_0, \sigma_1, \sigma_2 | \gamma) = p(\phi_0, \phi_1, \phi_2, \sigma_0, \sigma_1, \sigma_2 \mid \tau)$$

$$= \prod_{i=1}^{3} p(\phi_i \mid \sigma_i) p(\sigma_i),$$

with a normal-inverted gamma form for $p(\phi_i \mid \sigma_i) p(\sigma_i)$ (see, e.g., Judge, Griffiths, Hill, Lutkepohl, and Lee 1985, pp. 106–107 for further details about the normal-inverted gamma prior). Thus, conditional on σ_i^2 it is assumed that ϕ_i is multivariate normal with mean vector $\underline{\phi}$ and variance covariance matrix $\sigma_i^2 \underline{\mathbf{D}}$. The number of degrees of freedom of the inverted gamma distribution for the precision is $\underline{\nu}$, and the mean is $\underline{s}^{-2}$. It is assumed that the hyperparameters of the prior do not depend on γ or τ. This is a restriction that could be relaxed with little difficulty. The priors for γ and τ are discussed below.

Let us begin by considering the linear AR model. If we condition on p initial values of Y_t, then this model is identical to the standard normal regression model with a natural conjugate prior. The analytical expression for the marginal likelihood is well known (e.g., Judge, Griffiths, Hill, Lutkepohl, and Lee 1985, p. 129). The following sample information is required to calculate the marginal likelihood:

1. The sample size T.
2. The ordinary least squares estimates of the parameter vector ϕ:

$$\hat{\phi} = [X'X]^{-1}X'Y,$$

where $Y = [Y_{p+1}, \ldots, Y_{T+p}]'$ and

$$X = \begin{bmatrix} 1 & Y_p & \cdots & Y_1 \\ 1 & Y_{p+1} & \cdots & Y_2 \\ \vdots & \vdots & & \vdots \\ 1 & Y_{T+p-1} & \cdots & Y_T \end{bmatrix}.$$

3. The moment matrix $X'X$.
4. The sum of squared errors:

$$e^2 = (Y - X\hat{\phi})'(Y - X\hat{\phi}).$$

Combining the sample information with the prior information, the marginal likelihood is

$$\ell_{\text{LAR}}(Y) = \frac{\Gamma(\bar{\nu}/2)(\underline{\nu}\,\underline{s}^2)^{\underline{\nu}/2}}{\Gamma(\underline{\nu}/2)\pi^{T/2}} \frac{|\bar{D}|^{1/2}}{|\underline{D}|^{1/2}} (\bar{\nu}\bar{s}^2)^{-\bar{\nu}/2},$$

where:

1. One has

$$\bar{\nu} = \underline{\nu} + T.$$

2. One has

$$\bar{\nu}\bar{s}^2 = \underline{\nu}\,\underline{s}^2 + e^2 + (\bar{\phi} - \hat{\phi})'X'X(\bar{\phi} - \hat{\phi}) + (\bar{\phi} - \underline{\phi})'\underline{D}^{-1}(\bar{\phi} - \underline{\phi}),$$

where $\bar{\phi}$ is the posterior mean of ϕ and is given by

$$\bar{D}[\underline{D}^{-1}\underline{\phi} + X'y].$$

3. $\bar{D}$ combines the prior and sample information on the variance–covariance matrix of the lagged variables and is given by

$$\bar{D} = [\underline{D}^{-1} + X'X]^{-1}.$$

The homoscedastic structural break model conditional on τ and the homoscedastic TAR model conditional on γ can be written (using appropriate dummy variables) in the form of linear regression models. Furthermore, the heteroscedastic structural break model conditional on τ and the heteroscedastic TAR model conditional on γ both divide into three simple linear regression models, each of which has a known marginal likelihood. Hence, conditional on γ or τ, the marginal likelihood for these models can easily be calculated. The following sample information is required to calculate the marginal likelihood for the two regime model:

1. The sample size T_i for each of the regimes in the case of the heteroscedastic model.
2. The ordinary least squares estimates of the parameter vector ϕ_i:

$$\hat{\phi}_i = [X_i' X_i]^{-1} X_i' Y_i,$$

   where $Y_i = 1(I_t = i)Y$ and $X_i = 1(I_t = i)X$.
3. The moment matrices $X_i' X_i$.
4. The sum of squared errors within each regime:

$$e_i^2 = (Y_i - X_i \hat{\phi})'(Y_i - X_i \hat{\phi}).$$

5. In the case of the heteroscedastic model the three pieces of sample information are used to construct separate $\bar{\nu}_i \bar{s}_i^2$, $\Gamma(\bar{\nu}_i/2)$, and $|\bar{\mathbf{D}}_i|$ values. In the case of the homoscedastic model the information is combined in constructing the values of $\bar{\nu}\bar{s}^2$, $\Gamma(\bar{\nu}/2)$, and $|\bar{\mathbf{D}}|$.
6. Efficient algorithms can be constructed to calculate this sample information across all the possible splits of the data.

In the case of a two regime heteroscedastic model the marginal likelihood conditional on γ or τ is

$$\ell_{2\,\mathrm{REGIME}}(Y \mid \gamma) = \prod_{i=0}^{1} \frac{\Gamma(\bar{\nu}_i/2)(\underline{\nu}\,\underline{s}^2)^{\underline{\nu}/2}}{\Gamma(\underline{\nu}/2)\pi^{T_i/2}} \frac{|\bar{\mathbf{D}}_i|^{1/2}}{|\underline{\mathbf{D}}|^{1/2}} \left(\bar{\nu}_i \bar{s}_i^2\right)^{-\bar{\nu}_i/2}.$$

A comparison of the marginal likelihoods for the linear AR and the two regime heteroscedastic model shows that:

1. $\prod_{i=0}^{1}(\bar{\nu}_i \bar{s}_i^2)^{-\bar{\nu}_i/2}$ is likely to be larger than $(\bar{\nu}\bar{s}^2)^{-\bar{\nu}/2}$ even if

$$\bar{\nu}\bar{s}^2 \approx \bar{\nu}_0 \bar{s}_0^2 + \bar{\nu}_1 \bar{s}_1^2,$$

   but this is balanced by the fact that

$$\Gamma(\bar{\nu}/2) > \prod_{i=0}^{1} \Gamma(\bar{\nu}_i/2)$$

   when $T > 1$ and $T_i > 0$, $i = 0, 1$.

2. Parsimony in terms of the number of conditional mean parameters used in the model comes from the fact that it is very likely that $|\bar{\mathbf{D}}| > |\bar{\mathbf{D}}_0||\bar{\mathbf{D}}_1|$. For example, in the case of $p = 0$ and assuming that $\underline{\mathbf{D}}$ was 1, we would have $|\bar{\mathbf{D}}| = 1/(T + 1)$ and

$$|\bar{\mathbf{D}}_0||\bar{\mathbf{D}}_1| = \frac{1}{(T_0 + 1)(T_1 + 1)},$$

with $T_0 + T_1 = T$.

Similar considerations apply to the comparison of two regime models with three regime models and of three regime models with linear models.

For the comparison across two regime models some insight can be gained into the threshold versus structural break case. Suppose that the sample size of each regime is the same. Then the ratio of the marginal likelihoods will have two components: the relative size of $\prod_{i=0}^{1}(\bar{v}_i\bar{s}_i^2)$ and the relative size of $|\bar{\mathbf{D}}_0||\bar{\mathbf{D}}_1|$ between the two models. If we assume that the latter is approximately equal across models, then the comparison will mainly be determined by the relative fit of the two models. Recall, though, that this is still only a conditional comparison, and its effect on the overall Bayes factor for the two models will depend on the weight the relevant value of γ and τ receives.

For the structural break model, a discrete uniform prior over all possible sample breaks which imply at least 15% of the data lies in each regime is used for τ. Thus, we do not use any *a priori* information about the location of the break point. The 15% rule is intended to ensure that an adequate amount of data is available in each regime and is consistent with the notion that structural breaks are spaced out over time. The marginal posterior $p(\tau \mid Y)$ is easy to derive from these conditional marginal likelihoods by normalizing them to a probability measure. The marginal likelihood for the model is found by simply averaging the conditional marginal likelihoods across all the values of τ.

For the TAR we use a continuous uniform prior over r_1, r_2 again with the restriction that at least 15% of the data must be in each regime. This leads to the following prior: r_1 has a uniform distribution from the 30th percentile of the data to the 85th percentile; conditional on r_1, r_2 is uniformly distribution with lower support at the 15th percentile and upper support at the sample value with 15% of the data below r_1. The prior for d is discrete uniform over the integers $1, 2, \ldots, p$. This means that the possible values of d depend the autoregressive lag used. This is an assumption that could also be relaxed.

The calculation of the overall marginal likelihood in the nonlinear case is a little more difficult. Again for each different value of γ we have calculated $\ell(Y \mid \gamma)$. First, for each fixed pair of r_1, r_2 we can average out the discrete values of d. This gives us $\ell(Y \mid r_1, r_2)$. Since the likelihood is flat for values of thresholds between the data points, this is also true of the marginal likelihood.

Thus to integrate the conditional marginal likelihoods against the uniform prior on the thresholds one needs to weight the conditional marginal likelihoods by the size of the interval and divide by the height of the integrating constant of the prior. In the case of the three regime model this is achieved by first averaging out over r_2 conditional on r_1 (note that this requires a different integrating constant as r_1 varies). If the threshold effect is very obvious, that is, clear breaks can be observed in the path of the time series, then the conditional marginal likelihoods around the observed thresholds will receive additional weight over other points. This is because for the threshold effect to be visually obvious the marginal distribution of the data must be multimodal with large gaps between the modes.

2.3   *Models and hyperparameters*

The results of the previous paragraphs can be extended in the obvious way to cases where more than two regimes, a structural break, or an outlier occurs. This chapter limits itself to the following 11 classes of models (short form acronyms are given in parentheses):

1. Linear autoregressive (LAR).
2. Homoscedastic TAR with one threshold and hence two regimes (TAR2-hom).
3. Heteroscedastic TAR with one threshold and two regimes (TAR2-het).
4. Homoscedastic TAR with two thresholds and hence three regimes (TAR3-hom).
5. Heteroscedastic TAR with two thresholds (TAR3-het).
6. Homoscedastic structural break model with one break (Break1-hom).
7. Heteroscedastic structural break model with one break (Break1-het).
8. Homoscedastic model with two structural breaks (Break2-hom).
9. Heteroscedastic model with two structural breaks (Break2-het).
10. Outlier models with one outlier (Out1).
11. Outlier models with two outliers (Out2).

We also examine the models across various lag lengths. For example, in the case of the industrial production data, we let p take on values between 1 and 6. Hence, we compare 66 models in that case.

It remains to discuss the choice of the hyperparameters for the normal–inverted-gamma priors. We begin by eliciting a prior which is relatively noninformative but accords with our subjective prior beliefs. To simplify matters, we assume the prior means and covariances are zero for all the regression parameters in all models (i.e., the ϕ_i's are centered over zero). The prior variances for the regression coefficients are assumed to be the same for all parameters except the intercept(s) for all models (in the case of outlier models the prior variance is the same as the slope coefficients). The prior variance for the intercept is taken to be 10 times as large as that for the slope coefficients. In particular, we assume the marginal prior variance for ϕ_i is $E(\sigma_i^2)cA_{p+1}$, where A_{p+1} is a $(p+1) \times (p+1)$ diagonal matrix with $(1, 1)$th element 10 and all other diagonal elements 1. Hence in the notation above $\underline{\mathbf{D}} = cA_{p+1}$. The number of degrees of freedom $(\underline{v})$ for the inverted gamma priors is 3 for all models, which is very noninformative, but which allows for the first two marginal prior moments to exist for all parameters. The other hyperparameter of the inverted gamma prior is $\underline{s}$. This hyperparameter is defined so that $E(\sigma_i^2) = [\underline{v}/(\underline{v} - 2)]\underline{s}^2$.

For real GDP growth we choose $c = \frac{2}{3}$ and $\underline{s}^2 = \frac{1}{4}$. This implies a very flat prior for σ_i^2, but one that has mean 0.75. Since the data is measured as percentage changes (e.g., 1.0 implies a 1.0 percent change in GDP), this choice of $\underline{s}^2$ is sensible. The prior centers the AR coefficients over zero for all series,[3] but allows for great prior uncertainty. In particular, prior variances of the AR coefficients are 0.5, which implies very large prior standard deviations relative to the size of the stationary region for AR models.

For growth in real industrial production we expect the error variance to be much larger, since, this data is observed annually and industrial production tends to be more volatile than GDP. In particular, we choose $\underline{s}^2 = \frac{4}{3}$, which implies we expect the error standard deviation to be around 4. With this change, we set $c = 0.25$, which implies the prior variance for the AR coefficients is again 0.5.

3   Empirical results

US real GDP and UK industrial production are plotted in Figure 1(a) and 1(b) respectively. It can be seen that UK industrial production is much more volatile than US real GDP. For the former, annual increases/decreases of more than 10 percent occur almost 5 percent of the time, while for the latter increases/decreases of more than 2 percent are rare. This, of course, is mostly due to the fact that industrial production is observed annually, while GDP is observed quarterly. Further, we note that our prior reflects this difference in volatility. A visual inspection of Figure 1(a) indicates a possible decline in volatility since the early 1980s, and Figure 1(b) indicates periods of high volatility in the early 1700s and the 1920s and 1930s. However, it is hard to ascertain through visual inspection, which of the models is supported by the data, and so we turn to our Bayesian methods.

For both of the series we consider, we selected a maximum value of p which we felt was large enough to capture the dynamics of the data. For the real GDP growth series we chose a maximum value of $p = 4$, while for the industrial

---

[3] Since we are working with differenced data, this implies a prior centered over a random walk for the level of the series.

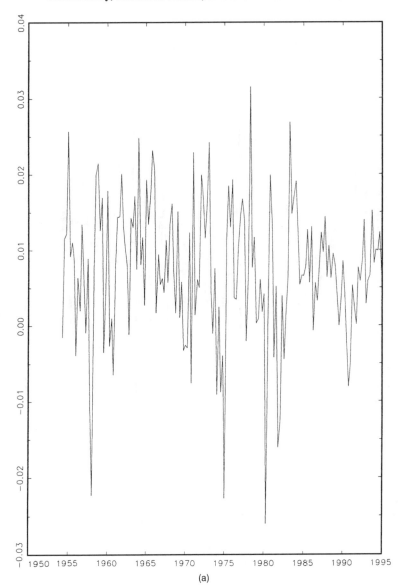

(a)

Figure 1. Growth of (a) US real GDP and (b) UK industrial production.

productions series we chose a maximum value of $p = 6$. Tables 1 and 2 present posterior model probabilities corresponding to each value of p and each class of models using the prior described in the previous section for US GDP and UK industrial production respectively.

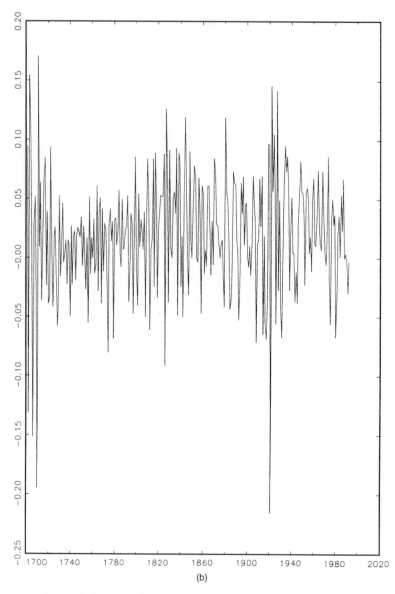

(b)

Figure 1. (continued)

Note first that the strong reward for parsimony reflected in Bayes factors with relatively flat priors implies that short lag lengths tend to be preferred for both series. Despite this, there is little evidence for the linear model (which, for a given lag length, is the most parsimonious model). This reinforces the

Table 1. *Posterior model probabilities for real US GDP data*

Model AR order	Probability			
	1	2	3	4
AR	0.0007	0.0001	0.0000	0.0000
TAR2-hom	0.0001	0.0000	0.0000	0.0000
TAR2-het	0.0001	0.0062	0.0001	0.0000
Break1-hom	0.0001	0.0000	0.0000	0.0000
Break1-het	0.8011	0.1306	0.0069	0.0003
TAR3-hom	0.0000	0.0000	0.0000	0.0000
TAR3-het	0.0000	0.0007	0.0000	0.0000
Break2-hom	0.0000	0.0000	0.0000	0.0000
Break2-het	0.0481	0.0026	0.0000	0.0000
Out1	0.0008	0.0002	0.0000	0.0000
Out2	0.0008	0.0003	0.0000	0.0000

Table 2. *Posterior model probabilities for UK industrial production data*

Model AR order	Probability					
	1	2	3	4	5	6
AR	0.0055	0.0039	0.0005	0.0000	0.0000	0.0000
TAR2-hom	0.1215	0.0176	0.0085	0.0003	0.0000	0.0000
TAR2-het	0.0000	0.0000	0.0000	0.0000	0.0000	0.0000
Break1-hom	0.0090	0.0070	0.0005	0.0001	0.0000	0.0000
Break1-het	0.0000	0.0000	0.0000	0.0000	0.0000	0.0000
TAR3-hom	0.1363	0.2797	0.0486	0.1922	0.0036	0.0001
TAR3-het	0.0000	0.0000	0.0000	0.0000	0.0000	0.0000
Break2-hom	0.0598	0.0839	0.0016	0.0000	0.0000	0.0000
Break2-het	0.0000	0.0000	0.0000	0.0000	0.0000	0.0000
Out1	0.0054	0.0038	0.0005	0.0000	0.0000	0.0000
Out2	0.0057	0.0039	0.0005	0.0000	0.0000	0.0000

common finding that departures from linearity are present in macroeconomic time series. However, just how this departure manifests itself seems to differ over series.

The real GDP growth series seems to be characterized by structural breaks in the variance. The heteroscedastic model with one break gets most of the posterior model probability, although substantial probability is allocated to the two-break model. Another point worth stressing is that the heteroscedastic TARs tend to be favored over the linear model (see, especially, TAR2-het with $p = 2$, which is roughly 10 times as likely as the linear AR with any choice of p). Hence, if a researcher were just looking for nonlinearity, he/she would likely conclude that it is present even though it is likely that this would merely reflect a structural break in the volatility of GDP growth. Estimates of the breakpoint indicate that volatility of the innovation to US GDP changed markedly in the early to mid-1980s. A similar estimate of a breakpoint in volatility is found in McConnell and Perez Quiros (1998) using classical methods. They show that the break emanates from a reduction in the volatility of durable goods production. Note also that there is little evidence for any nonlinearities in the conditional mean of the series (i.e., the homoscedastic models do not beat the linear AR model by much on limiting the comparison to these two groups). An additional interesting point is that, if we restrict attention to homoscedastic models, the outlier models are the most preferred (although they only slightly beat the linear AR model).

Results for the growth in UK industrial production series are markedly different. For this series there is absolutely no evidence for any type of heteroscedasticity, but there is strong evidence for departures from linearity in the conditional mean of the series. The posterior probability is spread out over a lot of different models, but overall the homoscedastic TAR models receive most support. If we sum across p to get the total probability for each class of models, then TAR3-hom receives 67 percent and TAR2-hom receives 15 percent of the posterior model probability. So there is surprisingly strong evidence that the departures from linearity observed in this series are not due to structural breaks (despite the length of the time series, which includes the industrial revolution), but rather by some endogenous process where changes in linear structure are predictable based on past information. However, we do not want push this story too hard, since there is some evidence for structural change (viz., the total probability allocated to the Break1-hom and Break2-hom classes of models is around 14 percent). Furthermore, our results do not completely rule out the possibility that the linear model adequately characterizes the data (this class gets around 2 percent of the posterior model probability).

These empirical results are meant to illustrate the practical usefulness of our techniques and show how great care needs to be taken when modeling departures from linearity. Even with these two series we can see that generalizations like: "structural breaks are present in macroeconomic time series" or "macroeconomic time series are nonlinear" are misleading. In a more extensive empirical exercise, a researcher would want to carry out a prior sensitivity analysis[4] and present other posterior features of interest. For reasons of brevity, we have not done so. Furthermore, in this chapter we have emphasized model choice. However, in practice, with Bayesian techniques the choice of one preferred model is

---

[4] We have informally experimented with changing the prior, and in practice the results seem to be robust to reasonable changes in our present prior.

not necessary. Rather, features of interest (e.g., impulse responses or forecasts) can be presented which average over all possible models where the weights are given by the posterior model probabilities.

4   Conclusions

This paper argues for the use of Bayesian "tests" as a complement to the myriad of classical tests now in use for testing for nonlinearities or structural breaks. Further, given that macroeconomists usually want to test a wide variety of specifications, it is important to develop tests that can be applied jointly across the specifications. The main methodological point of this paper has been to show how Bayesian methods for handling autoregressive models with unknown changes in regime can be used to construct such a joint test across specifications. Because these methods can be performed analytically, they are both computationally inexpensive and easy to use in practice.

In the empirical section of this chapter, our methods are implemented for two representative macroeconomic time series. For real GDP growth, our findings reinforce those of Stock and Watson (1996): structural breaks do indeed appear to occur. If this finding were to become widespread in macroeconomics, then a good portion of empirical macroeconomics would be called into question. If changes in structure are widespread and unpredictable, then it is difficult to imagine an appropriate forecasting strategy for use with macro data, and it may be difficult to develop economic theory to explain this. However, for our industrial production series, departures from linearity seem to be characterized by endogenous changes in structure of the sort captured by the TAR. If this behavior is widespread in macroeconomics, then a very productive research strategy exists involving development of more sophisticated nonlinear models and theories to explain the nonlinearities.

One obvious extension of our results would be to conduct a formal sensitivity analysis for the hyperparameters c, $\underline{s}^2$. Another important extension would involve examining the sensitivity of the results to the assumption of Gaussian innovations to the time series. Indeed, allowing for Student-t errors might make the outlier models considered in this paper redundant (see, e.g., Hoek, Lucas, and van Dijk 1995). However, this would involve some simulation. Thus, to reduce the computational burden it would make sense to focus on a subset of models suggested by an initial analysis based on the computationally simpler techniques described here.

REFERENCES

Albert, J., and S. Chib (1993), "Bayesian Inference via Gibbs Sampling of Autoregressive Time Series Subject to Markov Mean and Variance Shifts", *Journal of Business and Economic Statistics* 11: 1–15.

Andrews, D. W. K. (1993), "Tests for Parameter Instability and Structural Change with an Unknown Change Point", *Econometrica* 61: 821–56.

Andrews, D., and W. Ploberger (1994), "Optimal Tests When a Nuisance Parameter Is Present Only Under the Alternative", *Econometrica* 62: 1383–414.

Barry, D., and J. Hartigan (1993), "A Bayesian Analysis for Change Point Problems", *Journal of the American Statistical Association* 88: 309–19.

Beaudry, P., and G. Koop (1993), "Do Recessions Permanently Affect Output?"*Journal of Monetary Economics* 31: 149–63.

Carlin, B., A. Gelfand, and A. F. M. Smith (1992), "Hierarchical Bayesian Analysis of Changepoint Problems", *Applied Statistics* 41: 389–405.

Greasley, D., and L. Oxley (1994), "Rehabilitation Sustained: The Industrial Revolution as a Macroeconomic Epoch", *Economic History Review, 2nd Series* 47: 760–8.

Hamilton, J. (1989), "A New Approach to the Economic Analysis of Nonstationary Time Series and the Business Cycle", *Econometrica* 57: 357–84.

Hansen, B. (1996), "Inference When a Nuisance Parameter is Not Identified Under the Null Hypothesis", *Econometrica* 64: 413–30.

Hoek, H., A. Lucas, and H. van Dijk (1995), "Classical and Bayesian Aspects of Robust Unit Root Inference", *Journal of Econometrics* 69: 27–59.

Judge, G., W. E. Griffiths, R. C. Hill, H. Lutkepohl, and T. C. Lee (1985), *The Theory and Practice of Econometrics,* 2nd ed., New York: Wiley.

Koop, G., and S. Potter (1999), "Bayes Factors and Nonlinearity: Evidence from Economic Time Series", *Journal of Econometrics*, 88: 251–81.

McConnell, M. M., and G. Perez Quiros (1998), "Output Fluctuations in the United States: What has Changed since the Early 1980s?" Staff Report 41, Federal Reserve Bank of New York.

Mills, T., and N. Crafts (1996), "Trend Growth in British Industrial Output, 1700–1913: A Reappraisal", *Explorations in Economic History* 33: 277–95.

Perron, P. (1989), "The Great Crash, the Oil Price Shock, and the Unit Root Hypothesis", *Econometrica* 57: 1361–402.

Pesaran, M. H., and S. Potter (1997), "A Floor and Ceiling Model of US Output", *Journal of Economic Dynamics and Control* 21: 661–95.

Poirier, D. (1995), *Intermediate Statistics and Econometrics*. Cambridge, MA: The MIT Press.

Potter, S. (1995), "A Nonlinear Approach to US GNP", *Journal of Applied Econometrics* 10: 109–26.

Stephens, D. A. (1994), "Bayesian Retrospective Multiple-Changepoint Identification", *Applied Statistics* 43: 159–78.

Stock, J., and M. Watson (1996), "Evidence on Structural Instability in Macroeconomic Time Series Relations", *Journal of Business and Economic Statistics* 14: 11–30.

CHAPTER 5

Bayesian analysis of nonlinear time series models with a threshold

Michel Lubrano

1   Introduction

In a linear regression model, static or dynamic, the conditional expectation of the endogenous variable y_t is expressed as a *linear*[1] combination of the exogenous variables x_t with $E(y_t) = x_t'\beta$ or $y_t = x_t'\beta + e_t$. The type of models I want to analyze in this chapter illustrates a data generating process which is both dynamic and nonlinear and which can be expressed as

$$y_t = g(y_{t-1}, x_t, \beta, \tilde{\theta}) + e_t,\tag{1}$$

where g is a nonlinear function and e_t an additive error term following the usual hypothesis. This means that the conditional expectation of the endogenous variable is a *nonlinear* function of the exogenous variables and of lags of the endogenous variable with $E(y_t) = g(y_{t-1}, x_t, \beta, \tilde{\theta})$. Clearly this means that the data generation process itself is a nonlinear mechanism. Interest in this type of models started quite a long time ago with threshold and switching regression models.[2] See Quandt (1958) for the classical side and Bacon and Watts (1971) for the Bayesian side. In the recent literature, three books have had a great impact in popularising the topic of nonlinear modelling. Tong (1983) explores

This paper grew out of Chapter 8 of a book on Bayesian econometrics being written jointly with Luc Bauwens and Jean-François Richard. It was presented at the sixth meeting of the European Conference Series in Quantitative Economics and Econometrics, (EC)[2], entitled "Nonlinear modelling in economics" and held in Aarhus, December 1995. Useful discussions with the participants are gratefully acknowledged. Comments by Luc Bauwens, Peter Schotman, Timo Terasvirta, and an anonymous referee helped a lot in improving the presentation of the paper. Usual disclaimers apply.
[1] Sometimes a transformation is necessary to recover linearity in the parameters: an example is the Cobb–Douglass production function, which is a nonlinear model in the parameters. Taking the logs of the variables (with a suitable assumption on the error term) produces a model which becomes linear in the variables and the parameters.
[2] In this case g is a switching function.

the properties of the self-exciting autoregressive regression model (SETAR, a nonlinear AR model), and Tong (1990) develops the topic of nonlinear modelling from a dynamical system point of view. Granger and Teräsvirta (1993) represents the point of view of the econometrician. For an economist, interest in nonlinear time series comes from the fact that the economy has definite nonlinear features. Economic theory suggest models with floors and ceilings, switching regimes, quantity rationing, or nonlinear effects as in the Phillips curve. Particular interest has also been taken in asymmetry of behaviour of the business cycle in response to shocks, the first work being that of Keynes with his general theory. For applications to the US GNP see e.g., Hamilton (1989), Potter (1995), or Tiao and Tsay (1994).

In this chapter, I shall study a particular, but quite generic nonlinear dynamic model. In the linear regression model, the conditional expectation of the endogenous variable y_t is expressed as a linear combination of the elements of the exogenous or predetermined variables x_t. The regression coefficient β is fixed over time. One particularly interesting way of introducing nonlinearities is to suppose that the conditional expectation of the endogenous variable belongs to two or more regimes. In the simple two regime case, we have

$$E(y_t|x_t) = \begin{cases} x_t'\beta_1, & \text{first regime,} \\ x_t'\beta_2, & \text{second regime.} \end{cases} \tag{2}$$

The choice between the two regimes is represented by a nonlinear transition function $F(\tilde{z}_t, \tilde{\theta})$ which takes values between zero and one. We can write

$$y_t = (1 - F(\tilde{z}_t, \tilde{\theta})) x_t'\beta_1 + F(\tilde{z}_t, \tilde{\theta}) x_t'\beta_2 + \epsilon_t, \qquad \text{Var}(\epsilon_t) = \sigma^2, \tag{3}$$

with ϵ_t being normally distributed with zero mean and constant variance σ^2 over the two regimes. Extension to unequal variances would mean

$$y_t = [1 - F(\tilde{z}_t, \tilde{\theta})]x_t'\beta_1 + F(\tilde{z}_t, \tilde{\theta})x_t'\beta_2 + \epsilon_t,$$
$$\text{Var}(\epsilon_t) = [1 - F(\tilde{z}_t, \tilde{\theta})]\sigma_1^2 + F(\tilde{z}_t, \tilde{\theta})\sigma_2^2. \tag{4}$$

This model represents a transition regression model or a regression model with several regimes where the change of regimes is made according to the sign of a linear combination of the elements of $\tilde{z}_t$. This model covers a wide range of nonlinearities, but does not exhaust all the possible cases of nonlinearities in time series. In particular one can think of the mixture model introduced by Lindgren (1978) and later developed by Hamilton (1989), which corresponds to

$$y_t = \begin{cases} x_t'\beta_1 + \epsilon_{1t} & \text{with probability } p, \\ x_t'\beta_2 + \epsilon_{2t} & \text{with probability } 1 - p \end{cases} \tag{5}$$

and can alternatively be written

$$y_t = (x_t' \beta_1 + \epsilon_{1t})s_t + (x_t' \beta_2 + \epsilon_{2t})(1 - s_t), \qquad (6)$$

where s_t is a Bernouilli random variable being one with probability p and zero with probability $1 - p$. There are marked differences between (5) and (6). In model (5), the switching operates on the conditional expectation of y_t, whereas in model (6) it operates on the value of y_t itself. This makes a great difference in the likelihood function and in the identification properties. Moreover, in (5) the change of regime depends on the value of an exogenous variable z_t, whereas in (6) s_t is usually modelled as a Markov chain; see e.g., Lindgren (1978) or Hamilton (1989). Comparison with this model will no longer be pursued.

Starting from (3) or (5), I have then to specify a transition function $F(\cdot)$ and the switching variables contained in $\tilde{z}_t$. The various choices that can be made induce a large spectrum of models. First of all, $F(\cdot)$ can be a step or a smooth function, which means that the change of regime can occur abruptly or smoothly. Second, the switching variable $\tilde{z}_t$ can simply contain a time index and a constant. The time index can be replaced by a continuous exogenous variable or a lag of the endogenous variable. Let me review some of the possible combinations which have been used in the literature:

> Combining an abrupt transition function with a time index produces the structural break model where the breaking point is unknown. The Bayesian literature on change point models is quite large and includes Ferreira (1975) and DeJong (1996) for unit roots with a structural break in the US GNP, and Lubrano (1995) for cointegrating relationships applied to a wage equation. Introducing a smooth transition function allows some delay for the change of regime to occur.
>
> Threshold regression models replace the time index by a continuous variable, not necessarily growing over time. This means that we can have more than one change of regime over the sample period. Bacon and Watts (1971) started a long Bayesian literature on threshold regression models, which continued with e.g., Geweke and Terui (1993) and Chen and Lee (1995) and found applications in finance with e.g., Pfann, Schotman, and Tschernig (1996) and Forbes, Kalb, and Kofman (1997).

Nonlinearity has always been a delicate question for Bayesian computations, as it may imply numerical integration in high dimensions. What I show in the next sections is that the dimension of the numerical integration can be confined to one or two for most of the models I consider. This is an argument for using these models in empirical applications. Let me denote by β the vector of parameters indexing the regime equations. Conditionally on $\tilde{\theta}$, most of our models are linear in β, provided there is no element in common between β and $\tilde{\theta}$. The conditional posterior density of β is Student provided the prior density is noninformative or naturally conjugate. The marginal posterior density of $\tilde{\theta}$, on the contrary, has to be analysed numerically. The dimension of the numerical integration is the dimension of $\tilde{\theta}$ (minus one, because $\tilde{\theta}$ will be normalised). Posterior moments of β are found as a by-product of this integration.

The chapter is organised as follows. In Section 2, I present the general Bayesian analysis of two regime models, considering explicitly the identification problem introduced by the nonlinearity and the question of the integrability of the posterior density. Unequal variances of the error term are also treated. Section 3 is devoted to testing the null hypothesis of linearity in a Bayesian framework. Section 4 analyses models where the switching depends on a time trend. In Section 5, the time trend is replaced by a continuous variable, and various types of self-exciting threshold regression models are examined and their predictive densities computed. In both sections great attention is devoted to empirical applications concerning the French consumption function and the US business cycle. The last section concludes.

2   Bayesian analysis of the general two regime model

In this section, I propose a general Bayesian treatment for this type of model, considering first the case of equal variances, which is the most common one.[3] The case of unequal variances is treated at the end of the section. Nonlinear models display a specific identification problem when approaching linearity, which I shall solve by introducing a conditional prior information. Though the smooth transition function may appear at first sight to be more general than the step transition function, which may be viewed as a particular case, it in fact introduces a specific integrability problem which has to be solved also by the use of a specific prior information. Finally, when there is a constant term in $\tilde{z}_t$, a normalisation of $\tilde{\theta}$ gives birth to a special parameter, which is called the threshold. Its posterior density may have a very peculiar shape, depending on the nature of the transition function and of the switching variable z_t.

2.1   *Notation*

For practical computations it is useful to reparameterise the model so as to obtain

$$y_t = x_t'\beta_1 + F(\tilde{z}_t, \tilde{\theta})\, x_t'\delta + \epsilon_t \tag{7}$$

---

[3] For instance, Granger and Teräsvirta (1993) do not consider the threshold model with unequal variances.

with $\delta = \beta_2 - \beta_1$ being the contribution to the regression of considering a second regime. I shall suppose that the conditional expectation parameter β and the regime shift parameter $\tilde{\theta}$ have no element in common. Thus, it is possible to separate the parameter set into nonlinear parameters $\tilde{\theta}$ and linear parameters β. Moreover, the parameter $\tilde{\theta}$ may be partitioned into $\tilde{\theta} = [\gamma, \tilde{\theta}_2]$, where γ is a scalar parameter which measures the degree of smoothness of the transition. The change of regime itself depends on the sign of $z_t' \tilde{\theta}_2$, which shows that $\tilde{\theta}_2$ has to be normalised. In the simplest case, $\tilde{z}_t$ is composed of a variable and of a constant term, say $\tilde{z}_t = [z_t, 1]$. This implies that $\tilde{\theta}_2$ can be normalised as $\tilde{\theta}_2 = [1, -c]$. The switching is then determined by the sign of $z_t - c$, and c is referred to in the literature as the threshold parameter. Consequently the normalised nonlinear parameter is $\theta = [\gamma, c]$.

The separability between β and $\theta = [\gamma, c]$ makes it useful to consider the following reparameterisation:

$$x_t'(\gamma, c) = [x_t', F(\gamma, z_t - c) x_t'],$$
$$\beta' = [\beta_1', \beta_2' - \beta_1'], \tag{8}$$

so that the model can be written in a compact form:

$$y_t = x_t'(\gamma, c)\beta + \epsilon_t. \tag{9}$$

In the equal variance case, the variance of ϵ_t is σ^2. If I allow for different variances in each regime, I have an heteroskedastic model where

$$\text{Var}(\epsilon_t) = \sigma_1^2(1 - F(\cdot)) + \sigma_2^2 F(\cdot). \tag{10}$$

Finally let me note that the model is not fundamentally changed if I allow for a different set of regressors x_{it} for each regime. Supposing that k_1 and k_2 are the dimensions of these sets of regressors, I shall denote by $k = k_1 + k_2$ the total number of regressors in the model. Here k_1 and k_2 are the numbers of regressors in the two regimes, though one may have $k_1 = k_2$.

2.2   Likelihood function and posterior analysis

Under a normality assumption, the likelihood function of model (9) is

$$L(\beta, \gamma, c, \sigma^2; y) \propto \sigma^{-T} \exp\left(-\frac{1}{2\sigma^2} \sum_{t=1}^{T} (y_t - x_t'(\gamma, c)\beta)^2\right). \tag{11}$$

Following the distinction between linear and nonlinear parameters and the fact that in step transition models γ is dropped, it is convenient to decompose the prior into

$$\varphi(\beta, \sigma^2, \gamma, c) = \varphi(\beta, \sigma^2 | \gamma) \, \varphi(\gamma) \, \varphi(c). \tag{12}$$

The class of prior densities for γ or c is not restricted, as anyway these parameters will have to be integrated out *numerically*. Moreover, I want to be free to choose the prior on c according to the particular nonlinear model which is analysed. On the contrary, it is interesting to remain in the natural conjugate framework for the prior on β (or conditionally on γ in the natural conjugate framework), as I can integrate β out *analytically*. For ease of exposition, I shall choose, for the while, a noninformative prior on β and σ^2 with

$$\varphi(\beta, \sigma^2 | \gamma) \propto \sigma^{-2}. \tag{13}$$

This prior can be replaced by a normal inverted gamma-2 prior without fundamentally changing the next results.

Theorem 1. *Under a noninformative prior for β and σ^2, the conditional posterior densities of β and σ^2 are given by*

$$\begin{aligned}
\varphi(\beta | \gamma, c, y) &= f_t(\beta | \beta_*(\gamma, c), M_*(\gamma, c), s_*(\gamma, c), T - k), \\
\varphi(\sigma^2 | \gamma, c, y) &= f_{IG}(\sigma^2 | s_*(\gamma, c), T - k),
\end{aligned} \tag{14}$$

where

$$\begin{aligned}
M_*(\gamma, c) &= \sum x_t(\gamma, c) \, x_t'(\gamma, c), \\
\beta_*(\gamma, c) &= M_*^{-1}(\gamma, c) \sum x_t(\gamma, c) \, y_t, \\
s_*(\gamma, c) &= \sum y_t^2 - \beta_*'(\gamma, c) \, M_*(\gamma, c) \, \beta_*(\gamma, c).
\end{aligned} \tag{15}$$

The corresponding posterior density of γ and c is

$$\varphi(\gamma, c | y) \propto |s_*(\gamma, c)|^{-(T-k)/2} \, |M_*(\gamma, c)|^{-1/2} \, \varphi(\gamma) \, \varphi(c). \tag{16}$$

Proof: As the model is conditionally linear, the posterior densities of β and σ^2 result from the application of standard formulae of natural conjugate analysis due to Raiffa and Schlaifer (1961). The posterior density of γ and c is obtained as one over the integrating constant of the above Student density times the prior densities of γ and c □

The posterior density of γ and c has to be analysed numerically to compute its integrating constant and its moments, using a two dimensional numerical integration routine. The marginal posterior densities of β and σ^2 follow, with

$$\varphi(\beta | y) = \int \varphi(\beta | \gamma, c, y) \, \varphi(\gamma, c | y) \, d\gamma \, dc, \tag{17}$$

$$\varphi(\sigma^2 | y) = \int f_{IG}(\sigma^2 | s_*(\gamma, c), T - k) \, \varphi(\gamma, c | y) \, d\gamma \, dc. \tag{18}$$

The posterior density (16) has pathological behaviour in the directions of both γ and c.

2.3   Identification and nonlinearity

Model (7) becomes linear when $F(\gamma, z_t - c)$ is zero or constant. This is usually obtained for $\gamma = 0$. If $F(\cdot)$ is the logistic distribution detailed below, the model is reduced to

$$y_t = x_t'\beta_1 + 0.5\,x_t'\delta + \epsilon_t.$$

As a consequence, the matrix $M_*(\gamma = 0, c)$ becomes singular, which means that the joint posterior density of $\theta = [\gamma, c]$ is not integrable. The marginal posterior density of β is likewise not integrable, as when $\gamma = 0$, $\delta = \beta_2 - \beta_1$ may take any value and no longer enters the likelihood function. δ is not identified when $\gamma = 0$. Finally, c is also not identified, as $F(\gamma, z_t - c)$ is constant whatever the value of c when $\gamma = 0$.

The classical literature is concerned with that identification problem only when testing for linearity and not for inference. It is the problem of optimal testing when nuisance parameters are not identified under the null, treated for instance by Hansen (1996); see also the references cited therein. Hansen (1996) applied his methodology to the threshold regression model. He proposes a transformation of the threshold parameter c (his nuisance parameter) when testing for $\delta = 0$ which makes his test procedure feasible.

In the Bayesian framework, the problem is more serious and should be tackled also for inference, as Bayesians have to integrate over the whole domain of definition of γ. It is well known that numerical problems may begin well before the point zero in some samples. The Bayesian solution to this identification and numerical problem is to introduce an adequate prior information. We have two choices. One may imagine an informative prior density on γ which downweights the region near $\gamma = 0$, but that is not satisfactory. Instead we devise a prior information on δ. Consider the following conditional normal prior[4] for δ:

$$\varphi(\delta | \sigma^2, \gamma) = f_N\left(\delta | 0, \sigma^2 \exp(\gamma)\, N_0^{-1}\right). \tag{19}$$

This prior becomes informative when $\gamma \to 0$ and says that δ is zero in this case. This is coherent with the linearity hypothesis, which can be described either by $\gamma = 0$ or by $\delta = 0$. When $\gamma = 0$, the prior precision is equal to N_0/σ^2. The prior precision goes to zero when γ becomes positive, and the sample recovers all its freedom. As I have now an informative normal prior on a part of β, the

---

[4] See Schotman (1994) for a slightly different type of prior in the AR(1) model given by $(1 - \rho L)(y_t - \mu) = \epsilon_t$, where a similar identification problem occurs for the mean parameter μ when $\rho = 1$.

convolution formulae given in (15) have to be changed, and more precisely the expression of $|M_*(\gamma, c)|$, which becomes

$$|M_*(\gamma, c)| = M_0(\gamma) + \sum x_t(\gamma, c)x_t'(\gamma, c) \tag{20}$$

with

$$M_0(\gamma) = \begin{pmatrix} 0 & 0 \\ 0 & N_0/\exp(\gamma) \end{pmatrix}. \tag{21}$$

Usually it suffices to take N_0 as the identity matrix or a multiple of it.

Theorem 2. *The posterior density of γ and c is infinite at the point $\gamma = 0$ in the absence of a specific prior information, and finite under the partially informative normal prior (19).*

Proof: In (19), $s_*(\gamma, c)$ is always positive, as it is basically a sum of squared residuals. The crucial element is $M_*(\gamma, c)$. Expanding $M_*(\gamma = 0, c)$ given in (20) gives

$$M_*(\gamma = 0, c) = \begin{pmatrix} \sum x_t x_t' & 0 \\ 0 & N_0. \end{pmatrix}$$

This precision matrix is singular when $N_0 = 0$, and nonsingular if both $\sum x_t x_t' > 0$ and $N_0 > 0$. $\qquad\qquad \square$

2.4   *Integrability of the posterior density of γ*

An abrupt transition is obtained as the limiting case of a smooth transition for $\gamma \to \infty$. For certain sample configurations, γ is fairly large and smooth and abrupt transitions become observationally equivalent. As a consequence,

Theorem 3. *With a flat prior, the posterior density of γ is not integrable.*

Proof: For $\gamma \to \infty$, $F(\gamma, z_t - c)$ is $O(1)$ in γ. Consequently

$$\varphi(\gamma|y) = \int \varphi(\gamma, c|y)\,\varphi(c)\,dc$$

is finite and strictly positive for any positive value of γ. $\qquad\qquad \square$

A similar problem arises with regression models with Student errors, as underlined in Bauwens and Lubrano (1998). The conclusion is that a sufficient prior information is needed to force the posterior density to tend to zero quickly enough at its right tail in order to be integrable. The prior should at least be $O(\gamma^{1+\upsilon})$ with $\upsilon > 0$. A convenient prior is the truncated Cauchy density with [5]

$$\varphi(\gamma) = \begin{cases} (1 + \gamma^2)^{-1} & \text{if } \gamma > 0, \\ 0 & \text{otherwise.} \end{cases} \tag{22}$$

The scale of γ depends on the scale of the variable z_t. Thus it is wise to standardise the difference $z_t - c$ by the standard deviation of z_t. This makes the posterior expectation of γ interpretable.

2.5   The nature of the threshold

As the change of regime depends only on the sign of $z_t - c$, the domain of variation of c is determined by the minimum and the maximum of the sample values of z_t (if z_t is of dimension one). The interpretation of c then depends on the nature of z_t. If z_t is a time index, c represents a date and is usually denoted by τ. It is by nature a discrete parameter. [6] Its posterior density is discrete with equally spaced points, and the integration over τ should be done by direct summation. On the contrary, if z_t is a continuous variable, for instance the lagged endogenous variable as in self-exciting transition autoregressive (SETAR) models, then c is a continuous parameter with a posterior density which is generally continuous. However, if the transition function is a step function, the posterior density of c presents a series of discontinuities at a finite number of points. Let me sort the observations of the switching variable z_t. The continuous parameter c has to be compared with these discrete observations. Between two consecutive sorted values, c can vary without modifying the sample classification and consequently the likelihood function. Thus the marginalised likelihood function $l(c; y)$ in c is a step function which looks like an histogram with $T - 1$ cells. For $T \rightarrow \infty$, the length of the steps (or the cells of the histogram) tends to zero provided the distribution of z_t is absolutely continuous with respect to the Lebesgue measure over the real line. In this case, the posterior density $\varphi(c|y)$ tends to a continuous density. However, there is no reason why $l(c; y)$ or $\varphi(c|y)$ should approach a nice bell shape. A consequence is that a pure classical approach by maximum likelihood is difficult: the likelihood function is not differentiable and may be multimodal. Any classical measure of uncertainty for c seems infeasible. The Bayesian approach, on the contrary, causes no problem, as we simply have to average on the domain of variation of c.

---

[5] A flat prior on $1/\gamma$ yields $\varphi(\gamma) \propto 1/\gamma^2$. But this prior creates a singularity at $\gamma = 0$, which is a point of interest. The Cauchy prior simply translates the singularity outside of the region $\gamma > 0$.

[6] Even if the model is reparameterised by dividing z_t by the sample size T. In that case $\tau \in [0, 1]$ becomes a rational number, but still identifies a particular observation and a date in a discrete time process.

2.6   *The heteroskedastic case*

It is easy to generalise the previous model so as to allow for a different variance of the error term in each regime. I can factorise (10) into

$$\text{Var}(\epsilon_t) = \sigma^2[(1 - F(\gamma, z_t - c) + \phi\, F(\gamma, z_t - c))]$$

$$= \sigma^2 h_t(\gamma, c, \phi), \tag{23}$$

where $\sigma^2 = \sigma_1^2$, $\phi = \sigma_2^2/\sigma_1^2 \in [0, +\infty[$. The same analysis as before can be reproduced, provided I rescale the data by $\sqrt{h_t(\gamma, c, \phi)}$. Let me define

$$y_t(\gamma, c, \phi) = y_t/\sqrt{h_t(\gamma, c, \phi)},$$
$$x_t(\gamma, c, \phi) = x_t(\gamma, c)/\sqrt{h_t(\gamma, c, \phi)}. \tag{24}$$

Then the likelihood function of the model is

$$L(\beta, \gamma, c, \sigma^2, \phi; y) \propto \sigma^{-T} \left[\prod h_t(\gamma, c, \phi)\right]^{-1/2}$$

$$\times \exp\left(-\frac{1}{2\sigma^2}\sum_{t=1}^{T}(y_t(\gamma, c, \phi) - x_t'(\gamma, c, \phi)\beta)^2\right). \tag{25}$$

A noninformative prior for ϕ could be

$$\varphi(\phi) \propto 1/\phi. \tag{26}$$

Theorem 4. *The joint posterior density of γ, c, and ϕ is*

$$\varphi(\gamma, c, \phi|y) \propto \left[\prod h_t(\gamma, c, \phi)\right]^{-1/2}$$

$$\times |s_*(\gamma, c, \phi)|^{-(T-k)/2} |M_*(\gamma, c, \phi)|^{-1/2} \varphi(\gamma)\,\varphi(c)\,\varphi(\phi). \tag{27}$$

Proof: See Appendix A. □

Of course, compared to (16), I have increased the dimension of the numerical integration by one. Monte Carlo Markov chain algorithms may start to be competitive. The marginal posterior densities of β and σ^2 are given in Appendix A.

The natural question is now to know if an unequal variance model is really necessary. The answer is given by an inspection of the marginal posterior density of ϕ. If 1 belongs to a posterior confidence interval of ϕ, it means that the two variances σ_1^2 and σ_2^2 are likely to be equal.

Finally, let me note that the integration problem can be simplified when the transition function is a step function, as in this case the likelihood function becomes similar to that proposed for instance in Pole and Smith (1985) or Geweke and Terui (1993). This is explained in Appendix B, where separate posterior densities for β_i and σ_i^2 are obtained. This simplification reduces the dimension of the integration problem by one, but it has a price. Separability is made possible by the fact that conditionally on c, β_1 and β_2 are independent. But of course, they are not marginally independent. Separability implies that it is no longer possible to derive the posterior density of $\beta_2 - \beta_1$ or that of the ratio of the two error variances. This precludes some testing procedures for linearity and homoskedasticity.

3   Testing for linearity

The many possibilities among the different cases of nonlinearity induce some very natural questions which can be very hard to answer.

The first question is of course to know if a nonlinear model is necessary to give a correct account of the behaviour of the data. Some tests presented in the classical literature do not specify a precise model for the alternative (such as the RESET test, the bispectrum test, or tests based on Volterra expansions). But most of the tests, such as the ones presented for instance in Tsay (1989) or in Lin and Teräsvirta (1994), the data are supposed to follow a particular nonlinear model under the alternative. These classical tests should be more powerful, as they are based on a precise direction, even if they may be powerful against other directions.

The second question of interest concerns the determination of the switching variable z_t. There is no universal answer to this question. The economic problem under inspection of course gives some natural choices, which are very helpful and which the pure statistician terribly lacks (when analysing for instance the famous data sets of Wolfe's sunspot numbers and Canadian lynx captures). In the breaking point literature, for instance, z_t is a time trend. In the switching regime literature, z_t is a continuous variable which is usually suggested by economic theory. But z_t governs the switching with some delay, so that it appears as z_{t-d}, where d is called the delay parameter. It is a statistical question to determine the value of d. In the Bayesian literature (see e.g., Geweke and Terui 1993 or Koop and Potter 1995), d is treated as a parameter for which a posterior density is derived. In this case, d is an element of θ. We could also decide that it is part of the specification of the model, as for instance the lag length in autoregressive models. Then, d should be determined with an information criterion, as for instance Tong and Lim (1980) do in a classical framework. There should be no fixed doctrine in that respect.

3.1   Selecting linearity versus nonlinearity

The linear model

$$y_t = x_t'\beta + u_t \tag{28}$$

can be tested for various misspecification directions, one of which is the presence of nonlinearity in the conditional expectation. Following the tradition of the augmented regressions of Pagan (1984), reinterpreted in a Bayesian framework by Aprahamian, Lubrano, and Marimoutou (1994), I have to look for a regression direction, essentially a set of additional variables, which express nonlinearity. The linear model (28) becomes nonlinear in particular when an $x_t'\delta F(\gamma, z_{t-d} - c)$ term is added. A linearisation of $F(\cdot)$ around $\gamma = 0$ yields the additional variables looked for. Luukkonen, Saikkonen, and Teräsvirta (1988) suggest using a third order Taylor expansion, which means

$$F(\gamma, z_{t-d} - c) \simeq \alpha_0 + \alpha_1 z_{t-d} + \alpha_2 z_{t-d}^2 + \alpha_3 z_{t-d}^3. \tag{29}$$

The augmented regression for detecting a nonlinear direction becomes

$$y_t = x_t'\beta + x_t'\delta_1 z_{t-d} + x_t'\delta_2 z_{t-d}^2 + x_t'\delta_3\alpha_3 z_{t-d}^3 + \epsilon_t. \tag{30}$$

The choice between the linear model (28) and the nonlinear approximation (30) is made on the basis of a Schwarz criterion. The augmented regression is built for various values of d. The value of d which minimises the Schwarz criterion is selected. This sequence of model choices constitutes a tentative search strategy for specifying a nonlinear model in a Bayesian framework. See also Teräsvirta (1994) in a classical framework.

3.2   A linearity test based on the posterior density

Once a particular nonlinear model has been estimated, one may ask if the nonlinearity found is significant, i.e., if zero belongs to a posterior confidence interval of δ. Cook and Broemeling (1996) propose to evaluate such an interval by simulation. The marginal posterior density of δ, as given in the following theorem, is appropriate for testing linearity:

Theorem 5. *The marginal posterior distribution of δ,*

$$\varphi(\delta|y) = \int f_t(\delta|\delta_*(\gamma, c), M_{22.1}^*(\gamma, c), s_*(\gamma, c), T - k)\varphi(\gamma, c|y)\, d\gamma\, dc, \tag{31}$$

is a proper density, whatever the value of γ, when the prior for γ is the truncated Cauchy (22) and that of δ is the conditional normal (19).

Proof: The hyperparameters are conformable partitions of (15) and (20) with

$$\beta_*(\gamma, c) = \begin{pmatrix} \beta_1^*(\gamma, c) \\ \delta_*(\gamma, c) \end{pmatrix},$$

$$M_*(\gamma, c) = \begin{pmatrix} M_{11}^*(\gamma, c) & M_{12}^*(\gamma, c) \\ M_{21}^*(\gamma, c) & M_{22}^*(\gamma, c) \end{pmatrix}, \tag{32}$$

$$M_{22.1}^*(\gamma, c) = M_{22}^*(\gamma, c) - M_{21}^*(\gamma, c) M_{11}^{*-1}(\gamma, c) M_{12}^*(\gamma, c).$$

$M_*(\gamma = 0, c)$ is a regular matrix under prior (19). The posterior density $\varphi(\gamma, c|y)$ is fully integrable under priors (22) and (19). □

When δ is of dimension one, a posterior confidence interval is straightforward to build. In order to build a multivariate posterior confidence interval for δ, one usually considers a quadratic transformation $\xi(\gamma, c)$ of δ, defined by

$$\xi(\gamma, c) = (\delta - \delta^*(\gamma, c))' M_{22.1}^*(\gamma, c)(\delta - \delta^*(\gamma, c)) \frac{T - k}{k_2 \, s_*(\gamma, c)}. \tag{33}$$

Conditionally on γ and c, $\xi(\gamma, c)$ is distributed as an $F(k_2, T - k)$. The marginal posterior density of ξ is obtained by numerical integration of the conditional Fisher density

$$\varphi(\xi|y) \propto \int \xi(\gamma, c)^{(k_2-2)/2} \left(1 + \frac{T - k}{k_2} \xi(\gamma, c)\right)^{-(T+k+k_2)/2} \varphi(\gamma, c|y) \, d\gamma \, dc. \tag{34}$$

Define ξ_0 as

$$\xi_0 = \int \delta_*'(\gamma, c) \, M_{22.1}^*(\gamma, c) \, \delta_*(\gamma, c) \, \frac{T - k}{k_2 \, s_*(\gamma, c)} \, \varphi(\gamma, c|y) \, d\gamma \, dc. \tag{35}$$

The posterior probability that $\delta = 0$ is given by the partial integral

$$\int_0^{\xi_0} \varphi(\xi|y) \, d\xi. \tag{36}$$

4   Transition models based on a time index

This type of model is concerned with the analysis of structural breaks or segmented trends at an unknown point of time. The regression coefficient β changes according to an unknown date that I call τ and treat as a discrete parameter.[7] This model was first analysed by Quandt (1958) in a maximum likelihood context and gave birth to a huge literature, mainly classical, known as the change point or structural breaks literature, which can be divided mainly into two branches. Many authors are concerned with testing parameter constancy in linear (or even nonlinear) models and thus detecting model inadequacies. Examples go from Quandt (1960) to Andrews (1993) and Lin and Teräsvirta (1994). Interest may secondly lie in estimating a two regime model with a change of regime at an unknown point of time. In the classical econometric literature, Perron and Vogelsang (1992) test for unit roots in the presence of structural change occurring at an unknown date; Gregory and Hansen (1996) and Campos, Ericsson, and Hendry (1996) test for cointegration in the same context. Economists looked for economic slowdown after the oil crisis (Ben David and Papell 1995), break in inequality measures (Raj and Slottje 1994), inflation rates, and the stability of money demand equations (Baba, Hendry, and Starr 1992, Raj 1995, Lutkepohl 1993).

4.1   Step change

The change between the two regimes is made via a step transition function $\text{ID}(t - \tau)$ defined by

$$\text{ID}(t - \tau) = \begin{cases} 0 & \text{if} \quad t \geq \tau, \\ 1 & \text{if} \quad t < \tau. \end{cases} \tag{37}$$

$\text{ID}(\cdot)$ is thus an indicator function (or Heaviside function). Replacing $F(\gamma, z_t - c)$ by (37) in (7), the general model becomes

$$y_t = x_t'\beta_1 + \text{ID}(t - \tau) x_t'\delta + \epsilon_t \tag{38}$$

and

$$x_t'(\gamma, c) = x_t'(\tau) = [x_t', \text{ID}(t - \tau) x_t']. \tag{39}$$

Of course, the parameter γ and all the problems attached to its presence disappear. The simplest case I can think of is when the change point τ is known in advance, as then (38) is simply a model with multiplicative dummy variables. Perron (1989), for instance, used this type of model to test for unit roots in the presence of a break in the constant term and/or in the trend. With an unknown τ, we have access of course to a richer class of models. A first Bayesian treatment of this model was proposed by Ferreira (1975). Ohtani (1982) extended this

---

[7] This is the c of the previous section. I have adopted this notation to stress the fact that this is a discrete parameter representing a date and also to be coherent with a part of the literature concerning this model.

model to autocorrelated errors. Lubrano (1993) and DeJong (1996) test for unit roots in the presence of an endogenous break.

The prior on τ has to incorporate the fact that τ is a discrete parameter. A noninformative prior, for instance, would mean an equal mass on each integer $1, 2, \ldots, T$. There is an identification problem in the model whenever in regime i the number of observations (T_i) is lower than the number of variables (k_i) and the prior on β_i is noninformative. The matrix $M_*(\tau)$ is singular in this case. A prior giving a zero weight to infeasible regime classifications is one way to solve the identification problem:

$$\varphi(\tau) = \begin{cases} 0 & \text{if } \tau < k_1 \text{ or } \tau > T - k_2, \\ \dfrac{1}{T - k_1 - k_2} & \text{for each other point of the sample;} \end{cases} \tag{40}$$

while still being noninformative for the remaining regime classifications. We can note that the same type of problem occurs in a classical framework when Andrews (1993) tests for structural change. He recommends choosing the prior domain of τ so that $\tau/T \in [0.15, 0.85]$.

Ferreira (1975) uses another type of noninformative prior which is based on the expression of the posterior density of τ. In my notation and using (15), the two priors advocated by Ferreira are

$$\varphi(\tau) \propto [\tau \, (T - \tau)]^{1/2} \tag{41}$$

and

$$\varphi(\tau) \propto [\tau \, (T - \tau) \, M_*(\tau)]^{1/2}. \tag{42}$$

The weights in (41) could be interpreted as coming from a beta density on the interval $[0, T]$. The second prior (42) includes a term coming from the posterior density of τ [see (16)].

The general formulation of the posterior density (16) hides the fact that the posterior density of τ is now a discrete density. As a consequence τ is integrated out by a discrete summation with $d\tau = 1$:

$$\varphi(\beta|y) \propto \sum_{\tau=1}^{T} \varphi(\beta|\tau, y) \, \varphi(\tau|y) \tag{43}$$

Posterior results, as reported by Ferreira for an artificial sample, do not seem to be fundamentally affected by the shape of the prior density on τ. However, using (42) gives a posterior expectation for σ^2 which is slightly lower than in the other cases. As a matter of fact, this type of prior belongs to a wider class of prior densities which are liked by some Bayesians because they have the property of suppressing unimportant factors of the likelihood which sometimes may cause trouble. Here the prior (42) has the property of cancelling out the term $|M_*(\tau)|^{-1/2}$ from the posterior density of τ, which simplifies to

$$\varphi(\tau|y) \propto s_*(\tau)^{-(T-k)/2}[\tau\,(T - \tau)]^{1/2}. \tag{44}$$

As $|M_*(\tau)|$ does not vary very much in general (see for instance Bacon and Watts (1971), there is no major problem in using such a prior and posterior for τ. But there is also no good reason to do so, as here the term $|M_*(\tau)|$ does not cause a problem (contrary to the smooth transition case detailed below).

As the marginal posterior of τ is discrete, each point of this marginal can be interpreted as the probability that τ takes a particular value. So if no break point has a probability greater than a certain level, I can conclude with a posterior odds argument that there is no dominant break in the sample. However, the argument should be refined. There is no dominant break point if no point of the posterior of τ is higher than 0.5. But a gradual transition around a particular date may be the alternative. This is the case when the posterior of τ is concentrated around a particular value with a configuration like the one for instance presented in Figure 1 below. On the contrary, it is evidence of linearity when the probability is uniformly scattered along the domain of definition of τ.

The model with a single time break generates two regimes. It can be usefully generalised to two (or m) breaks with thus three (or $m + 1$) regimes. The algebra is not substantially changed. A particular three regime model may be of special interest. A structural break means that a long term relationship is modified under the influence of a shock or a change of habit, or a change in the composition of a macroeconomic aggregate. With a single break, no way is given to the model to return to its previous state of equilibrium. In a two regime model, the step function

$$\mathrm{ID}(\tau_1, \tau_2) = \mathrm{ID}(t - \tau_2) - \mathrm{ID}(t - \tau_1) \quad \text{with} \quad \tau_1 < \tau_2, \tag{45}$$

where $\mathrm{ID}(\cdot)$ is as defined in (37), is equal to zero when t is outside the interval $[\tau_1, \tau_2]$, and one in between. This step function says that one regime corresponds to the long term situation represented by $t < \tau_1$ or $t > \tau_2$ while the supplementary regime (regime 2) corresponds to the temporary disequilibrium situation which occurs between dates τ_1 and τ_2.

4.2   Smooth transition

A smooth transition between the two regimes is obtained by replacing the step indicator function (37) by a smooth, monotonically increasing, odd function $F(\cdot)$ with $F(-\infty) = 0$ and $F(\infty) = 1$. Many types of such functions are available (see the discussion in Bacon and Watts 1971). However, $F(\cdot)$ is usually chosen to be a cumulative distribution: the cumulative distribution of a normal distribution as in Goldfeld and Quandt (1973),[8] or more commonly the logistic function as advocated by Teräsvirta (1994):

$$F(\gamma, t - \tau) = \frac{1}{1 + e^{-\gamma(t-\tau)}},$$ (46)

or the hyperbolic tangent as in Tsurumi (1982). The parameter γ determines the degree of smoothness in the transition and is chosen to be positive in order to define the transition function uniquely. The step transition model (38) is changed to

$$y_t = x_t' \beta_1 + F(\gamma, t - \tau) x_t' \delta + \epsilon_t$$ (47)

and

$$x_t'(\gamma, c) = x_t'(\gamma, \tau) = [x_t', \ F(\gamma, t - \tau) x_t'].$$ (48)

There is now one extra parameter γ, which is restricted to be positive. As underlined in Section 2, this parameter, together with the form of the nonlinearity, causes identification and integrability problems which render the use of both the noninformative Cauchy prior (22) on γ and the normal prior (19) mandatory for a safe Bayesian analysis of this model.

As the transition is smooth, it is no longer possible to classify an observation as belonging to a single regime, and consequently β becomes apparently identified, whatever the value of τ. But for large values of γ, we go back to the previous configuration, so it may be safe to keep the same prior on τ as in the step transition model.

With these prior informations, the joint posterior density of τ and γ is well defined all over the parameter space including the point $\gamma = 0$. Its expression is

$$\varphi(\gamma, \tau | y) \propto |s_*(\gamma, \tau)|^{-(T-k)/2} |M_*(\gamma, \tau)|^{-1/2} \varphi(\tau) \varphi(\gamma).$$ (49)

This density is peculiar, as it is a mixture of a discrete and a continuous density. By a numerical integration over γ, I can compute the discrete probabilities attached to each of the $T - k$ possible values of τ. By a discrete summation over τ, I get the posterior density and moments of γ. The marginal posterior density of β is obtained by the weighted sum over τ of an integral over γ:

$$\varphi(\beta \mid y) \propto \sum_{\tau=1}^{T} \int_{\gamma>0} \varphi(\beta | \gamma, \tau, y) \varphi(\gamma, \tau | y) \, d\gamma$$ (50)

---

[8] They, however, make that assumption just for estimation purposes, to implement a maximum likelihood procedure as a step function is not differentiable and thus creates problems for a numerical optimisation routine.

The interpretation of the regression coefficients β may be difficult. When the change of regime is abrupt, the value of the function $\text{ID}(t - \tau)$ is either zero or one. So the regimes are clearly defined by a zero–one rule, and β_i is concerned only with the observations belonging to regime i. When the transition is smooth, the posterior expectations of the regression coefficients computed with (50) reflect the fact that the change of regime could be very gradual. Observation of one regime influences the regression coefficient of the other regime. So it may be rather difficult to compare the inference results on β between the two models when γ is small.

The double step function (45) can be generalised to the smooth transition case with

$$F(\gamma, \tau_1, \tau_2) = \frac{1}{1 + e^{\gamma(t - \tau_1)(t - \tau_2)}}. \tag{51}$$

This transition function was suggested as a generalisation of the exponential smooth transition function by Jansen and Teräsvirta (1996).

4.3   *Comparing step and smooth transition models*

Step and smooth transitions based on a time index are two alternative ways of modelling a structural break. The step transition model is simpler to manipulate in a Bayesian framework and is often used first. There are marked differences between the classical and the Bayesian approach. In a classical framework, τ is determined by a grid search, comparing a suite of conditional linear regressions. As a result, inference on β gives a conditional estimator, with a fixed sample separation in the step transition case. In the Bayesian approach, on the contrary, τ is integrated out, so $\text{E}(\beta|y)$ is a marginal estimator which depends not on a single sample separation, but on the most likely and averaged sample separations. So even with a step transition function, the Bayesian model can account for a rather smooth transition, depending on the sample configuration. This being said, it becomes less evident how to strictly separate the step and smooth transition models from a modelling point of view in a Bayesian framework.

The graph of the posterior density of τ in a step transition model gives direct intuitive results concerning the degree of abruptness of the switching. If most of the probability appears for one value of τ, this is the confirmation of an abrupt change. If on the contrary, most of the probability is scattered around one value of τ with a nice bell shape, this is evidence for a gradual transition.

In a smooth transition model, a high value of γ approximates the behaviour of the step transition model (38). When can $\text{E}(\gamma|y)$ be considered as high? *First,* γ must be made scale-free by an appropriate normalisation of $z_{t-d} - c$. Usual practice consists in dividing by the standard deviation of z_t. When z_t is a time index, γ is linked to the periodicity of the observations. The usual scaling is dividing t by $N \times S = T$, where N is the number of years and S the number of observations per year. In this case the standard deviation of the resulting time index is $\sqrt{0.083333 - 0.083333/T^2}$. Consequently it is preferable to divide the time index by $T \times \sqrt{0.083333}$. Second, the right tail of the posterior density of γ behaves like that of the truncated Cauchy prior. When the Cauchy prior is normalised to one on $[0, \infty]$, it has the property that the tail probability after 12.706 is equal to 5%, and after 63.66 to 1%. One can then conclude that $E(\gamma|y)$ is large if it is greater than 12 or greater than 66, respectively. However, a Schwarz criterion is certainly safer for choosing between the step and the smooth transition functions.

4.4   The French consumption function

A traditional consumption function, as reported for instance in Davidson, Hendry, Sbra, and Yeo (1978), relates the logarithm of real consumption to the logarithm of real disposable income. This formulation implies that in the long run the saving ratio is constant (provided the long run elasticity equals unity). Carruth and Henley (1990) pointed out that this ratio had declined a lot in the late eighties in England, leading to the predictive failure of many of the existing consumption functions.

The French economy seems to have also experienced a fall in its saving ratio; witness Figure 3 below, which displays the ratio between CP (real consumption) and RDP (real disposable income). This is the mean propensity to consume displayed for quarterly data covering the period 1963:1–1991:4, coming from the Laroque (INSEE) database. It remained fairly stationary till 1977, despite the first oil shock of 1974. As noted by Villa (1996), profits were first to be affected by the oil shock of 1974, which was transmitted only later to wages. The share of wages in real GNP rose till 1977 and eventually until 1982. Income distribution and consumption played a countercyclical role. Adjustments operated later with the conjugate effects of wage and fiscal policies.

This change in the structure of earnings is difficult to incorporate for modelling consumption. Hendry and von Ungern-Sternberg (1981) for the UK, Brodin and Nymoen (1992) for Norway, and Villa (1996) for France introduced personal wealth as an explanation for the rate of growth of consumption. A justification can be found in the life cycle theory (see e.g., Molana 1991). But the personal wealth series has a much greater variance than the usual disposable income series. Moreover, the theoretical life cycle model supposes that consumption and wealth are cointegrated (Molana 1991). This may fail to happen for some data sets (Pierse and Snell 1995). In this section, I prefer to investigate if allowing for a temporal change in the mean propensity to consume restores the stability of the traditional consumption function, and where the break occurs. If the model is expressed in logarithms, a break on the constant term is all I need.

I shall concentrate my efforts on estimating a consumption function of the following form:

$$\Delta \log CP = \delta + \delta_0 \, d69.2 + \gamma \Delta \log CP_{t-4} + \beta_0 \Delta \log RDP$$

$$+ (\alpha - 1)[\log CP_{t-1} - \nu \log RDP_{t-1}] + \epsilon_t, \qquad (52)$$

where d69.2 is a dummy variable which takes into account the effect of the important wage increases that followed the Matignon negotiations. In this type of consumption function, the steady state means $\log CP = \log K + \nu \log RDP$, K being the mean propensity to consume. In a dynamic equilibrium where $\Delta \log RDP = g_y$ and $\Delta \log CP = g_c = \nu g_y$, the log of the mean propensity to consume is expressed as

$$\log K = \frac{\delta}{1 - \alpha} + g_y \frac{\beta_0 - (1 - \gamma)\nu}{1 - \alpha}. \qquad (53)$$

A first estimation of this equation with a noninformative prior produces a result where the long term solution does not come in. The Schwarz criterion attached to this regression is -4.926. As I suspect a break in the constant term, I add a trend, a squared trend, and a cubic trend to build an augmented regression, coping with a possible nonlinearity in that direction. This augmented regression is preferred, as it gets a Schwarz criterion of -4.931.

Let me now consider a step transition function for the constant term. The model is reparameterised in such a way so as to allow for the possibility of a unit long term elasticity. Using a noninformative prior, Bayesian inference produces the following output

$$\Delta \log CP = \underset{[0.041]}{0.11} + \underset{[0.0035]}{0.013} \, ID(t - \tau) + \underset{[0.0067]}{0.021} \, d69.2$$

$$- \underset{[0.079]}{0.23} \, \Delta \log CP_{-4} + \underset{[0.067]}{0.25} \, \Delta \log RDP$$

$$- \underset{[0.040]}{0.16} \log \left(\frac{CP}{RDP} \right)_{-1} - \underset{[0.0031]}{0.011} \log RDP_{-1},$$

$$E(\sigma^2 | y) = 0.000423,$$

$$\text{Schwarz} = -5.030.$$

Note that this fully nonlinear model receives more evidence from the sample than the augmented regression, due to a lower value of its Schwarz criterion. The long term elasticity ν of consumption to income is lower than one – precisely, 0.93 (computed with posterior means) – while the marginal propensity to consume is $0.25/(1 + 0.23) = 0.20$.

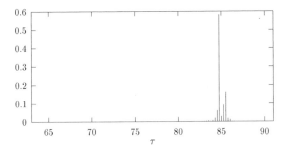

Figure 1. Posterior density of τ for the French consumption function with a step transition.

The expectation of τ is 1984.1, while the mode is 84.4. The probability of the posterior mode is 0.58. Figure 1 shows that the break point seems to be unique around this date, but 42% of the probability is scattered around this point. The shift in the constant term after 1984.1 means that the mean propensity to consume increased from $\exp[0.11/0.16 + g_x(0.25 - 0.93 \times 1.23)] = 1.99 \exp[-0.89gx]$ to $2.16 \exp[-0.89gx]$. These figures are unrealistic, whatever the value of g_y. They cast some doubt on the specification of the equation.

As an everlasting increase in the mean propensity to consume, as suggested by the single break model, is improbable, let me try the double break model (45), still with a diffuse prior. With this model, the unit elasticity hypothesis is accepted. The final results are

$$\Delta \log CP = -\underset{[0.00041]}{0.0088} - \underset{[0.0017]}{0.0071} \, ID(\tau_1, \tau_2) + \underset{[0.0067]}{0.019} \, d692$$

$$- \underset{[0.078]}{0.26} \, \Delta \log CP_{-4} + \underset{[0.069]}{0.23} \, \Delta \log RDP$$

$$- \underset{[0.022]}{0.11} \log \left(\frac{CP}{RDP} \right)_{-1},$$

$E(\sigma^2|y) = 0.000432,$

Schwarz $= -5.051.$

This model is slightly preferred to the one break model on the basis of the Schwarz criterion. Moreover, it has many remarkable features (Figure 2):

The long run elasticity is now equal to one.
There is one break with expectation at 1984.1, as before, but also a second break earlier, in 1973.3. These breaks are well marked, with posterior standard deviations of 3 quarters.

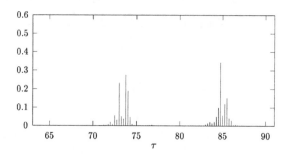

Figure 2. Double break for the French consumption function with a step transition.

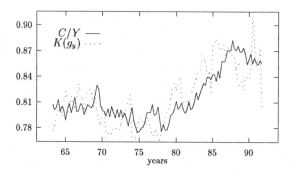

Figure 3. Mean propensity to consume: data and simulation with the double threshold model.

The mean propensity to consume is $0.92 \exp(-9.36 g_y)$ in the long term equilibrium, but seems lower, $0.87 \exp(-9.36 g_y)$, during the period of adjustment. With $g_y \in [0, 0.02]$, these figures are within the order of magnitude of the data. Figure 3 reproduces the data on C/Y and $K(g_y)$ with a g_y smoothed by a moving average of order 5. The two series follow similar paths. Even if the constant term is lower during the indermediate regime between 1973 and 1984, the resulting dynamic mean propensity to consume, $K(g_y)$, is slightly decreasing till 1975, but rises between 1978 and 1986, following the path shown by the ratio CP/RDP. The model thus manages to cope with the change in the structure of earning.

The final question is what will yield a smooth transition function. With the conditional normal and truncated Cauchy priors, the inference results are as

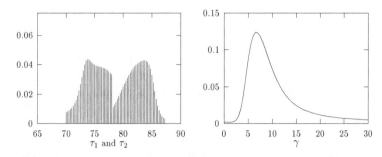

Figure 4. Double break for the French consumption function with a smooth transition.

follows:

$$\Delta \log CP = -\underset{[0.0047]}{0.0092} - \underset{[0.0043]}{0.012} \, ID(\tau_1, \tau_2) + \underset{[0.0069]}{0.019} \, d692$$

$$-\underset{[0.079]}{0.24} \, \Delta \log CP_{-4} + \underset{[0.070]}{0.26} \, \Delta \log RDP$$

$$-\underset{[0.025]}{0.10} \log \left(\frac{CP}{RDP} \right)_{-1},$$

$$E(\sigma^2|y) = 0.000453,$$

$$Schwarz = -5.003.$$

The posterior densities of τ_1 and τ_2 in Figure 4 indicate the same modes for the breaking dates. But the message becomes muddled, as the probability is regularly distributed around the modes. The jump comes from the fact that the posterior densities of τ_1 and τ_2 are presented on the same graph and that they both integrate to one. The integration ranges are the same as in the step function case. The posterior density of γ indicates that high values are not rejected with a mean of 10 and a posterior 95% confidence interval of [0, 23]. As a matter of fact, the Schwarz criterion would favour the step function.

5   Transition models based on a continuous variable

In the previous section, the change of regime was governed by a time index, comparing t with an unknown random date τ. The switching was thus exogenous and occurred at a single point of time as z_t was trending. I now study another type of model where the switching is done by comparing either the value of the lagged endogenous variable or the value of an exogenous variable (or a linear combination of exogenous variables) with an unknown random threshold c. The switching variable(s) contained in z_t is continuous, so the threshold c is no longer a discrete parameter. But the shape of its posterior density is greatly influenced by the nature of the transition function as explained in Section 2.

The models covered in this section are very popular in statistics and econometrics. One can distinguish TR (transition regression) models with exogenous variables and TAR (transition autoregressive) models. Generally the transition is based on exogenous variables in TR models and on past values of the endogenous variable in TAR models, which thus become SETAR models. SETAR models can be viewed as an alternative to linear ARMA models. They may introduce asymmetries in the dynamic path of the model. As the change of regime is self-exciting, these models are particularly interesting for prediction. Consequently I shall give in this section an algorithm to compute the predictive density for future observations.

For these models, one has also the choice between a step and a smooth transition. Apparently some authors do prefer the step transition, such as Potter (1995) in a classical framework or Pole and Smith (1985) and Geweke and Terui (1993) in a Bayesian framework. Other authors have a marked preference for smooth transitions such as Teräsvirta (1994) (classical side) or Bacon and Watts (1971) and Peguin-Feissolle (1994) (Bayesian side). I did not manage to find in these papers a discussion concerning the choice of a type of transition, step or smooth.

The economic literature has frequently exploited the possibilities created by the use of nonlinear models. I give here some examples concerning the business cycle, the neutrality of money, and wage formation. I could have also included nonlinear taxation. As underlined first by the NBER and then by Keynes in his general theory, the dynamics of the business cycle behave differently in expansion and in recession phases, which in general are shorter but violent. In a different domain, Keynes also stated that positive changes in the money supply are neutral, while negative changes have real effects on aggregate output. In their empirical study, Ravn and Sola (1996) found on the contrary that big changes in nominal demand are neutral, while small changes have a real effect, indicating the relevance of menu cost models and nominal rigidities. Bruno (1995) insists on a nonlinear trade-off between growth and inflation. In the domain of wage formation, Johansen (1995) rejects the Phillips curve but points out a nonlinear relation between real wages and the unemployment rate for Norway with a functional form of the type $WP = f(1/U^2)$, where WP is the level of the real wage and U the rate of unemployment. His economic model can be seen as an approximation of a truly nonlinear model. Finally, Burgess (1992) presents a model of variable adjustment costs for employment in Britain, where the adjustment cost is a function of the degree of tightness on the labour market. His model generates asymmetric dynamics.

5.1   *Step transition regression models*

I shall first review two step transition regression models which have been treated in the Bayesian literature and show that in fact the statistical treatment of these models is identical and enters the general framework developed in Section 2. Pole and Smith (1985) consider the following switching regression model:

$$y_t = \begin{cases} x'_{1t}\beta_1 + \epsilon_t & \text{if} \quad \tilde{z}'_t\theta \leq 0, \\ x'_{2t}\beta_2 + \epsilon_t & \text{if} \quad \tilde{z}'_t\theta > 0. \end{cases} \tag{54}$$

Geweke and Terui (1993) and Chen and Lee (1995) consider the SETAR model of Tong and Lim (1980):

$$y_t = \begin{cases} B_1(L)\, y_t + \epsilon_t & \text{if} \quad y_{t-d} \leq c, \\ B_2(L)\, y_t + \epsilon_t & \text{if} \quad y_{t-d} > c. \end{cases} \tag{55}$$

These models differ by the type of switching rule. But they can be made identical if

$$\begin{aligned} \tilde{z}'_t &= [y_{t-d}, 1], \\ \theta' &= [1, -c]. \end{aligned} \tag{56}$$

Consequently a quite generic model is

$$y_t = x'_t\beta_1 + \text{ID}(z_{t-d} - c)\, x'_t(\beta_2 - \beta_1) + \epsilon_t \tag{57}$$

with

$$\text{ID}(z_{t-d} - c) = \begin{cases} 0 & \text{if} \quad z_{t-d} - c \leq 0, \\ 1 & \text{otherwise} \end{cases} \tag{58}$$

and

$$x'_t(c) = [x'_t, \ \text{ID}(z_{t-d} - c)\, x'_t]. \tag{59}$$

The formulation of Pole and Smith of course allows for more than one variable to govern the switching. It does not explicitly consider that there is an unknown but constant threshold c. In their model, $\tilde{z}_t$ is composed of two or more continuous and nonconstant variables (there is no constant term in their $\tilde{z}_t$). The inequality $z'_t\theta < 0$ determines cones in the θ-space when the dimension of $\tilde{z}_t$ is two, and hypercones in the general case. Some data configurations may cause problems and lead in some cases to bimodal posterior densities in small samples. For more details, see their paper. In the sequel, I shall consider only the case where $\tilde{z}_t = [z_t, 1]$, i.e., models with a constant unknown threshold c. Note, however, that threshold models are sometime generalised by considering more than one variable in z_t. In the simple bivariate case, the inequality $z_{1t} + \lambda z_{2t} < c$ can induce some serious pathologies for the bivariate posterior density of c and λ which are hard to understand.

The important question treated in the model of Geweke and Terui (1993) is the determination of the value of d. Geweke and Terui consider it as an extra parameter called the delay parameter and simulate the joint posterior density of (d, c) in order to get information on the optimal value of d. Note that I have advocated in Section 2 that the choice of d can also be viewed as a model selection problem which can be solved using an information criterion.

As in the case where the transition variable was a time index, the model presents an identification problem. Basically, there must be enough observations per regime. Let me order the observations in z_t by increasing values and reorder the other variables accordingly. As the threshold c is compared with the ordered values of z_t, its domain of variation is determined by the minimum and maximum values of this vector. But in order to have at least k_i observations per regime, the integration range should be limited above the k_i smallest values of z_t and below the k_i greatest values of z_t. But in case of collinearity, this is not a sufficient condition. Let us denote by H the empirical distribution of z_t. As c belongs to R, a possible prior is

$$\varphi(c) = \begin{cases} 1 & \text{if} \quad c \in [c_a, c_b], \\ 0 & \text{otherwise} \end{cases} \tag{60}$$

with c_a and c_b such that $H(c_a) = 0.15$ and $H(c_b) = 0.85$. A similar suggestion can be found in the classical literature (Andrews 1993). Any type of informative prior could be used on the restricted domain of definition, as anyway this parameter is integrated out numerically.

Chen and Lee (1995) define a Gibbs sampler for TAR models, but the posterior analysis of this model enters the general formulation of Section 2. In the SETAR model, c is the sole nonlinear parameter, and thus inference is a one dimensional numerical integration exercise. As underlined in Section 2, the posterior density of c given by

$$\varphi(c|y) \propto |s_*(c)|^{-(T-k)/2} |M_*(c)|^{-1/2} \varphi(c) \tag{61}$$

may be extremely peculiar. It may be discontinuous for a finite number of points given by the sample values of z_t, but is continuous (and constant) between two consecutive values of z_t. Thus this parameter should be integrated out not by direct summation like τ in the previous section, but by a usual Simpson rule.

Pole and Smith (1985), Geweke and Terui (1993), and Chen and Lee (1995) consider explicitly the possibility of unequal variances between the two regimes, using the formulation given in Section 2. Various models with unequal variances are considered in Pfann, Schotman, and Tschernig (1996). It is easy to show that these models enter also the framework of Section 2. In particular, with

$$
y_t = \begin{cases} x_t'\beta_1 + w_t^\alpha \epsilon_t & \text{if } z_t \leq c, \\ x_t'\beta_2 + w_t^\alpha \epsilon_t & \text{if } z_t > c, \end{cases} \tag{62}
$$

the variance of the error term varies according to the variable w_t, independently of the regime switch. The authors use a Gibbs sampler to evaluate the posterior density of this model, but it can be rewritten as (57) with the variance of the error term being decomposed into

$$
\text{Var}(\epsilon_t) = \sigma^2 w_t^{2\alpha} = \sigma^2 h_t(\alpha). \tag{63}
$$

The second model they consider assumes that the variance of y_t can change according to a different threshold than the conditional expectation of y_t. More precisely,

$$
y_t = \begin{cases} x_t'\beta_1 + \epsilon_t & \text{if } z_t \leq c_1, \\ x_t'\beta_2 + \epsilon_t & \text{if } z_t > c_1 \end{cases} \tag{64}
$$

and

$$
\text{Var}(\epsilon_t) = \sigma^2[(1 - \text{ID}(z_t - c_2)) + \phi \, \text{ID}(z_t - c_2)] = \sigma^2 h_t(c_2, \phi). \tag{65}
$$

We have now a model with two thresholds c_1 and c_2 and a total of three nonlinear parameters. Note that c_1 and c_2 are identified by the restriction $c_1 < c_2$.

5.2   Smooth transition regression models

The step transition function is replaced in (57) by a smooth transition function. The generic model can thus be written

$$
y_t = x_t'\beta_1 + F(\gamma, z_{t-d} - c) x_t'(\beta_2 - \beta_1) + \epsilon_t \tag{66}
$$

and

$$
x_t' = [x't, \ F(\gamma, z_{t-d} - c) x_t']. \tag{67}
$$

c and γ are in general the sole nonlinear parameters of the model, leading to numerical integration of dimension two. As there is a smooth transition between the two regimes, the posterior density of c no longer presents the same discontinuities as in the step transition case.

The logistic transition function, already introduced in Section 4, is a generalisation of the step function (3). It is an odd function with $F(-\infty) = 0$, $F(+\infty) = 1$. In business cycle studies, it aims at detecting asymmetries between positive and negative shocks. On the contrary, an even function ($F(\pm\infty) = 1$, $F(0) = 0$), such as the exponential function used for instance by Haggan and

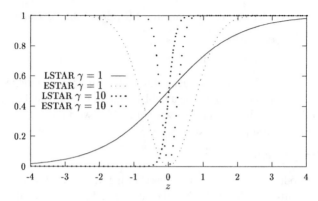

Figure 5. Two transition functions.

Ozaki (1981), aims at discriminating between the effects of big shocks compared to small ones. The exponential transition function is

$$F(\gamma, z_{t-d} - c) = 1 - e^{-\gamma (z_{t-d}-c)^2}. \tag{68}$$

See Bacon and Watts (1971) for other types of even functions. The difference of behaviour between the logistic and the exponential is well illustrated in Figure 5, where graphs of the logistic function and the exponential function are given for various values of γ. When $\gamma \to \infty$, the logistic function tends to the Heaviside function (58) and the model is still nonlinear. When $\gamma \to \infty$, the exponential function is always one except for the point $z_{t-d} = c$, which is of zero probability. Consequently, the model becomes linear when $\gamma = 0$, as was discussed in Sections 2 and 3, but also when $\gamma \to \infty$. The exponential transition function introduces a supplementary identification problem for $\delta = \beta_2 - \beta_1$. Note that the transition function (51), which was suggested as a generalisation of the exponential smooth transition function by Jansen and Teräsvirta (1996), does not present this problem.

Peguin-Feissolle (1994) has proposed a Bayesian analysis of the logistic smooth transition autoregressive (LSTAR) model, a simplified version of which is

$$y_t = \beta_0 + \beta_1 y_{t-1} + F(\gamma, y_{t-d} - c)(\delta_0 + \delta_1 y_{t-1}) + \epsilon_t. \tag{69}$$

She uses Monte Carlo integration by importance sampling to analyse the posterior and predictive densities of this model. But it is easy to show that once again this model enters the general framework developed in Section 2.

Remark: In the applications reported in Granger and Teräsvirta (1993, Chapter 9), it appears that the reported values of γ are fairly large for the LSTAR model, even when y_{t-d} is normalised. In the transition model based on a time index, there was presumably only one change at a particular point of time. When the transition variable is continuous and stationary, many changes of regime can occur in a given sample. All of these changes have to obey the same transition function, which has constant parameters. Consequently the threshold c is always the same and the speed of change γ is also always the same, irrespective of the size of the jump given by z_t which motivates the regime change. This restriction may explain why it is in general so difficult to make inference on γ in LSTAR models. However, when Teräsvirta and Anderson (1992) use the exponential transition function, the reported γ has midrange values, because both small and large values correspond to linearity, which is rejected by the sample.

5.3 Predictive density

STAR, SETAR, and LSTAR models are very attractive for prediction, as the change of regime becomes totally endogenous. Let me consider a simple version of these models with

$$y_t = \beta_1 y_{t-1} + F(\gamma, y_{t-d} - c)\delta y_{t-1} + \epsilon_t. \tag{70}$$

Denote by y^* the set of s future values of y (i.e., y_{T+1}^{T+s}), and consider $g(y^*)$, a function of the future observations. I want to compute the expected value of this function. For various choices of $g(\cdot)$, I shall get the expectation, variance, and density function of future observations. The posterior expectation of this function corresponds to

$$E[g(y^*)|y] = E_\zeta[E_{y^*}(g(y^*)|y, \zeta)]$$

$$= \int_\zeta \left[\int_{\mathbf{R}^s} g(y^*)\, p(y^*|y, \zeta)\, dy^* \right] \varphi(\zeta|y)\, d\zeta, \tag{71}$$

where $p(y^*|y, \zeta)$ is the density of future observations, and ζ represents all the parameters of the model: c, γ, β, and σ^2. To evaluate these integrals, I shall propose a method which is adapted from Geweke (1989). For a given drawing of $\epsilon^* = \epsilon_{T+1}^{T+s}$ and conditionally on β, γ, and c, it is fairly easy to generate y^* by recursion starting from

$$y_{T+1} = \beta_1 y_T + F(\gamma, y_{T-d+1} - c)\, \delta\, y_T + \epsilon_{T+1}. \tag{72}$$

Conditionally on γ and c, I know analytically the posterior densities of β and σ^2, which are respectively Student and inverted gamma2. They are indexed by $M_*^{-1}(\gamma, c)$, $\beta_*(\gamma, c)$, and $s_*(\gamma, c)$. Consequently, a random drawing of β can be obtained conditionally on σ^2 and c. Moreover, in order to take account of the uncertainty on σ^2, a random drawing of ϵ is obtained from a Student density of $T - k$ degrees of freedom, zero mean, and scale parameter equal to the conditional posterior mean of σ^2.

I have now all the needed ingredients to evaluate the predictive moments of y in the same numerical integration loops used for computing the posterior moments of the parameters. The algorithm is as follows:

> For each point on the integration grid of γ, c,
> compute the conditional expectation $E(\sigma^2|y, \gamma, c)$,
> compute the conditional moments of β,
> draw s values for ϵ from a $t(0, E(\sigma^2|y, \gamma, c), T - k)$,
> draw a β from its conditional Student posterior density,
> compute by recursion y^*,
> accumulate with the appropriate weights of the Simpson rule

$$g(y^*)\,\varphi(\gamma, c|y).$$

This is thus a mix between a deterministic integration rule and a direct Monte Carlo simulation for the future observations.

5.4   *Business cycle and nonlinear asymmetries*

There is a debate about the presence or absence of nonlinearities in the US GNP. Hamilton's (1989) model explains the changes in the log of real quarterly US GNP over 1951:2 to 1984:4 using a discrete shift between two regimes, where the shift obeys a Markov process. His model reproduces the usual turning points of the US business cycle as determined by the NBER. Boldin (1996) showed that Hamilton found only a local maximum of the likelihood function and that his results were not robust to an extension of the sample period. Potter (1995), using the same data but on the period 1947:1–1990:4, adjusts a two regime step transition TAR(5) model with unequal variances where $d = 2$ and $c = 0$. Though his model seems to fit the data correctly, the linearity tests he reports, based on Hansen's (1996) theoretical results, reject linearity at 10%, but not at 5%. Hansen (1996) himself concludes a statistical artifact for nonlinearity in the US GNP. However, tests presented in Tiao and Tsay (1994) do conclude nonlinearity of the US GNP. Teräsvirta and Anderson (1992) underline that an industrial production index (IPI) has a greater variance than the GNP series and thus is more capable of displaying nonlinearity.

I shall consider both the US real GNP and the IPI for the period 1960:1–1994:1. The data come from the OECD database. They are seasonally adjusted for the GNP. Monthly, not seasonally adjusted data are available for the IPI. I converted them to quarterly by averaging. I then took the fourth difference of these two series to stationarise them and to remove possible remaining seasonality. The resulting series are displayed in Figure 6.

Table 1. *Selecting linearity or nonlinearity for US real GNP and US industrial production*

	GNP		IPI	
d	Schwarz	P-value	Schwarz	P-value
—	−3.91	—	−2.63	—
1	−3.81	0.74	−2.52	0.89
2	−3.87	0.036	−2.62	0.004
3	−3.86	0.069	−2.66	0.001
4	−3.90	0.006	−2.55	0.25
5	−3.80	0.87	−2.53	0.68

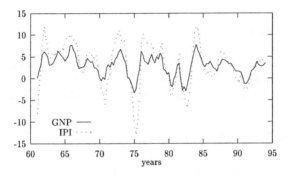

Figure 6. US business indicators.

I shall try first to verify the presence of nonlinearity by analysing a sequence of augmented regressions as described in Section 3. An AR(5) fits best for both series, but leads to very messy augmented regressions, as 15 supplementary regressors are needed. An AR(1) is probably biased, but nevertheless picks up a lot of the variance of the series. Let me now compare the linear AR(1) model with a sequence of augmented regressions where d varies between 1 and 5. The results are shown in Table 1.

The "Schwarz" columns give the Bayesian Schwarz criterion, and the "P-value" columns the classical P-value for the F-test of linearity. At the classical 5% level, linearity is rejected for the US GNP. The smallest P-value is obtained for $d = 4$. However, a Schwarz criterion selects the linear model. If I consider an AR(5) model, linearity is also no longer rejected by classical P-values. This is in agreement with some of the empirical results mentioned above.

For the US IPI, linearity is rejected at the classical level of 1%. The Schwarz criterion selects the nonlinear model with $d = 3$. This is in agreement with the results of Teräsvirta and Anderson (1992) on a smaller sample size. These results are confirmed with an AR(5).

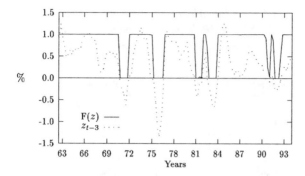

Figure 7. Transition function of the LSTAR model for US industrial production.

Let me now try to adjust an LSTAR model on the IPI data only, as there is not enough evidence in favour of nonlinearity in the GNP data. Let me call the transformed variable y_t. I shall use a restricted uniform prior on c, a Cauchy prior on γ, and a conditional normal prior on the second regime regression parameters as defined in (19). I started with a model including six lags of y_t and ended up with

$$y_t = \underset{[0.0051]}{0.016} + \underset{[0.11]}{1.06}\, y_{t-1} - \underset{[0.90]}{0.36}\, y_{t-2} - \underset{[0.19]}{1.06}\, y_{t-4} + \underset{[0.20]}{0.59}\, y_{t-5}$$

$$+ (1 + \exp(-\underset{[30]}{43}\, \underset{}{20.34}(y_{t-3} + \underset{[0.0020]}{0.0036})))^{-1}$$

$$\times (\underset{[0.0062]}{0.020} + \underset{[0.10]}{0.25}\, y_{t-1} + \underset{[0.20]}{0.97}\, y_{t-4}) - \underset{[0.21]}{0.59}\, y_{t-5},$$

$$E(\sigma^2 | y) = 0.000245,$$

$$\text{Schwarz} = -3.164.$$

As the transition variable was normalised by its standard deviation, the posterior expectation of γ can be directly interpreted. With a value of 43, the transition between expansions and recessions is quite abrupt; witness the corresponding graph of $F(z)$ in Figure 7, which is shown together with the rescaled series z_{t-3}. Note that with a shorter sample (ending in 1986), $E(\gamma|y)$ was much lower. This is in agreement with the remark made at the end of Section 5.2.

The posterior expectation of c is slightly negative at -0.0036, but with such a large standard deviation that it could have been taken equal to zero. The posterior density of c has not a regular shape (see Figure 8). In fact, it tends toward an irregular histogram as $\gamma \rightarrow \infty$. Note that despite the very high posterior expectation of γ, there is still a difference between the step transition and the logistic transition.

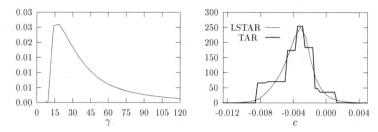

Figure 8. Posterior densities of γ and c of the LSTAR model for US industrial production.

The step transition model has a Schwarz criterion of -3.165, which is only marginally better than that of the smooth transition model. There is no major change in the regression coefficients. The posterior density of the threshold is displayed in Figure 8 on the same graph as for the smooth case. It shows how the shape of this posterior density is modified by the nature of the threshold. It is less scattered, but of course presents a series of points where it is not continuous. By the way, it is still centered on zero, which suggests again that the threshold could be pinpointed at this value. In this case the model is reduced to a linear regression model with multiplicative dummy variables.

The model is characterised by very different dynamic behaviour between the two regimes. Teräsvirta and Anderson (1992) suggest computing the dominant root of the lag polynomial for each regime. I have a dominant complex root of 1.124 for the recession regime and a dominant complex root of modulus 0.827 for the expansion regime. Consequently, the dynamics of the expansion is stationary, while the dynamics of the recession is explosive, suggesting that recessions are short-lived and violent. Out of 127 observations, 25 are in the recession regime, while 102 are in the expansion regime.

6   Conclusion

Bayesian methods are flexible enough to analyse nonlinear time series models. I have shown that they lead to simple calculations requiring simple integration in one, two, eventually three dimensions. In the case of step switching (switching rules based on inequalities), Bayesian methods appear to be superior to classical ones, as they are not in trouble with the nondifferentiability of the likelihood function. Bayesian methods proceed by simple averages over the parameter space. Osiewalski and Welfe (1998), for instance, analyse a wage–price two equation model where price indexation occurs only above an unknown threshold c. Their posterior density of c has no obvious maximum. It is a collection of pieces, the shape of which is determined by the presence of a Jacobian generated by the simultaneity of the model.

The simplicity of the calculations presented in this paper may seem at odds with the modern literature on the Gibbs sampler. Whenever possible, deterministic integrations rule must be used because they are quicker and more precise than sampling methods. However, much more complicated models have been treated in the Bayesian literature, and they all need Monte Carlo methods. Lubrano (1985) analyses a switching model where β and θ are identical. This model is linked to disequilibrium models. Inference is made by importance sampling. Koop and Potter (1995) study an autoregressive threshold model where the variable governing the switching is built recursively during inference, which is conducted using the Gibbs sampler. Ghysels, McCulloch, and Tsay (1994) propose a periodic switching regime model where the change of regime follows a seasonal Markov process. This model is also estimated using the Gibbs sampler. Related papers are Albert and Chib (1993) and McCulloch and Tsay (1993).

I shall say a final word concerning smooth and step transition functions. Bayesian methods are particularly at ease with step transition functions, whereas classical methods experience most of their difficulties in this area. On the contrary, a smooth transition is amenable to a classical treatment, as its likelihood function is differentiable. But as the posterior density of this type of model is not integrable under a flat prior, a Bayesian treatment is more delicate. As a matter of fact, in the empirical examples I presented, the step transition is preferred by the data most of the time. This is due to the fact that in a Bayesian framework, because of the averaging rule, a step transition function always allows for some progressivity in the transition.

Appendix A: Posterior analysis for the heteroskedastic case

Proof of Theorem 4:   With (23), σ^2 can still be integrated out analytically. The conditional posterior density of β is proportional to a Student density with

$$\varphi(\beta|\gamma, c, \phi, y) \propto \left[\prod h_t(\gamma, c, \phi)\right]^{-1/2}$$

$$\times f_t(\beta|\beta_*(\gamma, c, \phi), M_*(\gamma, c, \phi), s_*(\gamma, c, \phi), T - k), \quad (73)$$

and that of σ^2 is proportional to an inverted gamma2:

$$\varphi(\sigma^2|\gamma, c, \phi, y) \propto f_{IG}(\sigma^2|s_*(\gamma, c, \phi), T - k), \quad (74)$$

where:

$$M_*(\gamma, c, \phi) = \sum x_t(\gamma, c, \phi) x_t'(\gamma, c, \phi),$$

$$\beta_*(\gamma, c, \phi) = M_*^{-1}(\gamma, c, \phi) \sum x_t(\gamma, c, \phi) y_t(\gamma, c, \phi), \quad (75)$$

$$s_*(\gamma, c, \phi) = \sum y_t(\gamma, c, \phi)^2 - \beta_*'(\gamma, c, \phi) M_*(\gamma, c, \phi) \beta_*(\gamma, c, \phi).$$

The joint posterior density of γ, c, and ϕ is obtained as one over the integrating constant of the above Student density times its scaling factor times the prior densities of γ, c, and ϕ. $\square$

Corollary 1. *The marginal posterior density of β and σ^2 follows from*

$$\varphi(\beta|y) = \int f_t(\beta|\beta_*(\gamma, c, \phi), M_*(\gamma, c, \phi), s_*(\gamma, c, \phi), T - k)$$
$$\times \varphi(\gamma, c, \phi|y)\, d\gamma\, dc\, d\phi, \tag{76}$$

$$\varphi(\sigma^2|y) \propto \int f_{IG}(\sigma^2|s_*(\gamma, c, \phi), T - k)\, \varphi(\gamma, c, \phi|y)\, d\gamma\, dc\, d\phi. \tag{77}$$

The latter corresponds to the posterior density of the variance of the error term in the first regime. The posterior expectation of σ^2 is obtained by taking the expectation of the analytical conditional expectation of σ^2:

$$E(\sigma^2|y) = \int \frac{s_*(\gamma, c, \phi)}{T - k - 2}\varphi(\gamma, c, \phi|y)\, d\gamma\, dc\, d\phi. \tag{78}$$

In order to recover the posterior density of $\sigma_2^2 = \sigma^2\phi$, let me consider the one to one transformation $(\sigma^2, \phi) \longrightarrow (\sigma^2\phi, \phi)$ of the Jacobian ϕ. Starting from the joint posterior density of (σ^2, ϕ) given by

$$\varphi(\sigma^2, \phi|y) \propto \int \varphi(\sigma^2|\gamma, c, \phi, y)\, \varphi(\gamma, c, \phi|y)\, d\gamma\, dc. \tag{79}$$

I get

$$\varphi(\sigma_2^2|y) \propto \int \phi f_{IG}(\sigma_2^2|s_*(\gamma, c, \phi), T - k)\, \varphi(\gamma, c, \phi|y)\, d\gamma\, dc\, d\phi, \tag{80}$$

and the posterior expectation of σ_2^2 is given by

$$E(\sigma_2^2|y) = \int \phi \frac{s_*(\gamma, c, \phi)}{T - k - 2}\varphi(\gamma, c, \phi|y)\, d\gamma\, dc\, d\phi. \tag{81}$$

Appendix B: Simplification with a step transition function

The transition function $F(\cdot)$ becomes an indicator function $\mathrm{ID}_t(c)$ defined by

$$\mathrm{ID}_t(c) = \begin{cases} 0 & z_t - c \geq 0, \\ 1 & \text{otherwise.} \end{cases} \tag{82}$$

Each observation is now unambiguously classified in one regime or the other. For a given sample separation (a given value for c), the conditional likelihood function is

$$L\left(\beta, \sigma_1^2, \sigma_2^2; y, c\right) \propto \sigma_1^{-T_1(c)} \sigma_2^{-T_2(c)}$$

$$\times \exp\left(-\frac{1}{2\sigma_1^2}(y_1(c) - x_1'(c)\beta_1)'(y_1(c) - x_1'(c)\beta_1)\right)$$

$$\times \exp\left(-\frac{1}{2\sigma_2^2}(y_2(c) - x_2'(c)\beta_2)'(y_2(c) - x_2'(c)\beta_2)\right).$$

$$(83)$$

The quantities $y_i(c)$, $X_i(c)$, and $T_i(c)$ are defined by

$$\begin{aligned}
y_1(c) &= [y_t(1 - \mathrm{ID}_t(c))], & y_2(c) &= [y_t\,\mathrm{ID}_t(c)], \\
X_1(c) &= [X_t(1 - \mathrm{ID}_t(c))], & X_2(c) &= [X_t\,\mathrm{ID}_t(c)], \\
T_1(c) &= \sum_{t=1}^{T}(1 - \mathrm{ID}_t(c)), & T_2(c) &= T - T_1(c).
\end{aligned}$$

$$(84)$$

The subscripts indicate the regime. The symbol $[\cdot]$ indicates a complete matrix with generic element $\cdot$. The number of rows in $y_i(c)$ and $X_i(c)$ is equal to T, but some of the rows are rows of zeros. This type of parameterisation is different from the simple parameterisation (8) chosen in (11) and (25) and corresponds to the initial parameterisation given in (5).

The conditional posterior density of the regression parameters and of the two variances of the error terms can be decomposed into

$$\varphi\left(\beta, \sigma_1^2, \sigma_2^2|c, y\right) \propto \prod_{i=1}^{2} f_{IG}\left(\sigma_i^2\big|s_{*i}(c), T_i(c) - k_i\right) f_N\left(\beta_i\big|\sigma_i^2 M_{*i}^{-1}(c)\right) \quad (85)$$

with

$$\begin{aligned}
M_{*i}(c) &= X_i'(c)\,X_i(c), \\
\beta_{*i}(c) &= M_{*i}^{-1}(c)\,X_i'(c)\,y_i(c), \\
s_{*i}(c) &= y_i'(c)\,y_i(c) - \beta_{*i}'(c)\,M_{*i}(c)\,\beta_{*i}(c).
\end{aligned}$$

$$(86)$$

The posterior density of c is given by the integrating constants of the conditional posterior density of β, σ_1^2, and σ_2^2 times the prior density of c:

$$\varphi(c|y) \propto \varphi(c) \prod_{i=1}^{2} |s_{*i}(c)|^{-(T_i(c)-k_i)/2}\,|M_{*i}(c)|^{-1/2}. \quad (87)$$

The marginal posterior densities of β_i and σ_i^2 can all be analysed separately, and I have

$$\varphi(\beta_i|y) \propto \int f_t(\beta_i|\beta_{*i}(c), M_{*i}(c), s_{*i}(c), T_i(c) - k_i)\,\varphi(c|y)\,dc,$$

$$\varphi\left(\sigma_i^2|y\right) \propto \int f_{IG}\left(\sigma_i^2\big|s_{*i}(c), T_i(c) - k_i\right)\varphi(c|y)\,dc.$$

$$(88)$$

REFERENCES

Albert, J., and S. Chib (1993), "Bayesian Inference of Autoregressive Time Series with Mean and Variance Subject to Markov Jumps", *Journal of Business and Economic Statistics* 11: 1–15.

Andrews, D. W. K. (1993), "Tests for Parameter Instability and Structural Change with Unknown Change Point", *Econometrica* 61(4): 821–56.

Aprahamian, F., M. Lubrano, and V. Marimoutou (1994), "A Bayesian Approach to Misspecification Tests", Document de Travail GREQAM No. 94A06, Centre National de la Recheche Scientifique, Marseille.

Baba, Y., D. F. Hendry, and R. M. Starr (1992), "The Demand for Money in the U.S.A., 1960–1988", *Review of Economic Studies* 59(1): 25–61.

Bacon, D. W., and D. G. Watts (1971), "Estimating the Transition between Two Intersecting straight lines", *Biometrika* 58: 525–34.

Bauwens, L., and M. Lubrano (1998), "Bayesian Inference on GARCH Models Using the Gibbs Sampler", *The Econometric Journal*, 1: C23–46.

Ben David, D., and D. H. Papell (1995), "The Great Wars, the Great Crash and Steady State Growth: Some New Evidence about an Old Stylized Fact", *Journal of Monetary Economics* 36(3): 453–75.

Boldin, M. D. (1996), "A Check on the Robustness of Hamilton's Markov Switching Model Approach to the Economic Analysis of the Business Cycle", *Studies in Nonlinear Dynamics and Econometrics* 1(1): 35–46.

Brodin, P. A., and R. Nymoen (1992), "Wealth Effects and Exogeneity: The Norwegian Consumption Function 1966(1)–1989(4)", *Oxford Bulletin of Economics and Statistics* 54(3): 431–54.

Bruno, M. (1995), *Inflation, Growth and Monetary Control: Nonlinear Lessons from Crisis and Recovery*, Bank of Italy Paolo Baffi Lectures on Money and Finance. Rome: Edizioni dell'Elefante.

Burgess, S. M. (1992), "Nonlinear Dynamics in a Structural Model of Employment", *Journal of Applied Econometrics* 7(S): 101–18.

Campos, J., N. R. Ericsson, and D. F. Hendry (1996), "Cointegration Tests in the Presence of Structural Breaks", *Journal of Econometrics* 70(1): 187–220.

Carruth, A., and A. Henley (1990), "Can Existing Consumption Functions Forecast Consumer Spending in the Late 1980's?", *Oxford Bulletin of Economics and Statistics* 52: 211–22.

Chen, C. W. S., and J. C. Lee (1995), "Bayesian Inference of Threshold Autoregressive Models", *Journal of Time Series Analysis* 16: 483–92.

Cook, P., and L. D. Broemeling (1996), "Analyzing Threshold Autoregressions with a Bayesian Approach", in T. B. Fomby (ed.), *Advances in Econometrics: Bayesian Methods Applied to Time Series Data*, Vol. 11, Part B. Greenwich: JAI Press, pp. 89–107.

Davidson, J. E. H., D. F. Hendry, F. Sbra, and S. Yeo (1978), "Econometric Modelling of the Aggregate Time Series Relation between Consumption and Income in the UK", *The Economic Journal* 88: 661–92.

DeJong, D. N. (1996), "A Bayesian Search for Structural Breaks in U.S. GNP", in T. B. Fomby (ed.), *Advances in Econometrics: Bayesian Methods Applied to Time Series Data*, Vol. 11, Part B. Greenwich: JAI Press, pp. 109–46.

Ferreira, P. E. (1975), "A Bayesian Analysis of a Switching Regression Model: Known Number of Regimes", *Journal of the American Statistical Association* 70: 370–4.

Forbes, C. S., G. R. J. Kalb, and P. Kofman (1997), "Bayesian Arbitrage Threshold Analysis", Monash University Discussion Paper. Clayton, Australia.

Geweke, J. (1989), "Exact Predictive Densities for Linear Models with ARCH Disturbances", *Journal of Econometrics* 40: 63–86.

Geweke, J., and N. Terui (1993), "Bayesian Threshold Autoregressive Models for Nonlinear Time Series", *Journal of Time Series Analysis* 14: 441–54.

Ghysels, E., R. E. McCulloch, and R. S. Tsay (1994), "Bayesian Inference for Periodic Regime-Switching Models", Working Paper, CRDE, Université de Montréal.

Goldfeld, S. M., and R. E. Quandt (1973), "A Markov Model for Switching Regressions", *Journal of Econometrics* 1: 3–16.

Granger, C. W. J., and T. Teräsvirta (1993), *Modeling Nonlinear Economic Relationships*. Oxford: Oxford University Press.

Gregory, A. W., and B. E. Hansen (1996), "Tests for Cointegration in Models with Regime and Trend Shifts", *Oxford Bulletin of Economics and Statistics* 58(3): 555–60.

Haggan, V., and T. Ozaki (1981), "Modeling Nonlinear Random Vibrations Using an Amplitude-Dependent Autoregressive Time Series Model", *Biometrika* 68: 189–96.

Hamilton, G. D. (1989), "A New Approach to the Economic Analysis of Non-stationary Time Series and the Business Cycle", *Econometrica* 57: 357–84.

Hansen, B. E. (1996), "Inference When a Nuisance Parameter is Not Identified under the Null Hypothesis", *Econometrica* 64(2): 413–30.

Hendry, D. F., and T. von Ungern-Sternberg (1981), "Liquidity and Inflation Effects on Consumers' Expenditure", in A. Deaton (ed.), *Essays in the Theory and Measurement of Consumer Behaviour*. Cambridge: Cambridge University Press.

Jansen, E. S., and T. Teräsvirta (1996), "Testing Parameter Constancy and Super Exogeneity in Econometric Equations", *Oxford Bulletin of Economics and Statistics* 58(4): 735–63.

Johansen, K. (1995), "Norwegian Wage Curves", *Oxford Bulletin of Economics and Statistics* 57(2): 229–47.

Koop, G., and S. M. Potter (1995), "Bayesian Analysis of Endogenous Delay Threshold Models Using the Gibbs Sampler", Working Paper, Department of Economics, University of California, Los Angeles.

Lin, C. F., and T. Teräsvirta (1994), "Testing the Constancy of Regression Parameters against Continuous Change", *Journal of Econometrics* 62 (2): 211–28.

Lindgren, G. (1978), "Markov Regime Models for Mixed Distributions and Switching Regressions", *Scandinavian Journal of Statistics* 5: 81–91.

Lubrano, M. (1985), "Bayesian Analysis of Switching Regression Models", *Journal of Econometrics* 29: 69–95.

(1993), "Bayesian Unit Root Tests with Endogenous Break Point: An Application to the Analysis of Persistence in European Unemployment Rate Series", Mimeo, SPES Workshop on Wage and Price Formation, Maastricht, February.

(1995), "Bayesian Tests for Co-integration in the Case of Structural Breaks: An Application to the Analysis of Wage Moderation in France", *Recherches Economiques de Louvain* 61(4): 479–507.

Lutkepohl, H. (1993), "The Sources of the U.S. Money Demand Instability", *Empirical Economics* 18(4): 729–43.

Luukkonen, R., P. Saikkonen, and T. Teräsvirta (1988), "Testing Linearity against Smooth Transition Autoregression", *Biometrika* 75: 491–9.

McCulloch, R. E., and R. S. Tsay (1993), "Bayesian Inference and Prediction for Mean and Variance Shifts", *Journal of the American Statistical Association* 423: 968–78.

Molana, H. (1991), "The Time Series Consumption Function: Error Correction, Random Walk and the Steady-State", *The Economic Journal* 101: 382–403.

Ohtani, K. (1982), "Bayesian Estimation of the Switching Regression Model with Autocorrelated Errors", *Journal of Econometrics* 18: 251–61.

Osiewalski, J., and A. Welfe (1998), "The Wage-Price Mechanism: An Endogenous Switching Model", *European Economic Review* 42: 365–74.

Pagan, A. (1984), "Model Evaluation by Variable Addition", in D. F. Hendry and F. K. Wallis (eds.), *Econometrics and Quantitative Economics*. Oxford: Blackwell.

Peguin-Feissolle, A. (1994), "Bayesian Estimation and Forecasting in Nonlinear Models: Application to an LSTAR Model", *Economic Letters* 46: 187–94.

Perron, P. (1989), "The Great Crash, the Oil Price Shock and the Unit Root Hypothesis", *Econometrica* 57: 1361–401.

Perron, P., and T. J. Vogelsang (1992), "Nonstationarity and Level Shifts with an Application to Purchasing Power Parity", *Journal of Business and Economic Statistics* 10(3): 301–20.

Pfann, G. A., P. C. Schotman, and R. Tschernig (1996), "Nonlinear Interest Rate Dynamics and the Implications for the Term Structure", *Journal of Econometrics* 74: 149–76.

Pierse, R. G., and A. J. Snell (1995), "Temporal Aggregation and the Power of Tests for a Unit Root", *Journal of Econometrics* 65(2): 333–45.

Pole, A. M., and A. F. M. Smith (1985), "A Bayesian Analysis of Some Threshold Switching Models", *Journal of Econometrics* 29: 97–119.

Potter, S. M. (1995), "A Nonlinear Approach to US GNP", *Journal of Applied Econometrics* 10: 109–25.

Quandt, R. E. (1958), "The Estimation of Parameters of a Linear Regression System Obeying Two Separate Regimes", *Journal of the American Statistical Association* 53: 873–80.

(1960), "Tests of the Hypothesis That a Linear Regression System Obeys Two Separate Regimes", *Journal of the American Statistical Association* 55: 324–30.

Raiffa, H., and R. Schlaifer (1961), *Applied Statistical Decision Theory*. Boston: Harvard University Press.

Raj, B. (1995), "Institutional Hypothesis of the Long-Run Income Velocity of Money and Parameter Stability of the Equilibrium Relationship", *Journal of Applied Econometrics* 10(3): 233–53.

Raj, B., and D. J. Slottje (1994), "The Behaviour of Alternative Income Inequality Measures in the United States from 1947–1990 and the Structural Break", *Journal of Business and Economic Statistics* 12(4): 479–87.

Ravn, M., and M. Sola (1996), "A Reconsideration of the Empirical Evidence on the Asymmetric Effects of Money Supply Shocks: Positive versus Negative or Big versus Small?", Memo 1996/4, Department of Economics, University of Aarhus.

Schotman, P. C. (1994), "Priors for the AR(1) Model", *Econometric Theory* 10: 579–95.

Teräsvirta, T. (1994), "Specification, Estimation and Evaluation of Smooth Transition Autoregressive Models", *Journal of the American Statistical Association* 89: 208–18.

Teräsvirta, T., and H. Anderson (1992), "Modeling Nonlinearities in Business Cycles Using Smooth Transition Autoregressive Models", *Journal of Applied Econometrics* 7: S119–36.

Tiao, G. C., and R. S. Tsay (1994), "Some Advances in Nonlinear and Adaptive Modeling in Time Series", *Journal of Forecasting* 13: 109–31.

Tong, H. (1983), *Threshold Models in Nonlinear Time Series Analysis*. New York: Springer-Verlag.

(1990), *Nonlinear Time Series: A Dynamical System Approach*. Oxford: Oxford University Press.

Tong, H., and K. S. Lim (1980), "Threshold Autoregression, Limit Cycles and Cyclical Data", *Journal of the Royal Statistical Association* B42: 245–92.

Tsay, R. S. (1989), "Testing and Modeling Threshold Autoregressive Processes", *Journal of the American Statistical Association* 84: 231–40.

Tsurumi, H. (1982), "A Bayesian and Maximum Likelihood Analysis of a Gradual Switching Regression in a Simultaneous Equation Framework", *Journal of Econometrics* 19: 165–82.

Villa, P. (1996), "La Fonction de Consommation sur Longue Période en France", *Revue Economique* 47(1): 111–42.

CHAPTER 6

Nonlinear time series models: Consistency and asymptotic normality of NLS under new conditions

Santiago Mira & Alvaro Escribano

1   Introduction

The statistical properties of linear time series models are well known. There exist well-established diagnostic procedures in the strategy of identification, estimation, and testing. The class of ARIMA models is a clear example. However, there are serious limitations in a linear environment: all models are symmetric, sometimes it is difficult to find models that forecast better than a random walk, it is not easy to find models that are constant in the parameters, and many time series are subject to abrupt changes (outliers). Those cases have been analyzed with linear models but incorporating exogenous information (dummy variables, intervention analysis, etc.). Nonlinear time series models are interesting alternatives in those cases, since they can endogenize the interventions and thereby improve the forecasting accuracy.

The class of nonlinear models is too broad to be well defined, since obviously it includes any model that is not linear. In this paper we concentrate on two types of nonlinear models: nonlinear dynamic regression models (nonlinear autoregressive distributed lag models) and nonlinear autoregressive models. Examples of general nonlinear formulations are the state dependent models introduced by Priestley (1980) and the double stochastic models of Tjöstheim (1986). Consider the following nonlinear autoregressive distributed lag model:

$$y_t = g(y_{t-1}, \ldots, y_{t-p}, z_t, z_{t-1}, \ldots, z_{t-q}, \rho^*) + v_t, \qquad (1.1)$$

where the variable z_t is weakly exogenous and v_t is a random error term. This type of model is general enough to accommodate most of the asymmetric behavior that has been found in some economic time series of the United States and Europe (aggregate unemployment, sectorial unemployment, output, industrial production indices, etc.); see for example Neftcy (1984); Rothman (1991);

The authors thank Manuel Moreno, Benedikt Pötscher, and an anonymous referee for helpful comments. They acknowledge financial support from the Spanish DGICYT contract PB95-0298.

Teräsvirta and Anderson (1992); Burgess, Escribano, and Pfann (1993, 1997); Escribano and Pfann (1998); Escribano and Granger (1998); and Escribano and Jordá (1999).

In this paper we specify model (1.1) in an equivalent but more useful way. We first write the linear autoregressive distributed lag part and then add the nonlinear function, which will therefore measure departures from linearity:

$$y_t = \beta_0^* + \beta_1^* y_{t-1} + \cdots + \beta_p^* y_{t-p} + \delta_0^* z_t + \delta_1^* z_{t-1} + \cdots + \delta_q^* z_{t-q}$$

$$+ f(y_{t-1}, \ldots, y_{t-p}, z_t, z_{t-1}, \ldots, z_{t-q}, \gamma^*) + v_t. \tag{1.2}$$

This representation is useful in interpreting the nonlinear model as a model that is linear but with adjustable parameters. Changes in the parameters are explicitly modeled by particular nonlinear functions of interest. Therefore we can endogenize structural changes. The β's of (1.2) represent either an intrinsically linear part or the first term of a Taylor series expansion of $g(\cdot)$, with $f(\cdot)$ being the remainder. Furthermore, for theoretical reasons it is useful to explicitly study the conditions for estimation and inference in both formulations (1.1) and (1.2), since by doing that we are able to check in what sense those conditions differ from the well-known conditions of the linear models.

The class of nonlinear functions that we study in this paper is parametric. The main reason is that with macroeconomic time series the sample size is usually small, preventing us from using more general nonparametric procedures that require larger samples. Among the large class of parametric nonlinear models that are included in this framework we mention a few. Logistic smooth transition autoregressive (LSTAR) models are obtained from (1.2) by deleting the variables z_{t-i}, $i = 0, \ldots, q$, and by making

$$f(y_{t-1}, \ldots, y_{t-p}, \gamma^*) = (\phi_0^* + \phi_1^* y_{t-1} + \cdots + \phi_p^* y_{t-p})$$

$$\times (1 + \exp\{-\gamma_1^*(y_{t-d} - \gamma_2^*)\})^{-1}.$$

In the exponential smooth transition autoregressive (ESTAR) models the function $f(\cdot)$ is

$$f(y_{t-1}, \ldots, y_{t-p}, \gamma^*) = (\phi_0^* + \phi_1^* y_{t-1} + \cdots + \phi_p^* y_{t-p})$$

$$\times (1 - \exp\{-\gamma_1^*(y_{t-d} - \gamma_2^*)^2\}).$$

A similar class of models, proposed by Tong (1978), are the threshold autoregressive (TAR) models. These models can be thought of as representing the limit behavior of STAR models where the transition between the two regimes occurs instantaneously at point γ_2^*:

$$f(y_{t-1}, \ldots, y_{t-p}, \gamma^*) = (\psi_0^* + \psi_1^* y_{t-1} + \cdots + \psi_p^* y_{t-p}) I_{y_{t-d}},$$

where $I_{y_{t-d}} = 0$ if $y_{t-d} \leq \gamma_2^*$ and $I_{y_{t-d}} = 1$ if $y_{t-d} > \gamma_2^*$.

It is clear that those models are symmetric by construction. However, if some of the parameters of the two regimes are different or if we have more data in one part of the nonlinear function than in the other, the actual time series generating the data can be asymmetric and/or we can have extreme values (outliers) at particular dates; see Escribano and Jordá (1994). Teräsvirta (1994) discusses in detail how to estimate and evaluate these STAR models and proposes a decision rule for selecting between the logistic and the exponential function. Escribano and Jordá (1994, 1999) suggest a more natural decision rule and study its implications with Monte Carlo simulations and with some empirical examples. STAR models are usually estimated by NLS, but the assumptions under which NLS works are never checked in practice. In fact, convergence of NLS is one of the main problems one has to face when formulating this sort of models in empirical applications.

In this chapter, the specification procedure of nonlinear time series models is avoided, and the estimation and inference are attacked directly. The problem of estimation and inference in nonlinear time series has been approached by many authors using Markovian models; see for instance Tong (1990) and Teräsvirta (1994), based on Klimko and Nelson (1978), who gave sufficient conditions to ensure consistency and asymptotic normality of M-estimators. The application of these results requires stationary ergodicity of the Markov process implicit in the series. However, many economic time series are nonstationary, since they have mean, variance, covariance, or other moments changing through time (we do not consider unit root cases). The stationarity requirement seems to be too restrictive for these models, and we propose to relax the usual assumption of stationary ergodicity, or geometric ergodicity, to the condition of strong mixing, which allows for some heterogeneity. Recall that geometrically ergodic processes are strong-mixing; see Meyn and Tweedie (1992).

We establish a different set of conditions. Some of them, such as strong mixing, are extensively used in the time series literature. Although we cannot test directly for strong mixing, it is a fairly general condition to be satisfied by many economic variables, maybe after some transformations (differencing, etc.), and therefore it is not too restrictive; see Lo (1991) and Kwiatkowski, Phillips, Schmidt, and Shin (1992) for some indirect tests of mixing. However, the other conditions that we propose are easily checkable. The main advantage of our approach is that we are able to write explicit assumptions related to model (1.2), such as moment conditions and conditions on the nonlinear function $f(\cdot)$, to obtain consistency and asymptotic normality of NLS.

The structure of the chapter is the following. Section 2 briefly reviews the main results of the framework of Gallant and White (1988) that are useful for us and introduces concepts that will be used later on. Section 3 works out the details of the particular conditions that guarantee the consistency of NLS. Section 4 studies the extra conditions that are sufficient for asymptotic normality of NLS.

Section 5 makes those conditions explicit for two different ESTAR models and shows how they can be used in a more general framework. The conclusions are given in Section 6. Finally, all the proofs are in Appendices A, B, and C.

2   Estimation and inference in dynamic nonlinear models

The estimation of dynamic nonlinear models is usually done either by maximum likelihood (ML) or by nonlinear least squares (NLS). Conditions for NLS are presented here, although ML could also be studied with few modifications. The seminal paper on estimation with this approach is Klimko and Nelson (1978), and a clear exposition of the least squares method may be found in Tong (1990, Section 5.5). They prove consistency and asymptotic normality when the underlying process is a stationary ergodic process or geometrically ergodic with a unique stable distribution. Tong (1990) applies this result to the case of a STAR model with i.i.d. errors.

The estimation of nonlinear models with exogenous variables can be traced as early as Jennrich (1969). This work provides conditions for the consistency and asymptotic normality of nonlinear least squares with i.i.d. errors. After this important contribution, many others extended his results, for instance Hannan (1971) for time series, Robinson (1972) for systems of equations, Bierens (1981) for robust methods, and Burguete, Gallant, and Souza (1982) for implicit non-linear simultaneous systems of equations without dynamics. Gallant and White (1988) provide a unified theory which allows for dependent and heterogeneous variables, imposing only mixing conditions, and avoiding stationarity and ergodicity assumptions. An extensive exposition of more recent studies may be found in Potscher and Prucha (1991a,b) and in Wooldridge (1994).

The approach that is followed here is based on Gallant and White (1988). In what follows we review their basic results that we use in the rest of the paper.

The estimators considered in the previously mentioned literature are M-estimators, in particular NLS estimators, which are defined as solutions to an optimization problem. Suppose $\hat{\theta}_n$ is given as a function of the data such that

$$\hat{\theta}_n \in \arg \min_{\theta \in \Theta} Q_n(\theta).$$

For example, for the NLS estimation of the regression model $y_t = f(x_t, \theta) + u_t$ we have

$$Q_n(\theta) = n^{-1} \sum_{t=1}^{n} (y_t - f(x_t, \theta))^2$$

$$\equiv n^{-1} \sum_{t=1}^{n} q_t(\theta). \tag{2.1}$$

The idea behind the consistency result is that if $Q_n(\theta)$ converges almost surely (a.s.) to $\bar{Q}(\theta)$, a nonstochastic function, for every θ, and if $\bar{Q}(\theta)$ is minimized at the true parameter θ^0, then the limit of the minimum $\hat{\theta}_n$ should be the minimum θ^0 of the limit $\bar{Q}(\theta)$. On the other hand, the idea behind asymptotic normality is that in large samples, estimators are approximated by linear combinations of sample averages, and therefore the central limit theorem ensures the asymptotic normality of the estimators. For an intuitive discussion see Newey and McFadden (1994).

If $\hat{\theta}_n$ is interior to Θ, then the first order condition is $n^{-1} \sum_{t=1}^n \nabla_\theta q_t(\hat{\theta}_n) = 0$, and applying the mean value theorem,

$$n^{-1} \sum_{t=1}^n \nabla_\theta q_t(\theta^0) + \left[n^{-1} \sum_{t=1}^n \nabla_\theta^2 q_t(\ddot{\theta}) \right] (\hat{\theta}_n - \theta^0) = 0,$$

where $\ddot{\theta}$ is in the line joining $\hat{\theta}_n$ and θ^0 and can be different for each row of the matrix $\nabla_\theta^2 q_t(\ddot{\theta})$. Then multiplying by $\sqrt{n}$ and solving for $\sqrt{n}(\hat{\theta}_n - \theta^0)$, we obtain

$$\sqrt{n}(\hat{\theta}_n - \theta^0) = -\left[n^{-1} \sum_{t=1}^n \nabla_\theta^2 q_t(\ddot{\theta}) \right]^{-1} \left[n^{-1/2} \sum_{t=1}^n \nabla_\theta q_t(\theta^0) \right].$$

Under very general conditions (see White 1982) the score verifies $n^{-1/2} \sum_{t=1}^n \nabla_\theta q_t(\theta^0) \xrightarrow{d} \mathcal{N}(0, J)$ by the central limit theorem (CLT), and the hessian verifies $n^{-1} \sum_{t=1}^n \nabla_\theta^2 q_t(\ddot{\theta}) \xrightarrow{p} H$ by a law of large numbers (LLN), then from the Slutzky theorem

$$\sqrt{n}(\hat{\theta}_n - \theta^0) \xrightarrow{d} \mathcal{N}(0, H^{-1} J H^{-1}).$$

Sufficient conditions for the consistency of M-estimators are: (i) uniform convergence of $Q_n(\theta)$ to $\bar{Q}(\theta)$; (ii) the limit $\bar{Q}(\theta)$ is continuous; (iii) the parameter set Θ is compact; and (iv) $\bar{Q}(\theta)$ has a unique maximum at the true parameter θ^0.

The usual approach to ensure the uniform convergence and the continuity is a uniform law of large numbers (ULLN), which ensures that $Q_n(\theta) - E(Q_n(\theta)) \to 0$ a.s. uniformly on Θ, so that we can take $\bar{Q}_n(\theta) = E(Q_n(\theta))$ and $\bar{Q}(\theta) = \lim_{n \to \infty} \bar{Q}_n(\theta)$.

In what follows we introduce some primitive assumptions, which will be used to derive the consistency result, and we also present the main theorems that ensure the consistency and asymptotic normality of the estimators.

Definition 2.1 (Strong mixing). Let $\{v_t\}$ be a sequence of random variables. Let $\mathcal{F}_s^t \equiv \sigma(v_s, \ldots, v_t)$, and define the α-mixing coefficients

$$\alpha_m \equiv \sup_t \sup_{\{F \in \mathcal{F}_{-\infty}^t, \, G \in \mathcal{F}_{t+m}^\infty\}} |\mathbf{P}(G \cap F) - \mathbf{P}(G)\mathbf{P}(F)|.$$

The process $\{v_t\}$ is said to be strong mixing (also α-mixing) if $\alpha_m \to 0$ as $m \to \infty$.

The coefficients α_m measure the amount of dependence between events involving V_t separated by at least m time periods. Mixing processes are useful because they allow considerable time dependence without necessarily restricting the heterogeneity of the process. Let $\alpha_m = O(m^\lambda)$ for all $\lambda < -a$; then α_m is said to be of size $-a$.

Definition 2.2 (Near epoch dependence). Let $\{z_t : \Omega \to \mathfrak{R}\}$ be a sequence measurable-$(\mathcal{F}, \mathcal{B})$ with $E(z_t^2) < \infty$, for all t. Then $\{z_t\}$ is near epoch dependent on $\{v_t\}$ of size $-a$ if $\{\phi_m\}$ is of size $-a$, for ϕ_m given by

$$\phi_m \equiv \sup_t \left\| Z_t - E_{t-m}^{t+m}(Z_t) \right\|_{L2}$$

and $E_{t-m}^{t+m}(z_t) = E(z_t | v_{t-m}, \ldots, v_{t+m})$, where $\| \cdot \|_{L2}$ is the L_2 norm of a random variable.

Given the above definition, ϕ_m can be defined as the worst mean squared forecast error when z_t is predicted by $E_{t-m}^{t+m}(z_t)$. When ϕ_m tends to zero at an appropriate rate, then z_t depends essentially on the recent epoch of v_t. If z_t depends on a finite number of lags, then it is near epoch dependent of any size. Note that usually the forward values v_{t+r} $(r = 1, \ldots, m)$ are useless, but harmless, because in our case z_t is strictly exogenous and depends only on past values and not on future values.

Definition 2.3.

(a) (r-integrability) Let $D_t : \Omega \to \mathfrak{R}$ be a measurable-$(\mathcal{F}, \mathcal{B})$ sequence. Then D_t is r-integrable uniformly in t if $\|D_t\|_{Lr} \leq \Delta < \infty$ for $r > 0$, $t = 1, 2, \ldots$, where $\| \cdot \|_{Lr}$ is the usual L_r norm.

(b) (r-domination) $q_t(\omega, \theta)$ is r-dominated on Θ if there exists $D_t : \Omega \to \mathfrak{R}$ such that $|q_t(\theta)| \leq D_t$ for all $\theta \in \Theta$ and D_t is r-integrable uniformly in t.

Definition 2.4 (Lipschitz-L_1 a.s.). The sequence $\{q_t : \Omega \times \Theta \to \mathfrak{R}\}$ is defined to be Lipschitz-L_1 a.s. on Θ if $q_t(\cdot, \theta)$ is measurable-$(\mathcal{F}, \mathcal{B})$ for each θ, and for each θ^0 there exist (i) $\delta^0 > 0$, (ii) functions $L_t^0 : \Omega \to \mathfrak{R}^+$ measurable-$(\mathcal{F}, \mathcal{B}^+)$, and (iii) functions $a^0 : \mathfrak{R}^+ \to \mathfrak{R}^+$ such that:

(a) $a^0(\delta) \to 0$ as $\delta \to 0$,
(b) $\{n^{-1} \sum_{t=1}^n E(L_t^0)\}$ is $O(1)$, and
(c) for all θ with $\rho(\theta, \theta^0) \leq \delta^0$ it holds that

$$|q_t(\omega, \theta) - q_t(\omega, \theta^0)| \leq L_t^0 a^0(\rho(\theta, \theta^0)) \qquad \text{a.s.}$$

Notice that in Definition 2.3 we have that $\|D_t\|_{Lr} \le \|D_t\|_{Ls}$ if $r \le s$. The terminology of Definition 2.4 conveys the idea that the above Lipschitz condition holds almost surely, and that the Lipschitz functions L_t^0 satisfy a restriction on the average of their L_1 norms.

Definition 2.5 (Uniform convergence on Θ a.s.). Let $\{Q_n : \Omega \times \Theta \to \mathfrak{R}\}$ be a sequence of random functions. Let $\{\bar{Q}_n : \Theta \to \mathfrak{R}\}$ be a sequence of functions. Then $Q_n(\omega, \theta) - \bar{Q}_n(\theta) \to 0$ a.s. uniformly on Θ if there exists $F \in \mathcal{F}$ with $\mathbf{P}(F) = 1$ such that, $\forall \varepsilon > 0, \forall \omega \in F, \exists N(\omega, \varepsilon) < \infty$ such that $\forall n > N(\omega, \varepsilon)$

$$\sup_{\theta \in \Theta} |Q_n(\omega, \theta) - \bar{Q}_n(\theta)| < \varepsilon$$

as $n \to \infty$. The uniform convergence arises because $N(\omega, \varepsilon)$ does not depend on Θ.

Definition 2.6 (Identifiable uniqueness). Suppose that θ_n^0 minimizes $\bar{Q}_n(\theta)$ on Θ. Let $S_n^0(\varepsilon)$ be an open sphere in $\mathfrak{R}^k$ with radius ε centered at θ_n^0. Let $\eta_n^0(\varepsilon) = S_n^0(\varepsilon) \cap \Theta$, and define $\eta_n^0(\varepsilon)^c = S_n^0(\varepsilon)^c \cap \Theta$. The sequence $\{\theta_n^0\}$ is said to be identifiable unique minimizers on Θ if either (i) for all $\varepsilon > 0$, $\eta_n^0(\varepsilon)^c$ is empty, or (ii) for all $\varepsilon > 0$

$$\lim_{n\to\infty} \inf \left(\min_{\theta \in \eta_n^0(\varepsilon)^c} \bar{Q}_n(\theta) - \bar{Q}_n(\theta^0) \right) > 0.$$

This condition rules out the possibility that $\bar{Q}_n$ might be flatter and flatter in a neighborhood of θ_n^0 as $n \to \infty$, and also rules out the possibility that some other sequence, taking values in Θ, might yield values of the objective function approaching $\bar{Q}_n(\theta_n^0)$ arbitrarily close as $n \to \infty$.

The following assumptions are sufficient for consistency and asymptotic normality of NLS. Notice that some assumptions used for consistency are reinforced for normality by adding (in parenthesis) an extra requirement.

Assumption 2.1: The observed data are generated as a realization of a stochastic process $\{X_t\}$, with $X_t : \Omega \to \mathfrak{R}$, such that X_t is measurable-$(\mathcal{F}, \mathcal{B})$. The process X_t depends on an underlying process V_t such that $X_t = h_t(\ldots, V_{t-1}, V_t, V_{t+1}, \ldots)$. Both processes lie in a complete probability space $(\Omega, \mathcal{F}, \mathbf{P})$.

Assumption 2.2: Let $Q_n : \Omega \times \Theta \to \mathfrak{R}$ be the optimand $Q_n(\omega, \theta) \equiv n^{-1} \sum_{t=1}^n q_t(\omega, \theta)$, with $q_t : \Omega \times \Theta \to \mathfrak{R}$, where Θ is a compact subset of $\mathfrak{R}^k$ and (Θ, ρ) is a separable metric space. For each θ, $q_t(\cdot, \theta)$ is measurable-$(\mathcal{F}, \mathcal{B})$, and for each θ, $q_t(\omega, \cdot)$ is continuous on θ almost surely (continuously differentiable of order 2 on θ a.s., for normality).

Assumption 2.3: The sequence $\{\bar{Q}_n(\theta)\}$ has identifiable unique minimizers $\{\theta_n^0\}$ on Θ.

Assumption 2.4: The elements of $q_t(\theta)$ are r-dominated on Θ uniformly in t for $r \geq 2$, (for $r > 2$, for normality).

Assumption 2.5: $\{V_t\}$ is a strong mixing sequence such that α_m is of size $-r/(r-2)$ with $r > 2$ (of size $-2r/(r-2)$ with $r > 2$, for normality).

Assumption 2.6: The elements of $\{q_t(\omega, \theta)\}$ are near epoch dependent on $\{V_t\}$ of size $-\frac{1}{2}$ on (Θ, ρ) (of size -1 uniformly on (Θ, ρ), for normality).

Assumption 2.7: $\{q_t(\omega, \theta)\}$ is almost surely Lipschitz-L_1 on Θ.

The above conditions allow us to prove the following theorem, which states the main result on consistency.

Theorem 2.1 (Consistency). *Given* $(\Omega, \mathcal{F}, \mathbf{P})$ *and a compact set* $\Theta \in \mathfrak{R}^k$, *let* $Q_n : \Omega \times \Theta \to \mathfrak{R}$ *be a random function continuous on* Θ *a.s. Let* $\hat{\theta}_n$ *be a measurable solution to the problem* $\min_{\theta \in \Theta} Q_n(\theta)$. *Suppose there exists* $\{\bar{Q}_n : \Theta \to \mathfrak{R}\}$ *such that* $Q_n(\theta) - \bar{Q}_n(\theta) \to 0$ *a.s. uniformly on* Θ. *If* $\{\bar{Q}_n\}$ *has identifiable unique minimizers* $\{\theta_n^0\}$ *on* Θ, *then* $\hat{\theta}_n - \theta_n^0 \to 0$ *a.s.*

Proof: See Theorem 3.3 of Gallant and White (1988).   □

Note that in case $\bar{Q}_n \equiv \bar{Q}$, as for instance if $\bar{Q}_n \equiv \lim_{n \to \infty} E(Q_n(\omega, \theta))$, then the sequence $\{\theta_n^0\}$ is one point θ^0.

Some assumptions have to be added to the above set of assumptions to prove asymptotic normality.

Assumption 2.8: The elements of $\{\nabla_\theta q_t(\theta)\}$ are r-dominated on Θ uniformly in t, $r > 2$.

Assumption 2.9: The elements of $\{\nabla_\theta^2 q_t(\theta)\}$ are r-dominated on Θ uniformly in t, $r > 2$.

Assumption 2.10: The elements of $\{\nabla_\theta q_t(\theta)\}$ are near epoch dependent on $\{V_t\}$ of size -1 uniformly on (Θ, ρ).

Assumption 2.11: The elements of $\{\nabla_\theta^2 q_t(\theta)\}$ are near epoch dependent on $\{V_t\}$ of size $-\frac{1}{2}$ uniformly on (Θ, ρ).

Assumption 2.12: The elements of $\{\nabla_\theta q_t(\theta)\}$ are Lipschitz-L_1 a.s.

Assumption 2.13: The elements of $\{\nabla_\theta^2 q_t(\theta)\}$ are Lipschitz-L_1 a.s.

Assumption 2.14: The sequence $\{B_n^0\}$ is uniformly positive definite, with $\{B_n^0\}$ defined below.

Assumptions 2.8–2.14, together with Assumptions 2.1, 2.2, and 2.4–2.6, allow us to prove the following theorem.

Theorem 2.2 (Normality). *Given $(\Omega, \mathcal{F}, \mathbf{P})$, (Θ, ρ), and $Q_n : \Omega \times \Theta \to \Re$ as in the assumptions, with Q_n continuously differentiable of order 2 on Θ a.s., $n = 1, 2, \ldots$. Let $\hat{\theta}_n : \Omega \to \Theta$ be a function $(\mathcal{F}, \mathcal{B}^k)$-measurable $n = 1, 2, \ldots$ which solves $\min_\Theta Q_n(\theta)$ a.s., and suppose $\hat{\theta}_n - \theta_n^0 \to 0$ a.s. for $\theta_n^0 \in \arg\min_\Theta \bar{Q}_n(\theta)$ and θ_n^0 an interior point of Θ. Suppose there exists a nonstochastic sequence of $k \times k$ matrices $\{B_n^0\}$ which are $O(1)$ and uniformly positive definite such that*

$$\left(B_n^0\right)^{-1/2} n^{1/2} \nabla_\theta \bar{Q}\left(\theta_n^0\right)' \overset{A}{\sim} \mathcal{N}(0, I_k).$$

If there exists a nonstochastic sequence $\{A_n : \Theta \to \Re^{k \times k}\}$ such that $\{A_n\}$ is continuous on Θ uniformly in n, $\nabla_\theta^2 Q_n(\theta) - A_n(\theta) \to 0$ a.s. uniformly on Θ, and $\{A_n(\theta_n^0)\}$ is $O(1)$ and uniformly positive definite, then

$$\left(B_n^0\right)^{-1/2} A_n^0 n^{-1/2} \left(\hat{\theta}_n - \theta_n^0\right) \overset{A}{\sim} \mathcal{N}(0, I_k).$$

Proof: See Theorem 5.1 of Gallant and White (1988). □

In the next two sections we introduce sufficient conditions, which are not difficult to verify, for Theorem 2.1 and Theorem 2.2 to hold.

3 Consistency of NLS estimation

Without loss of generality, we write model (1.2) without the constant β_0^* in the linear part of the model to avoid tedious demonstrations, and without the contemporaneous values of the weakly exogenous variable z_t for convenience, although the results could be easily generalized to incorporate this possibility. In this case we have

$$y_t = \beta_1^* y_{t-1} + \cdots + \beta_p^* y_{t-p} + \delta_1^* z_{t-1} + \cdots + \delta_q^* z_{t-q}$$
$$+ f(y_{t-1}, \ldots, y_{t-p}, z_{t-1}, \ldots, z_{t-q}, \gamma^*) + v_t. \tag{3.1}$$

In this section we concentrate on the assumptions that are sufficient to guarantee the consistency of NLS. For convenience we specify the nonlinear autoregressive distributed lag model (3.1) in companion form. Let us define the vectors $(p \times 1)$ $Y_t = [y_t, \ldots, y_{t-p+1}]'$, $(q \times 1)$ $Z_t = [z_t, \ldots, z_{t-q+1}]'$, and $(p \times 1)$ $V_t = [v_t, 0, \ldots, 0]'$. Let us define the matrices $(p \times p)$ B^* and $(p \times q)$ D^* by

$$B^* = \begin{pmatrix} \beta_1^* & \beta_2^* & \cdots & \beta_{p-1}^* & \beta_p^* \\ 1 & 0 & \cdots & 0 & 0 \\ \vdots & \vdots & & \ddots & \vdots \\ 0 & 0 & \cdots & 1 & 0 \end{pmatrix}, \quad D^* = \begin{pmatrix} \delta_1^* & \delta_2^* & \cdots & \delta_{q-1}^* & \delta_q^* \\ 0 & 0 & \cdots & 0 & 0 \\ \vdots & \vdots & & \ddots & \vdots \\ 0 & 0 & \cdots & 0 & 0 \end{pmatrix}.$$

Define also the vector $(p \times 1)$ $F(Y_{t-1}, Z_{t-1}, \gamma^*) = [f(y_{t-1}, \ldots, y_{t-p}, z_{t-1}, \ldots, z_{t-q}, \gamma^*), 0, \ldots, 0]'$. Now our model (3.1) can be rewritten as

$$Y_t = B^* Y_{t-1} + D^* Z_{t-1} + F(Y_{t-1}, Z_{t-1}, \gamma^*) + V_t. \tag{3.2}$$

Notice that if there were a constant, then model (3.2) would be $Y_t = C^* + B^* Y_{t-1} + D^* Z_{t-1} + F(Y_{t-1}, Z_{t-1}, \gamma^*) + V_t$ for C^* a $p \times 1$ vector equal to $[\beta_0^*, 0, \ldots, 0]'$. Furthermore, if there were no linear part in the model, then $\beta_i^* = 0$ for all i and the companion form would be $Y_t = Q^* Y_{t-1} + F(Y_{t-1}, Z_{t-1}, \gamma^*) + V_t$ for a matrix Q^* equal to B^* replacing β_i^* by zeros. This comment will be useful later on when comparing models (1.1) and (1.2).

The parameter vector that has to be estimated in (3.1) is

$$\theta^* = [\beta_1^*, \ldots, \beta_p^*, \delta_1^*, \ldots, \delta_q^*, (\gamma^*)']',$$

which for $\gamma^* \in \Re^g$ belongs to $\Theta \in \Re^{p+q+g}$. The space Θ is determined by the assumptions below. Since we are interested in nonlinear least squares estimators, the optimand $Q_n : \Omega \times \Theta \to \Re$, which depends on the data $y_{-p+1}, \ldots, y_0, y_1, \ldots, y_n$ and $z_{-q+1}, \ldots, z_0, z_1, \ldots, z_n$, is given by

$$Q_n(\omega, \theta) \equiv n^{-1} \sum_{t=1}^{n} q_t(\omega, \theta)$$

and $\bar{Q}_n(\theta) = E(Q_n(\omega, \theta))$, where $q_t(\omega, \theta)$ is given by

$$\begin{aligned} q_t(\omega, \theta) &\equiv |y_t - \beta_1 y_{t-1} - \cdots - \beta_p y_{t-p} - \delta_1 z_{t-1} - \cdots - \delta_q z_{t-q} \\ &\quad - f(y_{t-1}, \ldots, y_{t-p}, z_{t-1}, \ldots, z_{t-q}, \gamma)|^2 \\ &\equiv \|Y_t - BY_{t-1} - DZ_{t-1} - F(Y_{t-1}, Z_{t-1}, \gamma)\|_2^2, \end{aligned}$$

where the 2-norm of a vector $X = [x_1, \ldots, x_p]$ is $\|X\|_2 = (x_1^2 + \cdots + x_p^2)^{1/2}$. Other norms can be used, for instance, $\|\cdot\|_r$ for $r \geq 2$.

For any vector norm $\|\cdot\|$ we can define a matrix norm $\|A\|$, which is a subordinate matrix norm, such that for any vector X

$$\|AX\| \leq \|A\| \|X\|.$$

The following theorem finds a suitable matrix norm which will be useful for our purposes.

Theorem 3.1. *For any given matrix A and any number $\delta > 0$, there exists at least one subordinate matrix norm $\|\cdot\|_S$ such that*

$$\|A\|_S \leq \rho(A) + \delta,$$

where $\rho(A)$ is the spectral radius of A, i.e., the largest eigenvalue of the matrix A.

Proof: See Appendix A. $\qquad\qquad\qquad\qquad\qquad\qquad\qquad\qquad\qquad\qquad$ $\square$

Definition 3.2. Let Y_t be a random vector. We define its Sr norm as $\|Y_t\|_{Sr} \equiv (E(\|Y_t\|_S^r))^{1/r} \equiv E^{1/r}(\|Y_t\|_S^r)$.

Note that this is usually called the L_p norm when random variables appear instead of random vectors and $\|\cdot\|_S$ is changed to the absolute value.

Lemma 3.3. *If W is a random vector, the function $N(\cdot)$ defined by $N(W) \equiv \|g\|_{Sr}$ is a norm.*

Proof: See Appendix A. $\qquad\qquad\qquad\qquad\qquad\qquad\qquad\qquad\qquad\qquad$ $\square$

The following assumptions are sufficient to prove that the nonlinear least squares estimator, $\hat{\theta}$, is consistent for θ^*. Note that the inequality between random variables holds a.s., and the values $\delta_{BY}, \delta_H, \dots$ are positive and finite real numbers used as upper bounds; the same can be said for $\Delta_Z^{(r)}$, etc. The indexing of the subsequent assumptions is made (in the spirit of Gallant and White 1988) for mnemonic purposes.

Assumption MD: Model (3.1) is the true model in the sense that

$$E(y_t | y_{t-1}, \dots, y_{t-p}, z_{t-1}, \dots, z_{t-q})$$
$$\equiv \beta_1^* y_{t-1} + \cdots + \beta_p^* y_{t-p} + \delta_1^* z_{t-1} + \cdots + \delta_q^* z_{t-q}$$
$$+ f(y_{t-1}, \dots, y_{t-p}, z_{t-1}, \dots, z_{t-q}, \gamma^*).$$

Assumption MX: The sequence $\{(V_t, Z_t)\}$ is strong mixing with $\{\alpha_n\}$ of size $-v/(v-2)$ with $v > 2$.

Assumption CT:

(i) For some fixed value $\varepsilon > 0$ and for all matrices $B\nabla F$ given by $B\nabla F \equiv B + \nabla_Y F(Y, Z, \gamma)$, with $\theta \in \Theta$, we have that

$$\rho(B\nabla F) < 1 - \varepsilon < 1 \,.$$

Notice that for each specific matrix $B\nabla F$, its associated norm $\|\cdot\|_S$ will verify that $\|B\nabla F\|_S \equiv \delta_{BY} < 1 - \varepsilon$.

(ii) $|f(0, \ldots, 0, z_{t-1}, \ldots, z_{t-q}, \gamma^*)| \le \delta_H \|Z_{t-1}\|_S$.

(iii) For the norms $\|\cdot\|_S$ and $\|\cdot\|_2$ we have $\|B\| \le \delta_{CB}$ and $\|D\| \le \delta_{CD}$.

(iv) For the norms $\|\cdot\|_S$ and $\|\cdot\|_2$ we have $\|\nabla_Y f(Y, Z, \gamma)\| \le \delta_{KY}$ and $\|\nabla_Z f(Y, Z, \gamma)\| \le \delta_{KZ}$.

(v) The compact parametric space Θ is such that the Jordan decomposition of the matrix $B\nabla F$ given in part (i), $J = M^{-1}(B\nabla F)M$, verifies $\|M^{-1}\|_\infty < \Delta^{-1}$ and $\|M\|_\infty < \Delta$ for some fixed values Δ and Δ^{-1}.

Assumption CN: $f(y_{t-1}, \ldots, y_{t-p}, z_{t-1}, \ldots, z_{t-q}, \gamma)$ is continuously differentiable in each argument, and its second order derivatives with respect to γ are continuous functions.

Assumption LR: For $r = 6$ we have

(i) $E\|Z_t\|_S^r \le \Delta_Z^{(r)}$;

(ii) $E\|V_t\|_S^r \le \Delta_V^{(r)}$;

(iii) $E\|Z_t\|_S^r\|Z_s\|_S^r \le \Delta_{ZZ}^{(r)}$;

(iv) $E\|V_t\|_S^r\|V_s\|_S^r \le \Delta_{VV}^{(r)}$;

(v) $E\|Z_t\|_S^r\|V_s\|_S^r \le \Delta_{ZV}^{(r)}$.

Assumption LN: For the norms $\|\cdot\|_S$ and $\|\cdot\|_2$,

(i) the following inequality holds a.s.:

$$\|F(Y_t, Z_t, \gamma)\|_S \le \delta_{CF}(\|Y_t\|_S + \|Z_t\|_S);$$

(ii) the following inequality holds a.s. :

$$\|\nabla_\gamma F(Y_{t-1}, Z_{t-1}, \gamma)\|_S^2$$
$$\le \|\nabla_\gamma f(y_{t-1}, \ldots, y_{t-p}, z_{t-1}, \ldots, z_{t-q}, \gamma)\|_S^2$$
$$\le \delta_L(\|Y_{t-1}\|_S + \|Z_{t-1}\|_S)^2.$$

Note in the last assumption that since $\|\cdot\|_S$ is a subordinate matrix norm, then $\|AX\|_S \le \|A\|_S\|X\|_S$. Note also that $\nabla_\gamma f(y_{t-1}, \ldots, y_{t-p}, z_{t-1}, \ldots, z_{t-q}, \gamma)$ is a $g \times 1$ vector and $\nabla_\gamma F(Y_{t-1}, Z_{t-1}, \gamma)$ is a $g \times g$ matrix with rows

$(\nabla_\gamma f', 0', \ldots, 0')$, and then $\nabla_\gamma F = M \nabla_\gamma f$ for a matrix M with a first row of 1's and zeros everywhere else, which has spectral radius equal to 1. Therefore the first inequality of (ii) is obvious.

The concept of short memory (or stability), defined in Granger and Teräsvirta (1993) (see also Granger 1995) is very close to the concept of α-mixing (see Mira 1996 for details). As Granger and Teräsvirta explain, there are a set of assumptions that ensure stability of the model $y_t = f(y_{t-1}, \gamma) + v_t$. In our approach a similar set of assumptions are given by CT and LN (notice that the assumptions in Granger and Teräsvirta can be rewritten in terms of $\|\cdot\|_{L2}$). Lemma 3.4 proves that under those assumptions y_t is near epoch dependent, which is a concept that captures the desirable properties of mixing, but avoids its strong requirements.

Lemma 3.4 and Theorem 3.5 prove the consistency of the NLS estimator.

Lemma 3.4. *Under Assumptions MD, MX, CT, CN, LR, and LN we have the following results:*

  (i) $\{Y_t\}$ *is near epoch dependent of any size on* $\{(Z_t, V_t)\}$*;*
  (ii) $\{Y_t\}$ *is r-integrable uniformly in t, for r given in Assumption LR;*
  (iii) $q_t(\omega, \theta)$ *is s-dominated for $s = r/2$;*
  (iv) $\{q_t(\omega, \theta)\}$ *is near epoch dependent on* $\{(Z_t, V_t)\}$ *of any size;*
  (v) $q_t(\omega, \theta)$ *is Lipschitz-L_1 a.s. on* Θ*.*

Proof: See Appendix B. □

Theorem 3.5 (Consistency). *Under Assumptions MD, MX, CT, CN, LR, and LN and the identification assumption stated below, the nonlinear least squares estimator for model (3.1), $\hat{\theta}$, converges a.s. to the true value θ^*.*

Proof: By Lemma 3.4 we can apply Theorem 2.1 if we prove that $\{\bar{Q}_n\}$ has identifiable unique minimizers. Since the mean square error has a unique minimum at the conditional mean, and since model (3.1) is the conditional mean from Assumption MD, then the identification condition is that

$$B^* Y_{t-1} + D^* Z_{t-1} + F(Y_{t-1}, Z_{t-1}, \gamma^*)$$
$$\neq B Y_{t-1} + D Z_{t-1} + F(Y_{t-1}, Z_{t-1}, \gamma)$$

for $(B^*, D^*, \gamma^*) \neq (B, D, \gamma)$. With this new assumption we can apply Theorem 2.1, and the consistency of NLS follows. □

3.1   *Comments on the assumptions*

In this section we explain the main implications of the assumptions and, when needed, use a simple nonlinear AR(1) model as example.

Assumption MD implies that the error term v_t is a martingale difference sequence.

Assumption CT(i) is analogous to the condition that the roots of the AR(p) model must be outside the unit circle. Consider a nonlinear example given by simple NAR(1),

$$y_t = g(y_{t-1}, z_{t-1}, \theta^*) + v_t. \tag{3.3}$$

Gallant and White (1988) demonstrate that under some regularity conditions y_t is near-epoch-dependent (NED) on $\{(z_t, v_t)\}$ if $g(\cdot, z, \gamma)$ is a contraction mapping, i.e., $|(\partial/\partial y)g(y, z, \gamma)| < 1$. If we write the NAR(1) as a linear AR(1) plus a nonlinear term, as in (2.1), Assumption CT means that

$$\left| \frac{\partial}{\partial y} g(y, z, \theta^*) \right| = \left| \beta_1^* + \frac{\partial}{\partial y} f(y, z, \gamma^*) \right| \equiv \delta_{BY} < 1.$$

Assumption CT(i) could be replaced by CTT given by

$$\left| \frac{\partial}{\partial y_j} f(y_1, \ldots, y_p, z_1, \ldots, z_q, \gamma^*) \right| \le \delta_{y_j};$$

$$\text{and} \quad \|B^*\|_S + \sqrt{p}\delta_{My}| \equiv |\delta_{BY}| < 1,$$

where $\delta_{My} = \max_j\{\delta_{y_j}\}$. CTT is more restrictive, essentially because $|a + b| \le \|a\| + |b\|$, although sometimes it is more easily interpretable, since it means that $f(y_{t-1}, \ldots, y_{t-p}, z_{t-1}, \ldots, z_{t-q}, \gamma)$ is a contraction mapping in each of the y_j arguments.

It is interesting to note that not any norm can be used in Assumption CT(i). If we replace $\|\cdot\|_S$ by $\|\cdot\|_2$, we have that $\|M\|_2$ is equal to the maximum eigenvalue of $(M'M)$, and for the matrix $M = (B^* + \nabla_Y F(Y, Z, \gamma^*)$ we have that the maximum eigenvalue of $(M'M)$ is greater than or equal to 1, no matter what the spectral radious of B^* is. Therefore the norm $\|\cdot\|_S$ is the appropriate norm and not $\|\cdot\|_2$.

Assumption CT(ii) can be expressed for the simple case (3.3) as

$$|g(0, z_{t-1}, \theta^*)| \le \delta_H |z_{t-1}|,$$

and it imposes a linearly growing upper bound on the growth of $f(\cdot)$ with respect to the exogenous variables z_t.

Assumption CT(iii) restricts the space Θ.

Note that *Assumption CT(iv)* implies Assumption LN(i). The reason is essentially the following. Consider the simplest model with $|(\partial/\partial y)f(y)| \le K$; then

$|f(y)| \leq C|y| + M$, and the proof can be done by taking a partition of the interval $[0, y]$ and considering that for ε small enough $|(f(x + \varepsilon) - f(x))/\varepsilon| \leq K$. Nevertheless, CT(iv) is not implied by LN(i), as the following counterexample shows. Take $f(x) = x \sin(1/x)$; then $f(x)$ verifies LN(i) but $(d/dx)f(x)$ does not verify CT(iv). For the simple case (3.3) we have

$$\left| \frac{\partial}{\partial y} g(y_{t-1}, z_{t-1}, \theta) \right| < \delta_{KY},$$

$$\left| \frac{\partial}{\partial z} g(y_{t-1}, z_{t-1}, \theta) \right| < \delta_{KZ}.$$

Assumption CN allows us to take a mean value expansion of the function $f(\cdot)$ as in Jennrich (1969).

Assumption LR are restrictions as moment conditions on Z_t and V_t. Note for example that the relationship between the norms $\|\cdot\|_S$, defined for some matrix A, and $\|\cdot\|_\infty$ gives

$$\|Z_t\|_S^r \leq C(A)^r \|Z_t\|_\infty^r$$
$$= C(A)^r \left(\max_{j=0,\ldots,q-1} |z_{t-j}| \right)^r$$
$$= C(A)^r \max_{j=0,\ldots,q-1} |z_{t-j}|^r,$$

and since $E(\max_i |z_i|) < \infty$ if $\max_i E|z_i| < \infty$, then $E\|Z_t\|_S^r < \Delta_Z^{(r)}$ is implied by the existence of moments of order r in $\{z_t\}$.

Assumption LN(i) implies that the growth of the function $F(\cdot, \cdot, \gamma)$ is at most linear. In fact, a linear bound would be $\|F(Y_t, Z_t, \gamma)\|_S^2 \leq C + CF(\|Y_t\|_S + \|Z_t\|_S)$, but maintaining the constant C makes the proofs more messy without changing the main argument. Assumption LN(i) can be replaced by $\|F(Y_t, Z_t, \gamma)\|_S^2 \leq CF_1\|Y_t\|_S + CF_2\|Z_t\|_S + CF_3\|Y_t\|_S\|Z_t\|_S$ or similar ones. For the simple case (3.3) we have

$$|g(y_{t-1}, z_{t-1}, \theta)| \leq \delta_{CF}(|y_{t-1}| + |z_{t-1}|).$$

Assumption LN(ii) implies linear behavior for $\nabla_\gamma f$. Note that $\nabla_\gamma f(y_1, \ldots, y_t, z_1, \ldots, z_q, \gamma)$ is a $g \times 1$ vector and $\nabla_\gamma F(Y_1, Z_1, \gamma)$ is a $g \times p$ matrix, with only the first row different from zero and equal to $\nabla_\gamma f(\cdot)$. It has to be noted that $\|\nabla_\gamma f(Y, Z, \gamma)\|_S^2 \leq L(\|Y\|_S + \|Z\|_S)^2$ implies $|(\partial/\partial \gamma_i)f(Y, Z, \gamma)| \leq L(\|Y\|_S + \|Z\|_S)^2$. For the simple case (3.2) we have

$$\left| \frac{\partial}{\partial \theta} g(y_{t-1}, z_{t-1}, \gamma) \right| \leq \delta_L(|y_{t-1}| + |z_{t-1}|).$$

In the next section we look for explicit conditions for the asymptotic normality of NLS that are not difficult to verify.

4   Asymptotic normality of NLS estimation

As in the previous section, our model is given by (3.1), and the optimand is given by (2.1). Define the score by the $(p + q + g) \times 1$ vector $\nabla_\theta q_t$ given by

$$\nabla_\theta q_t(\omega, \theta) = \left[\frac{\partial}{\partial \beta_1} q_t, \ldots, \frac{\partial}{\partial \beta_p} q_t, \frac{\partial}{\partial \delta_1} q_t, \ldots, \frac{\partial}{\partial \delta_q} q_t, \frac{\partial}{\partial \gamma_1} q_t, \ldots, \frac{\partial}{\partial \gamma_g} q_t \right]'.$$

Since each derivative for $i = 1, \ldots, p$ (and analogously for δ_j for $j = 1, \ldots, q$ is equal to

$$\frac{\partial}{\partial \beta_i} q_t(\omega, \theta) = -2 y_{t-i}(y_t - \beta_1 y_{t-1} - \cdots - \beta_p y_{t-p} - \delta_1 z_{t-1} - \cdots - \delta_q z_{t-q}$$

$$- f(y_{t-1}, \ldots, y_{t-p}, z_{t-1}, \ldots, z_{t-q}, \gamma)),$$

and for $i = 1, \ldots, g$ is equal to

$$\frac{\partial}{\partial \gamma_i} q_t(\omega, \theta) = -2 \frac{\partial}{\partial \gamma_i} f(y_{t-1}, \ldots, z_{t-q}, \gamma)$$

$$\times (y_t - \beta_1 y_{t-1} - \cdots - \beta_p y_{t-p} - \delta_1 z_{t-1} - \cdots - \delta_q z_{t-q}$$

$$- f(y_{t-1}, \ldots, y_{t-p}, z_{t-1}, \ldots, z_{t-q}, \gamma)),$$

then we can write the absolute value of the score by

$$|\nabla_\theta q_t(\omega, \theta)| = 2|(y_t - \beta_1 y_{t-1} - \cdots - \beta_p y_{t-p} - \delta_1 z_{t-1} - \cdots - \delta_q z_{t-q}$$

$$- f(y_{t-1}, \ldots, y_{t-p}, z_{t-1}, \ldots, z_{t-q}, \gamma)|$$

$$\times \left| \left[y_{t-1}, \ldots, y_{t-p}, z_{t-1}, \ldots, z_{t-q}, \frac{\partial}{\partial \gamma_1} \right. \right.$$

$$\left. \left. \times f(Y_{t-1}, Z_{t-1}, \gamma), \ldots, \frac{\partial}{\partial \gamma_g} f(Y_{t-1}, Z_{t-1}, \gamma) \right]' \right|$$

$$= 2\|Y_t - B Y_{t-1} - D Z_{t-1} - F(Y_{t-1}, Z_{t-1}, \gamma)\|_2$$

$$\times \left| \left[y_{t-1}, \ldots, y_{t-p}, z_{t-1}, \ldots, z_{t-q}, \frac{\partial}{\partial \gamma_1} \right. \right.$$

$$\left. \left. \times f(Y_{t-1}, Z_{t-1}, \gamma), \ldots, \frac{\partial}{\partial \gamma_g} f(Y_{t-1}, Z_{t-1}, \gamma) \right]' \right|$$

$$\equiv 2\|Y_t - B Y_{t-1} - D Z_{t-1} - F(Y_{t-1}, Z_{t-1}, \gamma)\|_2 \times |DV|$$

for the vector $DV \equiv [y_{t-1}, \ldots, y_{t-p}, z_{t-1}, \ldots, z_{t-q}, (\partial/\partial \gamma_1) f(Y_{t-1}, Z_{t-1}, \gamma), \ldots, (\partial/\partial \gamma_g) f(Y_{t-1}, Z_{t-1}, \gamma)]'$. Analogously define the hessian by the

$(p + q + g) \times (p + q + g)$ matrix $\nabla_\theta^2 q_t$ given by

$$
\nabla_\theta^2 q_t(\omega, \theta) = \begin{pmatrix}
\frac{\partial^2}{\partial \beta_1^2} q_t(\omega, \theta) & \frac{\partial^2}{\partial \beta_2 \partial \beta_1} q_t(\omega, \theta) & \cdots & \frac{\partial^2}{\partial \gamma_g \partial \beta_1} q_t(\omega, \theta) \\
\frac{\partial^2}{\partial \beta_1 \partial \beta_2} q_t(\omega, \theta) & \frac{\partial^2}{\partial \beta_2^2} q_t(\omega, \theta) & \cdots & \frac{\partial^2}{\partial \gamma_g \partial \beta_2} q_t(\omega, \theta) \\
\vdots & \vdots & \ddots & \vdots \\
\frac{\partial^2}{\partial \beta_1 \partial \gamma_g} q_t(\omega, \theta) & \frac{\partial^2}{\partial \beta_2 \partial \gamma_g} q_t(\omega, \theta) & \cdots & \frac{\partial^2}{\partial \gamma_g^2} q_t(\omega, \theta)
\end{pmatrix}.
$$

This matrix includes six types of elements, given by

$$
A : \frac{\partial^2}{\partial \beta_i \partial \beta_j} q_t(\omega, \theta) \equiv 2 y_{t-i} y_{t-j},
$$

$$
B : \frac{\partial^2}{\partial \beta_i \partial \delta_j} q_t(\omega, \theta) \equiv 2 y_{t-i} z_{t-j},
$$

$$
C : \frac{\partial^2}{\partial \beta_i \partial \gamma_j} q_t(\omega, \theta) \equiv 2 y_{t-i} \frac{\partial}{\partial \gamma_j} f(Y_{t-1}, Z_{t-1}, \gamma),
$$

$$
D : \frac{\partial^2}{\partial \delta_i \partial \delta_j} q_t(\omega, \theta) \equiv 2 z_{t-i} z_{t-j},
$$

$$
E : \frac{\partial^2}{\partial \delta_i \partial \gamma_j} q_t(\omega, \theta) \equiv 2 z_{t-i} \frac{\partial}{\partial \gamma_j} f(Y_{t-1}, Z_{t-1}, \gamma),
$$

$$
F : \frac{\partial^2}{\partial \gamma_i \partial \gamma_j} q_t(\omega, \theta) \equiv -2 \left[\frac{\partial^2}{\partial \gamma_i \gamma_j} f(Y_{t-1}, Z_{t-1}, \gamma) \right] \times \| Y_t - B Y_{t-1}
$$
$$
- D Z_{t-1} - F(Y_{t-1}, Z_{t-1}, \gamma) \|_2
$$
$$
+ 2 \left[\frac{\partial}{\partial \gamma_i} f(Y_{t-1}, Z_{t-1}, \gamma) \right]
$$
$$
\times \left[\frac{\partial}{\partial \gamma_j} f(Y_{t-1}, Z_{t-1}, \gamma) \right].
$$

We will study each type separately.

Some of the assumptions used for consistency have to be reinforced for normality. Again the values $\Delta_{ij}^{(f)}$, and δ_{Li} are positive and finite real numbers that are used as upper bounds.

Assumption LF. For $f = 8 + \varepsilon$ we have

(i) $E \| Z_t \|_S^f \leq \Delta_Z^{(f)}$;

(ii) $E \| V_t \|_S^f \leq \Delta_V^{(f)}$;

(iii) $E\|Z_t\|_S^f\|Z_s\|_S^f \le \Delta_{ZZ}^{(f)};$
(iv) $E\|V_t\|_S^f\|V_s\|_S^f \le \Delta_{VV}^{(f)};$
(v) $E\|Z_t\|_S^f\|V_s\|_S^f \le \Delta_{ZV}^{(f)}.$

Assumption LN′. For the norms $\|\cdot\|_2$ and $\|\cdot\|_S$ we have the following inequalities:

(i)   for $j = 1, \ldots, g$,

$$\left\|\nabla_\gamma \frac{\partial}{\partial\gamma_j} f(Y, Z, \gamma)\right\|_S^2 \le \delta_{L1}(\|Y\|_S + \|Z\|_S)^2;$$

(ii)   for $i, j = 1, \ldots, g$,

$$\left\|\nabla_\gamma \frac{\partial^2}{\partial\gamma_i\partial\gamma_j} f(Y, Z, \gamma)\right\|_S^2 \le \delta_{L2}(\|Y\|_S + \|Z\|_S)^2.$$

Assumption C T′. For the norms $\|\cdot\|_2$ and $\|\cdot\|_S$ we have the inequalities

(i) $\|(\partial/\partial\gamma_j)\nabla_Y f(Y, Z, \gamma)\| \le \delta_{KG};$
(ii) $\|(\partial^2/\partial\gamma_j\partial\gamma_i)\nabla_Y f(Y, Z, \gamma)\| \le \delta_{KGG}.$

The asymptotic normality is based on Theorem 2.2. The following lemma allows us to apply that theorem.

Lemma 4.1. *Under the set of assumptions MD, MX, CT, CN, LF, LN′, and CT′, the following results hold:*

(i)   *the elements of $\nabla_\theta q_t(\omega, \theta)$ are g-dominated on Θ uniformly in t for $g > 2$;*

(ii)   *the elements of $\nabla_{th}^2 q_t(\omega, \theta)$ are g-dominated on Θ uniformly in t for $g > 2$;*

(iii)   *the elements of $\nabla_\theta q_t(\omega, \theta)$ are near epoch dependent on $\{(Z_t, V_t)\}$ of any size uniformly on (Θ, ρ);*

(iv)   *the elements of $\nabla_\theta^2 q_t(\omega, \theta)$ are near epoch dependent on $\{(Z_t, V_t)\}$ of any size uniformly on (Θ, ρ);*

(v)   *$\nabla_\theta q_t(\omega, \theta)$ is Lipschitz-L_1 a.s. on Θ;*

(vi)   *$\nabla_\theta^2 q_t(\omega, \theta)$ is Lipschitz-L_1 a.s. on Θ.*

Proof: See Appendix C.   □

Theorem 4.2 (Asymptotic normality). *Under the set of assumptions MD, MX, CT, CN, LF, LN′, and CT′, the NLS estimator $\hat\theta$ of θ for model (3.1) is* *asymptotically distributed as*

$$\left(B_n^0\right)^{-1/2} A_n\left(\theta_n^0\right) \sqrt{n}\left(\hat{\theta}_n - \theta_n^0\right) \overset{A}{\sim} \mathcal{N}(0, I_k)$$

for $B_n^0 \equiv \text{Var}(n^{-1/2} \sum_{t=1}^n M_{nt}^0)$, $M_{nt}^0 \equiv \nabla_\theta q_t(\theta_n^0)$, *and* $A_n(\theta) \equiv \nabla_\theta^2 \bar{Q}_n(\theta)$.

Proof: From Lemma 4.1 we can apply Theorem 2.2, and the results follow. $\square$

4.1   *Comments on the assumptions*

This subsection presents the implications of the assumptions for the simple model (3.3).

Assumption LN′ is a strengthening of Assumption LN. For model (3.3), Assumption LN′ is

(i) $|(\partial^2/\partial\theta^2)g(y_{t-1}, z_{t-1}, \theta)| \le \delta_{L1}(|y_{t-1}| + |z_{t-1}|)$;

(ii) $|(\partial^3/\partial\theta^3)g(y_{t-1}, z_{t-1}, \theta)| \le \delta_{L2}(|y_{t-1}| + |z_{t-1}|)$.

Analogously, *Assumption CT′* is a strengthening of Assumption CT. For model (3.3), Assumption CT′ is

(i) $|(\partial/\partial\theta)(\partial/\partial y)g(y_{t-1}, z_{t-1}, \theta)| \le \delta_{KG}$;

(ii) $|(\partial^2/\partial\theta^2)(\partial/\partial y)g(y_{t-1}, z_{t-1}, \theta)| \le \delta_{KGG}$.

5   **Examples and extensions**

5.1   *Consistency of ESTAR models*

As an example of how our assumptions can be used, we analyze in detail the consistency and asymptotic normality of several ESTAR models. The simplest case of an ESTAR model is

$$\begin{aligned}
y_t &= \beta_1 y_{t-1} + \delta_1 y_{t-1}(1 - \exp\{-\gamma_1(y_{t-1} - \gamma_2)^2\}) + v_t \\
&\equiv \beta_1 y_{t-1} + f(y_{t-1}, \delta_1, \gamma_1, \gamma_2) + v_t \\
&\equiv \beta_1 y_{t-1} + f(y_{t-1}, \gamma) + v_t,
\end{aligned} \tag{5.1}$$

where $\gamma = [\delta_1, \gamma_1, \gamma_2]$. To check Assumption CT we take the partial derivative

$$\frac{\partial}{\partial y_{t-1}} f(y_{t-1}, \gamma) = \delta_1 + (\exp\{-\gamma_1(y_{t-1} - \gamma_2)^2\})$$
$$\times(-\delta_1 + 2\delta_1\gamma_1 y_{t-1}(y_{t-1} - \gamma_2)). \tag{5.2}$$

This function is bounded, because the function $g(z) = \exp\{-z^2\} z^2$ is clearly bounded, since it is continuous and $g(z) \to 0$ as $z \to \infty$, $g(z) \to 0$ as

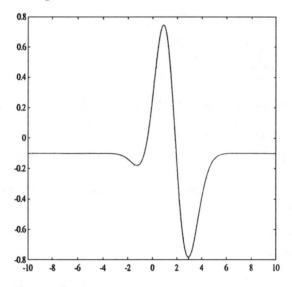

Figure 1. Case 1

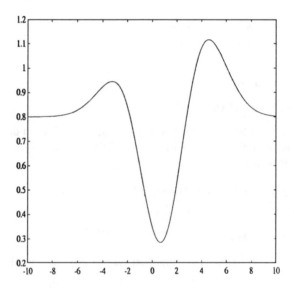

Figure 2. Case 2

$z \to 0$. The maximum value of $(\partial/\partial y_{t-1})f(y_{t-1}, \gamma)$ is obtained as a function of the parameters $\delta_1, \gamma_1, \gamma_2$. A grid of values for y_{t-1} provides a set of values for $(\partial/\partial y_{t-1})f(y_{t-1}, \gamma)$, and by inspection we can see if the maximum of $|\beta_1 + (\partial/\partial y_{t-1})f(y_{t-1}, \gamma)|$ is above or below 1. For instance, the graphs of figure 1 and 2 show the curve of the function $f(y_{t-1}) = \beta_1 + (\partial/\partial y_{t-1})f(y_{t-1}, \gamma)$

for values of the parameters $(\beta_1, \delta_1, \gamma_1, \gamma_2)$ given by $(0.6, -0.7, 0.4, 1.3)$ and $(0.3, 0.5, 0.1, 1)$ respectively.

For model (5.1) we have that Assumptions MD, MX, and CN are immediately satisfied; Assumption CT(i) can be checked by inspection as we have shown; Assumption CT(ii) is also immediately satisfied; Assumption CT(iii) restricts the parameter space; Assumption CT(iv) follows once Assumption CT(i) is satisfied. Furthermore, Assumption LN(i) is clearly satisfied, since $h(y_{t-1}, \gamma_1, \gamma_2) = 1 - \exp\{-\gamma_1(y_{t-1} - \gamma_2)^2\}$ verifies that $h(y_{t-1}, \gamma_1, \gamma_2) \to 0$ as $y_{t-1} \to 0$, and $h(y_{t-1}, \gamma_1, \gamma_2) \to 1$ as $y_{t-1} \to \infty$; therefore

$$|f(y_{t-1}, \delta_1, \gamma_1, \gamma_2)| \leq \delta_{CF}|y_{t-1}|$$

for $\delta_{CF} = \delta_1 + 1$. To study Assumption LN(ii) consider the score vector

$$\nabla_\gamma f(y_{t-1}, \gamma) = \left[\frac{\partial}{\partial \delta_1} f(y_{t-1}, \gamma), \frac{\partial}{\partial \gamma_1} f(y_{t-1}, \gamma), \frac{\partial}{\partial \gamma_2} f(y_{t-1}, \gamma) \right]$$

with

$$\frac{\partial}{\partial \delta_1} f(y_{t-1}, \gamma) \equiv y_{t-1}(1 - \exp\{-\gamma_1(y_{t-1} - \gamma_2)^2\}), \tag{5.3}$$

$$\frac{\partial}{\partial \gamma_1} f(y_{t-1}, \gamma) \equiv \delta_1 y_{t-1}(y_{t-1} - \gamma_2)^2 \exp\{-\gamma_1(y_{t-1} - \gamma_2)^2\}, \tag{5.4}$$

$$\frac{\partial}{\partial \gamma_2} f(y_{t-1}, \gamma) \equiv \delta_1 y_{t-1}\gamma_1(y_{t-1} - \gamma_2) \exp\{-\gamma_1(y_{t-1} - \gamma_2)^2\}. \tag{5.5}$$

Therefore the score behaves like

$$\nabla_\gamma f(y_{t-1}, \gamma) \sim \left[y_{t-1} \exp\{-y_{t-1}^2\}, y_{t-1}^3 \exp\{-y_{t-1}^2\}, y_{t-1}^2 \exp\{-y_{t-1}^2\} \right],$$

which has a finite bound.

From equations (5.2), (5.3), (5.4), and (5.5) it is clear that further derivatives are bounded. Therefore Assumptions LN'(i), LN'(ii), CT'(i), and CT'(ii) are also satisfied for this model.

Following the same procedure allows us to examine more general nonlinear dynamic models. Take for instance Assumption CT(i) in the following nonlinear autoregressive distributed lag model of order 2 (NARDL(2)):

$$\begin{aligned}
y_t &= \beta_1 y_{t-1} + \beta_2 y_{t-2} + \delta_1 z_{t-1} + \delta_2 z_{t-z} \\
&\quad + (\pi_1 y_{t-1} + \pi_2 y_{t-2} + \theta_1 z_{t-1} + \theta_2 z_{t-z}) \\
&\quad \times (1 - \exp\{-\gamma_1(y_{t-1} - \gamma_2)^2\}) + v_t \\
&\equiv \beta_1 y_{t-1} + \beta_2 y_{t-2} + \delta_1 z_{t-1} + \delta_2 z_{t-z} \\
&\quad + f(y_{t-1}, y_{t-2}, z_{t-1}, z_{t-z}, \gamma) + v_t, \tag{5.6}
\end{aligned}$$

where $\gamma' = [\pi_1, \pi_2, \theta_1, \theta_2, \gamma_1, \gamma_2]$. In this case we have to study the spectral radius of the matrix $B\nabla F \equiv B + \nabla F$ given by

$$B = \begin{pmatrix} \beta_1 & \beta_2 \\ 1 & 0 \end{pmatrix} \quad \text{and} \quad \nabla F = \begin{pmatrix} \xi_1 & \xi_2 \\ 0 & 0 \end{pmatrix},$$

where

$$\xi_1 = \frac{\partial}{\partial y_{t-1}} f(y_{t-1}, y_{t-2}, z_{t-1}, z_{t-2}, \gamma) \quad \text{and}$$

$$\xi_2 = \frac{\partial}{\partial y_{t-2}} f(y_{t-1}, y_{t-2}, z_{t-1}, z_{t-2}, \gamma).$$

Since the characteristic polynomial of the matrix $B\nabla F$ is $\mathrm{CP}(x) \equiv x^2 - (\beta_1 + \xi_1)x - (\beta_2 + \xi_2)$, then to study the spectral radius it is enough to make a grid of the eigenvalues λ_1 and λ_2 given as the solutions $\mathrm{CP}(\lambda_i) = 0$. Therefore, in general we need to make k grids if we have a NARDL(k) model.

If model (5.6) does not have an autoregressive endogenous linear part, i.e. $\beta_1 = \beta_2 = 0$, then the elements of matrix B, except the element $(1, 2)$, which is equal to 1, are all null. The general rule for this case is explained in Section 3.

5.2 *Extensions*

Consider the case of a TAR model given by

$$y_t = \beta^* y_{t-1} + \lambda^* y_{t-1} I(y_{t-1} < 0),$$

where $I(y_{t-1} < 0)$ is the characteristic function for the set $\{y_{t-1} < 0\}$. In this case the slope of the model is either β^* or $\beta^* + \lambda^*$. Then taking the derivative where possible, we have that Assumption CT(i) is $|\beta^* + \lambda^* I(y_{t-1} < 0)| < 1$, which means that in each region the function is "stable" in the sense that its slope is smaller than 1. In this case Assumption CN is violated, but the proof of each lemma can be done for each subset $y_{t-1} > \gamma_1$ and $y_{t-1} \leq \gamma_1$, and then the consistency result applies again.

The restriction of a linear growth can be weakened to a linear growth by intervals, for instance,

$$f(y, \gamma) \leq \begin{cases} C_1 + D_1 y & \text{for} \quad y > \gamma_1, \\ C_2 + D_2 y & \text{for} \quad \leq \gamma_1. \end{cases}$$

In this case the proof of each lemma can be done for each interval, and the consistency result follows. This allows us to deal with polynomial functions of any order, by truncating the polynomial at certain bounds. Thus, for a fourth order polynomial we have

$$
f(y, \gamma) = \begin{cases} C_0 + D_0 y + E_0 y^2 + F_0 y^3 + G_0 y^4 & \text{for} \quad B_1 \leq y \leq B_2, \\ C_1 + D_1 y & \text{for} \quad y < B_1, \\ C_2 + D_2 y & \text{for} \quad y > B_2. \end{cases}
$$

Notice that these polynomial functions are not required to be bounded by a constant, but by a linearly growth term. This extension allows us to consider nonlinear models in the form of a nonlinear error correction; see Escribano (1986, 1996) and Escribano and Mira (1997). The nonlinear term is the lagged error from a cointegration relationship. The results of this chapter include this class of models as long as the cointegrating vector is known and not estimated.

6   Conclusions

In this chapter we have given explicit sufficient conditions for consistency and asymptotic normality of the NLS estimator of the parameters of nonlinear dynamic regression models. In these models we allow for some heterogeneity (nonstationarity) as long as the underlying stochastic process is mixing. The basic statistical framework is based on Gallant and White (1988).

We show how, under some simple and explicit conditions on the linear and nonlinear terms of the model and under some moment conditions, the NLS estimator is consistent and asymptotically normal. These conditions are in sharp contrast with the conditions previously available, since they were highly technical and difficult to check in practice.

The analysis was based on general nonlinear autoregressive distributed lag models, but written as partially linear time series models. This is not restrictive, and on the other hand it allowed us to clarify the relationship with the well-known conditions for estimation and inference in linear time series models. This parameterization includes as particular cases the smooth transition regression (STR) models and the smooth transition autoregressive (STAR) models. These classes of models are becoming very popular in modeling the behavior of asymmetric time series variables; see for example Teräsvirta (1994).

However, the set of assumptions we have proposed allowed us to consider more general nonlinear functions, since they do not require the function to be bounded by a constant but only to be bounded by a linearly growing term. This important extension opens the opportunity to include a broader class of models, such as the nonlinear error correction models introduced by Escribano (1986).

Appendix A: Norms and matrices

A.1   *The S-norm*

Given a matrix A of size $n \times n$, let

$$A = MJM^{-1}$$

be its Jordan decomposition such that J is a diagonal matrix with boxes in its diagonal. The boxes are of the form

$$J = \begin{pmatrix} J_1 & 0 & \cdots & 0 \\ 0 & J_2 & \cdots & 0 \\ \vdots & \vdots & \ddots & \vdots \\ 0 & 0 & \cdots & J_s \end{pmatrix}, \quad \text{where} \quad J_i = \begin{pmatrix} \lambda_i & 1 & 0 & \cdots & 0 \\ 0 & \lambda_i & 1 & \cdots & 0 \\ \vdots & \vdots & \vdots & \ddots & \vdots \\ 0 & 0 & 0 & \cdots & 1 \\ 0 & 0 & 0 & \cdots & \lambda_i \end{pmatrix}.$$

Let us define now the matrix D_δ as

$$D_\delta = \begin{pmatrix} 1 & 0 & \cdots & 0 \\ 0 & \delta & \cdots & 0 \\ \vdots & \vdots & \ddots & \vdots \\ 0 & 0 & \cdots & \delta^{n-1} \end{pmatrix}.$$

If we take the matrix norm $\|\cdot\|_S$ as

$$\|A\|_S \equiv \|(MD_\delta)^{-1}A(MD_\delta)\|_\infty,$$

then it is clear that $\|A\|_S \leq \rho(A) + \delta$, because $(MD_\delta)^{-1}A(MD_\delta)$ is equal to the matrix J where the boxes J_i are replaced by boxes J_i^* of the form

$$J_i^* = \begin{pmatrix} \lambda_i & \delta & 0 & \cdots & 0 \\ 0 & \lambda_i & \delta & \cdots & 0 \\ \vdots & \vdots & \vdots & \ddots & \vdots \\ 0 & 0 & 0 & \cdots & \delta \\ 0 & 0 & 0 & \cdots & \lambda_i \end{pmatrix}.$$

In this case the vector norm is

$$\|Y\|_S = \|(MD_\delta)Y\|_\infty,$$

which means that the matrix norm is a subordinate vector norm. A very similar definition is given in Ciarlet (1989).

With the above definitions and Assumption CT we have

$$\|Y\|_S = \|Y^*\|_\infty$$
$$= \|(MD_\delta)^{-1}Y\|_\infty$$
$$\le \|D_\delta^{-1}\|_\infty \|M^{-1}\|_\infty \|Y\|_\infty$$
$$\le \Delta^{-1}\delta^{-1}\|Y\|_\infty,$$
$$\|Y\|_\infty = \|(MD_\delta)Y^*\|_\infty$$
$$\le \|M\|_\infty \|D_\delta\|_\infty \|Y^*\|_\infty$$
$$\le \Delta\|Y\|_S,$$

because δ is smaller than 1. Given that $\|\cdot\|_2$ and $\|\cdot\|_\infty$ are equivalent norms, then there exist inequalities similar to the above ones for $\|\cdot\|_S$ and $\|\cdot\|_2$, and we will make use of them in the following.

A.2 *Proof of Lemma 3.3*

(i) Triangular inequality: By the Minkowsky inequality we have

$$E\left(\|g+h\|_S^r\right) = E\left(\|g+h\|_S\|g+h\|_S^{r-1}\right)$$
$$\le E\left((\|g\|_S + \|h\|_S)\left(\|g+h\|_S^{r-1}\right)\right)$$
$$= E\left(\|g\|_S\|g+h\|_S^{r-1}\right) + E\left(\|h\|_S\|g+h\|_S^{r-1}\right).$$

The Hölder inequality states that $E(|XY|) \le E^{1/p}|X|^p E^{1/q}|Y|^q$ for $1/p + 1/q = 1$. Taking $p = r, q = (r-1)/r$, $|X| = \|g\|_S$, and $|Y| = \|g+h\|_S$ (and analogously for the second term in the summation), we have

$$E\left(\|g+h\|_S^r\right) \le E^{1/r}\|g\|_S^r E^{(r-1)/r}\left(\|g+h\|_S^{r-1}\right)^{r/(r-1)}$$
$$+ E^{1/r}\|h\|_S^r E^{(r-1)/r}\left(\|g+h\|_S^{r-1}\right)^{r/(r-1)}$$
$$= \left(E^{1/r}\|g\|_S^r + E^{1/r}\|h\|_S^r\right)E^{(r-1)/r}\left(\|g+h\|_S^{r-1}\right)^{r/(r-1)}$$
$$= \left(E^{1/r}\|g\|_S^r + E^{1/r}\|h\|_S^r\right)E^{1-1/r}\left(\|g+h\|_S^r\right);$$

therefore $1 \le (E^{1/r}\|g\|_S^r + E^{1/r}\|h\|_S^r)E^{-1/r}(\|g+h\|_S^r)$ or $E^{1/r}(\|g+h\|_S^r) \le (E^{1/r}\|g\|_S^r + E^{1/r}\|h\|_S^r)$, as we want.

(ii) Scalar multiplication:

$$\|\alpha X\|_{Sr} = E^{1/r}\|\alpha X\|_S^r = E^{1/r}|\alpha|^r\|X\|_S^r = |\alpha|E^{1/r}\|X\|_S^r = |\alpha|\|X\|_{Sr}.$$

Appendix B: Proof of Lemma 3.4

B.1   *Proof of Lemma 3.4(i)*

We will prove that Y_t is NED on $\{(V_t, Z_t)\}$ in the sense of the norm $\|\cdot\|_S$, i.e., $\|Y_t - E_{t-m}(Y_t)\|_{S2} \to 0$ when $m \to \infty$, for $E_{t-m}(Y_t) \equiv E(Y_t | v_t, z_{t-1}, v_{t-1}, \ldots, z_{t-m}, v_{t-m})$. Let us define the sequence of predictors $\{\bar{y}_t\}$ as

$$\bar{y}_t = \begin{cases} \beta_1^* \bar{y}_{t-1} + \cdots + \beta_p^* \bar{y}_{t-p} + f(\bar{y}_{t-1}, \ldots, \bar{y}_{t-p}, 0, \ldots, 0, \gamma^*) & \text{for } t > 0, \\ 0 & \text{for } t \leq 0; \end{cases}$$

therefore $\bar{Y}_t = [\bar{y}_t, \ldots, \bar{y}_{t-p+1}]'$, and also $\bar{Y}_t = B^* \bar{Y}_{t-1} + F(\bar{Y}_{t-1}, 0, \gamma^*)$. Let us define the sequence of predictors $\{\tilde{y}_{t,s}^m\}$ for $m \geq q$ as

$$\tilde{y}_{t,s}^m = $$

$$\begin{cases} \beta_1^* \tilde{y}_{t-1,s+1}^m + \cdots + \beta_p^* \tilde{y}_{t-p,s+p}^m + \delta_1^* z_{t-1} + \cdots + \delta_q^* z_{t-q} \\ \quad + f(\tilde{y}_{t-1,s+1}^m, \ldots, \tilde{y}_{t-p,s+p}^m, z_{t-1}, \ldots, z_{t-q}, \gamma^*) + v_t & \text{for } s + q \leq m, \\ \bar{y}_t & \text{for } s + q > m. \end{cases}$$

Therefore $\tilde{Y}_{t,s}^m = [\tilde{y}_{t,s}^m, \ldots, \tilde{y}_{t-p+1,s-p+1}^m]'$, and also $\tilde{Y}_{t,s}^m = B^* \tilde{Y}_{t-1,s+1}^m + D^* Z_{t-1} + F(\tilde{Y}_{t-1,s+1}^m, Z_{t-1}, \gamma^*) + V_t$. With this definition $\tilde{y}_{t,0}^m$ depends directly on $v_t, z_{t-1}, \ldots, z_{t-q}$ and indirectly on $v_{t-1}, \ldots, z_{t-m}, v_{t-m}$; therefore by definition $\tilde{y}_{t,s}^m$ is $\sigma(v_t, z_{t-1}, v_{t-1}, \ldots, z_{t-m}, v_{t-m})$-measurable. The difference between Y_t and its predictor $\bar{Y}_t$ is

$$\|Y_t - \bar{Y}_t\|_S \leq \|B^* Y_{t-1} + D^* Z_{t-1} + F(Y_{t-1}, Z_{t-1}, \gamma^*) + V_t$$
$$- B^* \bar{Y}_{t-1} + F(\bar{Y}_{t-1}, 0, \gamma^*)\|_S$$
$$\leq \|B^*(Y_{t-1} - \bar{Y}_{t-1}) + F(Y_{t-1}, Z_{t-1}, \gamma^*) - F(\bar{Y}_{t-1}, 0, \gamma^*)\|_S$$
$$+ \|D^*\|_S \|Z_{t-1}\|_S + \|V_t\|_S.$$

Using the mean value theorem for random variables and Assumption CN, we have

$$F(Y_{t-1}, Z_{t-1}, \gamma^*) - F(\bar{Y}_{t-1}, 0, \gamma^*)$$
$$= [f(y_{t-1}, \ldots, y_{t-p}, z_{t-1}, \ldots, z_{t-q}, \gamma^*)$$
$$- f(\bar{y}_{t-1}, \ldots, \bar{y}_{t-p}, 0, \ldots, 0, \gamma^*), 0, \ldots, 0]'$$
$$= \left[\frac{\partial}{\partial y_1} f(\ddot{y}_{t-1}, \ldots, \ddot{y}_{t-p}, \ddot{z}_{t-1}, \ldots, \ddot{z}_{t-q}, \gamma^*)(y_{t-1} - \bar{y}_{t-1}) + \cdots \right.$$
$$+ \frac{\partial}{\partial y_p} f(\ddot{y}_{t-1}, \ldots, \ddot{y}_{t-p}, \ddot{z}_{t-1}, \ldots, \ddot{z}_{t-q}, \gamma^*)(y_{t-p} - \bar{y}_{t-p})$$
$$+ \frac{\partial}{\partial z_1} f(\ddot{y}_{t-1}, \ldots, \ddot{y}_{t-p}, \ddot{z}_{t-1}, \ldots, \ddot{z}_{t-q}, \gamma^*)(z_{t-1}) + \cdots$$

$$+ \frac{\partial}{\partial z_q} f(\ddot{y}_{t-1}, \ldots, \ddot{y}_{t-p}, \ddot{z}_{t-1}, \ldots, \ddot{z}_{t-q}, \gamma^*)(z_{t-q}), 0, \ldots, 0\Big]'$$

$$= [\nabla_Y f(\ddot{Y}_{t-1}, \ddot{Z}_{t-1}, \gamma^*)(Y_{t-1} - \bar{Y}_{t-1})$$
$$+ \nabla_Z f(\ddot{Y}_{t-1}, \ddot{Z}_{t-1}, \gamma^*)Z_{t-1}, 0, \ldots, 0]'$$
$$= \nabla_Y F(\ddot{Y}_{t-1}, \ddot{Z}_{t-1}, \gamma^*)(Y_{t-1} - \bar{Y}_{t-1})$$
$$+ \nabla_Z F(\ddot{Y}_{t-1}, \ddot{Z}_{t-1}, \gamma^*)Z_{t-1},$$

where $\nabla_Y f(\ddot{Y}_{t-1}, \ddot{Z}_{t-1}, \gamma^*)$ is a $p \times 1$ vector and $\nabla_Y F(\ddot{Y}_{t-1}, \ddot{Z}_{t-1}, \gamma^*)$ is a $p \times p$ matrix with rows $(\nabla_Y f', 0', \ldots, 0')'$; and the third inequality represents a couple of inner products. Now we have

$$\|Y_t - \bar{Y}_t\|_S \le \|B^* + \nabla_Y F(\ddot{Y}, \ddot{Z}, \gamma^*)\|_S \|Y_{t-1} - \bar{Y}_{t-1}\|_S$$
$$+ (\|D^*\|_S + \|\nabla_Z F(\ddot{Y}, \ddot{Z}, \gamma^*)\|_S)\|Z_{t-1}\|_S + \|V_t\|_S.$$

Since $x_t = \alpha x_{t-1} + \beta_{t-1}$ then $x_t = \alpha^t x_0 + \sum_{i=0}^{t-1} \alpha^i \beta_{t-1-i}$, and so

$$\|Y_t - \bar{Y}_t\|_S \le (\|B^* + \nabla_Y F(\ddot{Y}, \ddot{Z}, \gamma^*)\|_S)^t \|Y_0 - \bar{Y}_0\|_S$$
$$+ \sum_{i=0}^{t-1} ((\|B^* + \nabla_Y F(\ddot{Y}, \ddot{Z}, \gamma^*)\|_S)^i [\|V_{t-i}\|_S$$
$$+ (\|D^*\|_S + \|\nabla_Z F(\ddot{Y}, \ddot{Z}, \gamma^*)\|_S)\|Z_{t-1-i}\|_S]) .$$

Now since $Y_0 = \bar{Y}_0 = 0$ and $\|B^* + \nabla_Y f(\ddot{Y}, \ddot{Z}, \gamma^*)\|_S \equiv \delta_{BY} < 1$ and $(\|D^*\|_S + \|\nabla_Z F(\ddot{Y}, \ddot{Z}, \gamma^*)\|_S) \equiv \delta_{DZ}$, then

$$\|Y_t - \bar{Y}_t\|_S \le \sum_{i=0}^{t-1} \delta_{BY}^i \|V_{t-i}\|_S + \sum_{i=0}^{t-1} \delta_{BY}^i \delta_{DZ}\|Z_{t-1-i}\|_S.$$

Therefore

$$E\|Y_t - \bar{Y}_t\|_S^2 \le E\left(\sum_{i=0}^{t-1} \delta_{BY}^i \|V_{t-i}\|_S + \sum_{i=0}^{t-1} \delta_{BY}^i \delta_{DZ}\|Z_{t-1-i}\|_S\right)^2$$
$$= E\left(\sum_{i=0}^{t-1} \delta_{BY}^i (\|V_{t-i}\|_S + \delta_{DZ}\|Z_{t-1-i}\|_S)\right)^2$$
$$= E\left(\sum_{i=0}^{t-1} \delta_{BY}^{2i} (\|V_{t-i}\|_S + \delta_{DZ}\|Z_{t-1-i}\|_S)^2\right.$$
$$\left. + \sum_{i=0}^{t-1}\sum_{i\ne j}^{t-1} \delta_{BY}^{i+j} (\|V_{t-i}\|_S + \delta_{DZ}\|Z_{t-1-j}\|_S)^2\right).$$

Now from Assumption LR we know that $E\|V_t\|_S^2 \le \Delta_V^{(2)}$, $E\|Z_t\|_S^2 \le \Delta_Z^{(2)}$, and $E\|V_t\|_S\|Z_t\|_S \le \Delta_{VZ}^{(1)}$, and then for some constants δ_{DVZ}

$$
E\|Y_t - \bar{Y}_t\|_S^2 \le \sum_{i=0}^{t-1} \delta_{BY}^{2i}\left(\Delta_V^{(2)} + \delta_{DZ}\Delta_Z^{(2)} + 2\delta_{DZ}\Delta_{ZV}^{(1)}\right)
$$

$$
+ \sum_{i=0}^{t-1}\sum_{i\ne j}^{t-1} \delta_{BY}^{i+j}\left(\Delta_V^{(2)} + \delta_{DZ}\Delta_Z^{(2)} + 2\delta_{DZ}\Delta_{ZV}^{(1)}\right)
$$

$$
\le \delta_{DVZ}\sum_{i=0}^{t-1} \delta_{BY}^{2i} + \delta_{DVZ}\sum_{i=0}^{t-1}\sum_{i\ne j}^{t-1} \delta_{BY}^{i+j}
$$

$$
= \delta_{DVZ}\frac{1}{\delta_{BY}^2} + \delta_{DVZ}\frac{1}{(1-\delta_{BY})^2}
$$

$$
\equiv \Delta_{Y-\bar{Y}}^{(2)},
$$

where the last inequality follows from

$$
\sum_{i=0}^{t-1}\sum_{i\ne j}^{t-1} \delta_{BY}^{i+j} = \sum_{i=0}^{t-1}\sum_{j=0}^{t-1} \delta_{BY}^{i+j} - \sum_{i=0}^{t-1} \delta_{BY}^{2i}
$$

$$
= \sum_{i=0}^{t-1} \delta_{BY}^{i} \times \sum_{j=0}^{t-1} \delta_{BY}^{j} - \sum_{i=0}^{t-1} \delta_{BY}^{2i}
$$

$$
= \left(\frac{1-\delta_{BY}^t}{1-\delta_{BY}}\right)^2 - \frac{1-\delta_{BY}^{2t}}{1-\delta_{BY}^2}
$$

$$
\le \left(\frac{1}{(1-\delta_{BY})}\right)^2 - \frac{1-\delta_{BY}^{2t}}{1-\delta_{BY}^2}
$$

$$
\le \left(\frac{1}{1-\delta_{BY}}\right)^2.
$$

Now the difference between Y_t and its predictor $\tilde{Y}_{t,0}^m$ is

$$
\left\|Y_t - \tilde{Y}_{t,0}^m\right\|_S \le \|B^*Y_{t-1} + D^*Z_{t-1} + F(Y_{t-1}, Z_{t-1}, \gamma^*) + V_t
$$

$$
- B^*\tilde{Y}_{t-1,1}^m - D^*Z_{t-1}F(\tilde{Y}_{t-1,1}, Z_{t-1}, \gamma^*) + V_t\|_S
$$

$$
\le \left\|B^*\left(Y_{t-1} - \tilde{Y}_{t-1,1}^m\right) + F(Y_{t-1}, Z_{t-1}, \gamma^*)\right.
$$

$$
\left. - F\left(\tilde{Y}_{t-1,1}^m, Z_{t-1}, \gamma^*\right)\right\|_S.
$$

Making use of the mean value theorem again, we have

$$\left\| Y_t - \tilde{Y}_{t,0}^m \right\|_S \leq \| B^* + \nabla_Y F(\ddot{Y}, Z, \gamma^*) \|_S \left\| Y_{t-1} - \tilde{Y}_{t-1,1}^m \right\|_S,$$

and by iteration we obtain

$$\left\| Y_t - \tilde{Y}_{t,0}^m \right\|_S \leq \| B^* + \nabla_Y F(\ddot{Y}, Z, \gamma^*) \|_S^m \| Y_{t-m} - \bar{Y}_{t-m} \|_S$$

$$\equiv \delta_{BY}^m \| Y_{t-m} - \bar{Y}_{t-m} \|_S,$$

where $|\delta_{BY}| < 1$. Taking expectations in $\| Y_t - \tilde{Y}_{t,0}^m \|_S^2$, we have

$$E \left\| Y_t - \tilde{Y}_{t,0}^m \right\|_S^2 \leq \delta_{BY}^{2m} E \| Y_{t-m} - \bar{Y}_{t-m} \|_S^2$$

$$= \delta_{BY}^{2m} \Delta_{Y-\bar{Y}}^{(2)},$$

so that we obtain a bound for $E \| Y_t - \tilde{Y}_{t,0}^m \|_S^2$, say $\Delta_{Y-\bar{Y}}^{(2)}$. Note that $E \| Y_t - \tilde{Y}_{t,0}^m \|_S^2 \to 0$ as $m \to \infty$ because $|\delta_{BY}| < 1$.

Now, given $E_{t-m}(Y_t) \equiv E(Y_t | v_t, z_{t-1}, v_{t-1}, \dots, z_{t-m}, v_{t-m})$, we can find a bound for $\| Y_t - E_{t-m}(Y_t) \|_{S2}$. Recall that $\| \cdot \|_{S2} = E^{1/2} \| \cdot \|_S^2$. Since $\tilde{Y}_{t,0}^m$ is $\sigma(v_t, \dots, z_{t-m}, v_{t+m})$-measurable, then

$$\| Y_t - E_{t-m}(Y_t) \|_{S2} \leq \left\| Y_t - \tilde{Y}_{t,0}^m \right\|_{S2}$$

$$= E^{1/2} \left\| Y_t - \tilde{Y}_{t,0}^m \right\|_S^2$$

$$\leq \left(\Delta_{Y-\bar{Y}}^{(2)} \right)^{1/2} \equiv \Delta_{Y-\bar{Y}}^2,$$

where the first inequality is valid up to a multiplicative constant and represents the generalization of the well-known fact in the scalar case that the least mean square error is given by the conditional expectation. Therefore $\{Y_t\}$ is NED in $\{(Z_t, V_t)\}$, and since $E \| Y_t - \tilde{Y}_{t,0}^m \|_S^2 \to 0$ at an exponential rate, then it is NED of any size.

B.2   *Proof of Lemma 3.4(ii)*

Recall the following property. For any matrix M and vector X we have

$$\| MX \|_{Sr} = E^{1/r} \| MX \|_S^r \leq E^{1/r} \| M \|_S^r \| X \|_S^r$$

$$= \| M \|_S E^{1/r} \| X \|_S^r = \| M \|_S \| X \|_{Sr}.$$

The Minkowsky inequality states $\| X + W \|_{Sr} \leq \| X \|_{Sr} + \| W \|_{Sr}$, and then we have

$$\| Y_t \|_{Sr} = \| B^* Y_{t-1} + D^* Z_{t-1} + F(Y_{t-1}, Z_{t-1}, \gamma^*) + V_t \|_{Sr}$$

$$\leq \| B^* Y_{t-1} + F(Y_{t-1}, Z_{t-1}, \gamma^*) \|_{Sr} + \| D^* \|_S \| Z_{t-1} \|_{Sr} + \| V_t \|_{Sr}.$$

Let us study the summation above. From the mean value theorem we have

$$F(Y_{t-1}, Z_{t-1}, \gamma^*)$$

$$= \left[f(0, \dots, 0, z_{t-1}, \dots, z_{t-q}, \gamma^*) \right.$$

$$\left. + \sum_{i=1}^{p} \frac{\partial}{\partial y_i} f(\ddot{y}_{t-1}, \dots, \ddot{y}_{t-p}, z_{t-1}, \dots, z_{t-q}, \gamma^*)(y_{t-i}), 0, \dots, 0 \right]$$

$$= F(0, Z_{t-1}, \gamma^*) + \nabla_Y F(\ddot{Y}, Z_{t-1}, \gamma^*) Y_{t-1},$$

where the last inequality follows from the same steps as in Lemma 3.1(i). Taking expectations and recalling that $\|Z_t\|_{Sr} = E^{1/r}\|Z\|_S^r \le (\Delta_Z^{(r)})^{1/r} \equiv \Delta_Z^r$ and $\|V_t\|_{Sr} \le (\Delta_V^{(r)})^{1/r} \equiv \Delta_V^r$, we have by Assumption CT(iii) and for some constant δ_{HH} that

$$\|Y_t\|_{Sr} \le \|B^* Y_{t-1} + F(0, Z_{t-1}, \gamma^*) + \nabla_Y F(\ddot{Y}, Z_{t-1}, \gamma^*) Y_{t-1}\|_{Sr}$$

$$+ \|D^*\|_S \|Z_{t-1}\|_{Sr} + \|V_t\|_{Sr}$$

$$\equiv \|B^* + \nabla_Y F(\ddot{Y}, Z_{t-1}, \gamma^*)\|_S \|Y_{t-1}\|_{Sr} + \delta_{HH}$$

$$\le \delta_{BY} \|Y_{t-1}\|_{Sr} + \delta_{HH}.$$

Now by iteration in $\|Y_t\|_{Sr} \le \delta_{BY}\|Y_{t-1}\|_{Sr} + \delta_{HH}$, the fact that $\|Y_0\|_{Sr} = 0$, and Assumption CT we obtain that $\|Y_t\|_{Sr} \le \Delta_Y^r$.

B.3   *Proof of Lemma 3.4(iii)*

By Assumption LN we have

$$|q_t(\omega, \theta)| = \|Y_t - BY_{t-1} - DZ_{t-1} - F(Y_{t-1}, Z_{t-1}, \gamma)\|_2^2$$

$$\le \|Y_t\|_2^2 + 2\|Y_t\|_2 \|BY_{t-1} - DZ_{t-1} - F(Y_{t-1}, Z_{t-1}, \gamma)\|_2$$

$$+ \|BY_{t-1} - DZ_{t-1} - F(Y_{t-1}, Z_{t-1}, \gamma)\|_2^2$$

$$\le \|Y_t\|_2^2 + 2\|Y_t\|_2 [\delta_{CB}\|Y_{t-1}\|_2 + \delta_{CD}\|Z_{t-1}\|_2$$

$$+ \delta_{CF}(\|Y_{t-1}\|_2 + \|Z_{t-1}\|_2)] + [\delta_{CB}\|Y_{t-1}\|_2$$

$$+ \delta_{CD}\|Z_{t-1}\|_2 + \delta_{CF}(\|Y_{t-1}\|_2 + \|Z_{t-1}\|_2)]^2$$

$$= \|Y_t\|_2^2 + 2\|Y_t\|_2 [(\delta_{CB} + \delta_{CF})\|Y_{t-1}\|_2$$

$$+ (\delta_{CD} + \delta_{CF})\|Z_{t-1}\|_2] + [(\delta_{CB} + \delta_{CF})^2\|Y_{t-1}\|_2^2$$

$$+ (\delta_{CD} + \delta_{CF})^2\|Z_{t-1}\|_2^2 + 2(\delta_{CB} + \delta_{CF})$$

$$\times (\delta_{CD} + \delta_{CF})\|Y_{t-1}\|_2\|Z_{t-1}\|_2]$$

$$\equiv D_t.$$

Since s is integer, we have that $E^{1/s}|D_t|^s$ is the expectation (sth root) of a summation of terms of the form $(\|Y_t\|_2^2\|Z_t\|_2^2)^c$ for $c \leq s$ (see Section B.4 below). From Assumption LR, since $s = r/2$ and $\|Y_t\|_2 \leq \Delta\|Y_t\|_S$ for some constant Δ, we conclude that $E^{1/s}|D_t|^s \leq \Delta_D^s$ for some constant Δ_D^s.

B.4   *Proof of Lemma 3.4(iv)*

We use Theorem 4.5 from Gallant and White (1988). Define

$$X_t = (Y_t, Y_{t-1}, Z_{t-1}),$$
$$\tilde{X}_t^m = \left(\tilde{Y}_{t,0}^m, \tilde{Y}_{t-1,1}^m, Z_{t-1}\right).$$

Define $b(X_t, \theta) = q_t(\omega, \theta)$. As we saw in Lemma 3.4(iii), $b(X_t, \theta)$ is s-dominated for $s = r/2$. Since $r > 4$, we impose that $b(X_t, \theta)$ is at least 2-dominated.

Recall that $\|W - V\|_2 \geq |\|W\|_2 - \|V\|_2|$ and $\|W + V\|_2 \leq \|W\|_2 + \|V\|_2$, then we obtain

$$
\begin{aligned}
\left|b(X_t, \theta) - b\left(\tilde{X}_t^m, \theta\right)\right| &= \left|\|Y_t - BY_{t-1} - DZ_{t-1} - F(Y_{t-1}, Z_{t-1}, \gamma)\|_2^2\right. \\
&\quad \left. - \left\|\tilde{Y}_{t,0}^m - B\tilde{Y}_{t-1,1}^m - DZ_{t-1} - F\left(\tilde{Y}_{t-1,1}^m, Z_{t-1}, \gamma\right)\right\|_2^2\right| \\
&= \left|\left(\|Y_t - BY_{t-1} - DZ_{t-1} - F(Y_{t-1}, Z_{t-1}, \gamma)\|_2\right.\right. \\
&\quad \left.+ \left\|\tilde{Y}_{t,0}^m - B\tilde{Y}_{t-1,1}^m - DZ_{t-1} - F\left(\tilde{Y}_{t-1,1}^m, Z_{t-1}, \gamma\right)\right\|_2\right) \\
&\quad \times \left(\|Y_t - BY_{t-1} - DZ_{t-1} - F(Y_{t-1}, Z_{t-1}, \gamma)\|_2\right. \\
&\quad \left.\left.- \left\|\tilde{Y}_{t,0}^m - B\tilde{Y}_{t-1,1}^m - DZ_{t-1} - F\left(\tilde{Y}_{t-1,1}^m, Z_{t-1}, \gamma\right)\right\|_2\right)\right| \\
&\leq \left|\|Y_t\|_2 + \|B\|_2\|Y_{t-1}\|_2 + \|D\|_2\|Z_{t-1}\|_2\right. \\
&\quad + \|F(Y_{t-1}, Z_{t-1}, \gamma)\|_2 + \left\|\tilde{Y}_{t,0}^m\right\|_2 + \|B\|_2\left\|\tilde{Y}_{t-1,1}^m\right\|_2 \\
&\quad \left.+ \|D\|_2\|Z_{t-1}\|_2 + \left\|F\left(\tilde{Y}_{t-1,1}^m, Z_{t-1}, \gamma\right)\right\|_2\right| \\
&\quad \times \left|\left\|Y_t - \tilde{Y}_{t,0}^m\right\|_2 + \|B\|_2\left\|Y_{t-1} - \tilde{Y}_{t-1,1}^m\right\|_2\right. \\
&\quad \left.+ \left\|F(Y_{t-1}, Z_{t-1}, \gamma) - F\left(\tilde{Y}_{t-1,1}^m, Z_{t-1}, \gamma\right)\right\|_2\right| \\
&\leq \left|\|Y_t\|_S + \|B\|_2\|Y_{t-1}\|_S + \|D\|_2\|Z_{t-1}\|_S\right. \\
&\quad + \|F(Y_{t-1}, Z_{t-1}, \gamma)\|_S + \left\|\tilde{Y}_{t,0}^m\right\|_S + \|B\|_2\left\|\tilde{Y}_{t-1,1}^m\right\|_S \\
&\quad \left.+ \|D\|_2\|Z_{t-1}\|_S + \left\|F\left(\tilde{Y}_{t-1,1}^m, Z_{t-1}, \gamma\right)\right\|_S\right| \\
&\quad \times \left\|Y_t - \tilde{Y}_{t,0}^m\right\|_S + \|B\|_2\left\|Y_{t-1} - \tilde{Y}_{t-1,1}^m\right\|_S \\
&\quad + \left\|F(Y_{t-1}, Z_{t-1}, \gamma) - F\left(\tilde{Y}_{t-1,1}^m, Z_{t-1}, \gamma\right)\right\|_S| \\
&\equiv B(X, \tilde{X}, \theta)\left|\left\|Y_t - \tilde{Y}_{t,0}^m\right\|_S + \|B\|_2\left\|Y_{t-1} - \tilde{Y}_{t-1,1}^m\right\|_S\right. \\
&\quad \left.+ \left\|F(Y_{t-1}, Z_{t-1}, \gamma) - F\left(\tilde{Y}_{t-1,1}^m, Z_{t-1}, \gamma\right)\right\|_S\right|.
\end{aligned}
$$

Note that in fact the second inequality should be $\|X\|_2 \leq \Delta\|X\|_S$, but the multiplicative constant Δ has been removed for ease of reading. Recall that $\|F(Y, Z, \gamma)\|_S = \|f(Y, Z, \gamma)\|_S$, because the left hand side is a vector with only its first component different from zero. Let us denote $\|\nabla_Y f(Y, Z, \gamma)\|_S \equiv \delta_{KY}$, as in Assumption CT. As we saw above,

$$\|F(Y, Z, \gamma)\|_S = \|F(0, Z, \gamma) + \nabla_Y F(\ddot{Y}, Z, \gamma) \cdot Y\|_S \leq \delta_H \|Z\|_S + \delta_{KY}\|Y\|_S,$$

$$\|F(\tilde{Y}, Z, \gamma)\|_S = \|F(0, Z, \gamma) + \nabla_Y F(\ddot{Y}, Z, \gamma) \cdot \tilde{Y}\|_S \leq \delta_H \|Z\|_S + \delta_{KY}\|\tilde{Y}\|_S,$$

$$\left\|F(Y_{t-1}, Z_{t-1}, \gamma) - F\left(\tilde{Y}^m_{t-1,1}, Z_{t-1}, \gamma\right)\right\|_S \leq \delta_{KY}\left\|Y_{t-1} - \tilde{Y}^m_{t-1,1}\right\|_S,$$

and applying this result to $B(X_t, \tilde{X}^m_t, \theta)$, we obtain

$$
\begin{aligned}
\left|b(X_t, \theta) - b\left(\tilde{X}^m_t, \theta\right)\right| &\leq B\left(X_t, \tilde{X}^m_t, \theta\right) \\
&\quad \times \left[\left\|Y_t - \tilde{Y}^m_{t,0}\right\|_S + \|B\|_2\left\|Y_{t-1} - \tilde{Y}^m_{t-1,1}\right\|_S \right. \\
&\quad \left. + \left\|F(Y_{t-1}, Z_{t-1}, \gamma) - F\left(\tilde{Y}^m_{t-1,1}, Z_{t-1}, \gamma\right)\right\|_S\right] \\
&\leq B\left(X_t, \tilde{X}^m_t, \theta\right) \\
&\quad \times \left[\left\|Y_t - \tilde{Y}^m_{t,0}\right\|_S + (\delta_{CB} + \delta_{KY})\left\|Y_{t-1} - \tilde{Y}^m_{t-1,1}\right\|_S\right] \\
&\leq \left[\|Y_t\|_S + \left\|\tilde{Y}^m_{t,0}\right\|_S + \delta_{CB}\left(\|Y_{t-1}\|_S + \left\|\tilde{Y}^m_{t-1,1}\right\|_S\right)\right. \\
&\quad + 2\delta_{CD}\|Z_{t-1}\|_S + 2\delta_H\|Z_{t-1}\|_S + \delta_{KY} \\
&\quad \left. \times \left(\|Y_{t-1}\|_S + \left\|\tilde{Y}^m_{t-1,1}\right\|_S\right)\right] \\
&\quad \times \left[\left\|Y_t - \tilde{Y}^m_{t,0}\right\|_S + (\delta_{CB} + \delta_{KY})\left\|Y_{t-1} - \tilde{Y}^m_{t-1,1}\right\|_S\right] \\
&\equiv \bar{B}\left(X_t, \tilde{X}^m_t\right) \times d\left(X_t, \tilde{X}^m_t\right).
\end{aligned}
$$

Taking $p = q = 2$, we have that $B(X_t, \tilde{X}^m_t, \theta)$ is q-dominated uniformly, $B(X_t, \tilde{X}^m_t, \theta)d(X_t, \tilde{X}^m_t)$ is s-dominated uniformly, and $\sup_t \|d(X_t, \tilde{X}^m_t)\|_{Sp}$ is nonzero.

In order to prove that $B(X_t, \tilde{X}^m_t, \theta)$ is q-dominated for $q = 2$, we begin by noting that $B(X_t, \tilde{X}^m_t) \leq \bar{B}(X_t, \tilde{X}^m_t)$. Now

$$
\begin{aligned}
\left\|\bar{B}\left(X_t, \tilde{X}^m_t\right)\right\|_{Sq} &\leq \|Y_t\|_{Sq} + \left\|\tilde{Y}^m_{t,0}\right\|_{Sq} + \delta_{CB}\left(\|Y_{t-1}\|_{Sq} + \left\|\tilde{Y}^m_{t-1,1}\right\|_{Sq}\right) \\
&\quad + 2\delta_{CD}\|Z_{t-1}\|_{Sq} + 2\delta_H\|Z_{t-1}\|_{Sq} \\
&\quad + \delta_{KY}\left(\|Y_{t-1}\|_{Sq} + \left\|\tilde{Y}^m_{t-1,1}\right\|_{Sq}\right).
\end{aligned}
$$

Also $\|\tilde{Y}^m_t\|_{Sr} \leq \Delta^r_{\tilde{Y}}$ by the same reasoning used to see that $\|Y_t\|_{Sr} \leq \Delta^r_Y$. Since $\|Y_t\|_{Sr} \leq \Delta^r_Y$ and $\|Z_t\|_{Sr} \leq (\Delta^{(r)}_Z)^{1/r} \equiv \Delta^r_Z$, and $\|X\|_{Sq} \leq \|X\|_{Sr}$ (since

$q = 2 < r = 4 + \varepsilon$), then

$$\left\| \bar{B}\left(X_t, \tilde{X}_t^m\right) \right\|_{Sq} \leq \|Y_t\|_{Sr} + \left\| \tilde{Y}_{t,0}^m \right\|_{Sr} + \delta_{CB}\left(\|Y_{t-1}\|_{Sr} + \left\| \tilde{Y}_{t-1,1}^m \right\|_{Sr} \right)$$
$$+ 2\delta_{CD}\|Z_{t-1}\|_{Sr} + 2\delta_H\|Z_{t-1}\|_{Sq}$$
$$+ \delta_{KY}\left(\|Y_{t-1}\|_{Sr} + \left\| \tilde{Y}_{t-1,1}^m \right\|_{Sr} \right)$$
$$\leq \Delta_Y^r + \Delta_{\tilde{Y}}^r + \delta_{CB}\left(\Delta_Y^r + \Delta_{\tilde{Y}}^r \right)$$
$$+ 2\delta_H + \delta_{KY}\left(\Delta_Y^r + \Delta_{\tilde{Y}}^r \right)$$
$$\equiv \Delta_B^r.$$

We have seen that $\|\bar{B}(X_t, \tilde{X}_t^m)\|_{Sr}$ is bounded by Δ_B^r. Now since $q = 2$, then $d(X_t, \tilde{X}_t^m)$ is q-dominated, because

$$\left\| d\left(X_t, \tilde{X}_t^m\right) \right\|_{Sq} \leq \left\| Y_t - \tilde{Y}_{t,0}^m \right\|_{Sq} + (\delta_{CB} + \delta_{KY})\left\| Y_{t-1} - \tilde{Y}_{t-1,1}^m \right\|_{Sq}$$
$$= \left\| Y_t - \tilde{Y}_{t,0}^m \right\|_{S2} + (\delta_{CB} + \delta_{KY})\left\| Y_{t-1} - \tilde{Y}_{t-1,1}^m \right\|_{S2}$$
$$\leq \Delta_{Y-\tilde{Y}}^2 + (\delta_{CB} + \delta_{KY})\Delta_{Y-\tilde{Y}}^2.$$

Note that from Lemma 3.4(i) $E\|Y_t - \tilde{Y}_{t,0}^m\|_S^2 \leq \Delta_{Y-\tilde{Y}}^{(2)} \to 0$ as $m \to \infty$; then as $m \to \infty$, since $p = 2$, we have

$$\sup_t \left\| d\left(X_t, \tilde{X}_t^m\right) \right\|_{Sp} \to 0$$

at exponential rate, so that it can be made arbitrarily small.

The only point that remains is to prove that $\|d(X_t, \tilde{X}_t^m)\|_{Sr}$ is bounded. This norm is as the norm used in Lemma 3.4(i), except that then r was 2, and now it is Sr. The same reasoning as in Lemma 3.4(i) may be now used, but we have to prove that $E\|Y_t - \bar{Y}_t\|_S^r$ is bounded. Since $r = 6$, then

$$\left(\sum_{i=1}^n a_i \right)^6 = \sum_{\{(b_1,\dots,b_n)|b_1+\cdots+b_n=6\}} K(b_1, \dots, b_n)a_1^{b_1}a_2^{b_2}\cdots a_n^{b_n},$$

where the constants K depend on $(b_1, \dots, b_n)$. The formula above can be proved by induction. Taking expectations,

$$E\|Y_t - \bar{Y}_t\|_S^6 \leq E\left(\sum_{i=0}^{t-1} \delta_{BY}^i(\|V_{t-i}\|_S + \delta_{DZ}\|Z_{t-1-i}\|_S) \right)^6$$

$$\equiv \sum_{\{(b_1,\dots,b_n)|b_1+\cdots+b_n=6\}} K(b_1, \dots, b_n)E\left(a_1^{b_1}a_2^{b_2}\cdots a_n^{b_n} \right)$$

for $n = t$, and $a_j = BY^{i-1}(\|V_{t-j-1}\|_S + DZ\|Z_{t-j}\|_S)$ for $j = 1, \ldots, n$; then each term is bounded by Assumption LR with $r = 6$, because $E\|Z\|_S^i\|V\|_S^j \le E\|Z\|_S^6\|V\|_S^6$, for $i, j \le 6$ if $\|Z\|_S > 1$, and $E\|Z\|_S^i\|V\|_S^j \le E\|V\|_S^6$, if $\|Z\|_S \le 1$. Then we obtain for $r = 6$ that $E\|Y_t - \bar{Y}_t\|_S^r \le \Delta_{Y_t - \bar{Y}_t}^r$.

Recalling the definition of $d(X, \tilde{X})$ we can say $\|d(X, \tilde{X})\|_{Sr} \le \Delta_d^r$ for $r = 5$. Therefore $\|XY\|_{Ss} = \|X\|_{S2s}\|Y\|_{S2s}$ and $\|X\|_{L2S} = \|X\|_{Sr}$; then for $s = 2 + \varepsilon$,

$$
\begin{aligned}
\left\| B(X_t, \tilde{X}_t^m, \theta) d(X_t, \tilde{X}_t^m) \right\|_{Ss} &\le \left\| \bar{B}(X_t, \tilde{X}_t^m) d(X_t, \tilde{X}_t^m) \right\|_{Ss} \\
&\le \left\| \bar{B}(X_t, \tilde{X}_t^m) \right\|_{Sr} \left\| d(X_t, \tilde{X}_t^m) \right\|_{Sr} \\
&\le \Delta_B^r \Delta_d^r.
\end{aligned}
$$

We have obtained that $q_t(\omega, \theta) \equiv b(X_t, \theta)$ is 2-dominated, $d(X_t, \tilde{X}_t^m)$ is p-dominated, $B(X_t, \tilde{X}_t^m, \theta)$ is q-dominated for $p = q = 2$, and $B(X_t, \tilde{X}_t^m, \theta) d(X_t, \tilde{X}_t^m)$ is s-dominated for $s = 2 + \varepsilon$; then Theorem 4.5 of Gallant and White can be applied, and the desired result follows.

B.5   *Proof of Lemma 3.4(v)*

Recall that $\|W - V\| \ge |\|W\| - \|V\||$. For the first inequality we can write

$$
\begin{aligned}
&|q_t(\omega, \theta^1) - q_t(\omega, \theta^2)| \\
&= |\|Y_t - B^1 Y_{t-1} - D^1 Z_{t-1} - F(Y_{t-1}, Z_{t-1}, \gamma^1)\|_2^2 \\
&\quad - \|Y_t - B^2 Y_{t-1} - D^2 Z_{t-1} - F(Y_{t-1}, Z_{t-1}, \gamma^2)\|_2^2| \\
&= |\|Y_t - B^1 Y_{t-1} - D^1 Z_{t-1} - F(Y_{t-1}, Z_{t-1}, \gamma^1)\|_2 \\
&\quad + \|Y_t - B^2 Y_{t-1} - D^2 Z_{t-1} - F(Y_{t-1}, Z_{t-1}, \gamma^2)\|_2| \\
&\quad \times |\|Y_t - B^1 Y_{t-1} - D^1 Z_{t-1} - F(Y_{t-1}, Z_{t-1}, \gamma^1)\|_2 \\
&\quad - \|Y_t - B^2 Y_{t-1} - D^2 Z_{t-1} - F(Y_{t-1}, Z_{t-1}, \gamma^2)\|_2| \\
&\le |\|Y_t\|_2 + \|B^1\|_2\|Y_{t-1}\|_2 + \|D^1\|_2\|Z_{t-1}\|_2 \\
&\quad + \|F(Y_{t-1}, Z_{t-1}, \gamma^1)\|_2 + \|Y_t\|_2 + \|B^2\|_2\|Y_{t-1}\|_2 \\
&\quad + \|D^2\|_2\|Z_{t-1}\|_2 + \|F(Y_{t-1}, Z_{t-1}, \gamma^2)\|_2| \\
&\quad \times \|B^1 - B^2\|_2\|Y_{t-1}\|_2 + \|D^1 - D^2\|_2\|Z_{t-1}\|_2 + \|F(Y_{t-1}, Z_{t-1}, \gamma^2) \\
&\quad - F(Y_{t-1}, Z_{t-1}, \gamma^1)\|_2 \\
&\le |2\|Y_t\|_2 + 2\delta_{CB}\|Y_{t-1}\|_2 + 2\delta_{CD}\|Z_{t-1}\|_2 + 2\delta_{CF}(\|Y_{t-1}\|_2 + \|Z_{t-1}\|_2)| \\
&\quad \times \|B^1 - B^2\|_2\|Y_{t-1}\|_2 + \|D^1 \\
&\quad - D^2\|_2\|Z_{t-1}\|_2 + \|F(Y_{t-1}, Z_{t-1}, \gamma^2) \\
&\quad - F(Y_{t-1}, Z_{t-1}, \gamma^1)\|_2.
\end{aligned}
$$

Recalling Assumption LN, we have

$$
\begin{aligned}
\| & F(Y, Z, \gamma^2) - F(Y, Z, \gamma^1)\|_2 \\
&\leq \|\nabla_\gamma f(Y, Z, \ddot{\gamma}) \cdot (\gamma^2 - \gamma^1)\|_2 \\
&\leq \|\nabla_\gamma f(Y, Z, \ddot{\gamma})\|_2 \|\gamma^2 - \gamma^1\|_2 \\
&\leq \sqrt{\delta_L}(\|Y\|_2 + \|Z\|_2)\|\gamma^2 - \gamma^1\|_2;
\end{aligned}
$$

then

$$
\begin{aligned}
|q_t&(\omega, \theta^1) - q_t(\omega, \theta^2)| \\
&\leq |2\|Y_t\|_2 + 2\delta_{CB}\|Y_{t-1}\|_2 + 2\delta_{CD}\|Z_{t-1}\|_2 + 2\delta_{CF}(\|Y_{t-1}\|_2 + \|Z_{t-1}\|_2)| \\
&\quad \times |\|B^1 - B^2\|_2 \|Y_{t-1}\|_2 + \|D^1 - D^2\|_2 \|Z_{t-1}\|_2 \\
&\quad + \sqrt{L}(\|Y_{t-1}\|_2 + \|Z_{t-1}\|_2)\|\gamma^2 - \gamma^1\|_2| \\
&\leq |2\|Y_t\|_2 + 2\delta_{CB}\|Y_{t-1}\|_2 + 2\delta_{CD}\|Z_{t-1}\|_2 + 2\delta_{CF}(\|Y_{t-1}\|_2 + \|Z_{t-1}\|_2)| \\
&\quad \times |\|Y_{t-1}\|_2 (\|B^2 - B^1\|_2 + \sqrt{L}\|\gamma^2 - \gamma^1\|_2) \\
&\quad \times Z_{t-1}\|_2 (\|D^2 - D^1\|_2 + \sqrt{L}\|\gamma^2 - \gamma^1\|_2)| \\
&\leq |2\|Y_t\|_2 + 2\delta_{CB}\|Y_{t-1}\|_2 + 2\delta_{CD}\|Z_{t-1}\|_2 + 2\delta_{CF}(\|Y_{t-1}\|_2 + \|Z_{t-1}\|_2)| \\
&\quad \times |\|Y_{t-1}\|_2 (\|B^2 - B^1\|_2 + \|D^2 - D^1\|_2 + \sqrt{L}\|\gamma^2 - \gamma^1\|_2) \\
&\quad \times \|Z_{t-1}\|_2 (\|B^2 - B^1\|_2 + \|D^2 - D^1\|_2 + \sqrt{L}\|\gamma^2 - \gamma^1\|_2)| \\
&= |2\|Y_t\|_2 + 2\delta_{CB}\|Y_{t-1}\|_2 + 2\delta_{CD}\|Z_{t-1}\|_2 + 2\delta_{CF}(\|Y_{t-1}\|_2 + \|Z_{t-1}\|_2)| \\
&\quad \times (\|Y_{t-1}\|_2 + \|Z_{t-1}\|_2)(\|B^2 - B^1\|_2 + \|D^2 - D^1\|_2 + \sqrt{L}\|\gamma^2 - \gamma^1\|_2) \\
&\leq |2\|Y_t\|_S + 2\delta_{CB}\|Y_{t-1}\|_S + 2\delta_{CD}\|Z_{t-1}\|_S + 2\delta_{CF}(\|Y_{t-1}\|_S + \|Z_{t-1}\|_S)| \\
&\quad \times (\|Y_{t-1}\|_S + \|Z_{t-1}\|_S)(\|B^2 - B^1\|_2 + \|D^2 - D^1\|_2 + \sqrt{L}\|\gamma^2 - \gamma^1\|_2) \\
&\equiv L_t^0 \times a^0(\rho(\theta^2, \theta^1))
\end{aligned}
$$

(note that again in the last inequality a multiplicative constant has been removed), where

$$
\begin{aligned}
L_t^0 = {}& |2\|Y_t\|_S + 2\delta_{CB}\|Y_{t-1}\|_S + 2\delta_{CD}\|Z_{t-1}\|_S \\
& + 2\delta_{CF}(\|Y_{t-1}\|_S + \|Z_{t-1}\|_S)| \\
& \times |\|Y_{t-1}\|_S + \|Z_{t-1}\|_S|.
\end{aligned}
$$

Define a^0 as the identity function and $\rho(\theta^2, \theta^1)$ as

$$
\rho(\theta^2, \theta^1) = (\|B^2 - B^1\|_2 + \|D^2 - D^1\|_2 + \sqrt{L}\|\gamma^2 - \gamma^1\|_2)
$$

where $\theta^1 = [\beta_1^1, \ldots, \beta_p^1, \delta_1^1, \ldots, \delta_q^1, \gamma_1^1, \ldots, \gamma_g^1]' \equiv [\theta_B^1, \theta_D^1, \theta_G^1]'$. With this definition ρ is a distance, because $\|B^1 - B^2\|_2$ is a norm for $\theta_B^1 - \theta_B^2$, although the dependence between θ_B^1 and B^1 implies that $\|B^1\|_2$ is different from $\|\theta_B^1\|_2$. Now by Assumption LR and the previous lemmas we have $\|Y_t\|_{L4} \leq \Delta_Y^4$,

$\|Z_t\|_{L4} \le \Delta_Z^4$, and as we saw in Lemma 3.4(iii), $\|XY\|_{Ss} \le \|X\|_{S2s}\|Y\|_{S2s}$. Then

$$
\begin{aligned}
E\left(L_t^0\right) = E\{2\|Y_t\|_S &+ 2\delta_{CB}\|Y_{t-1}\|_S + 2\delta_{CD}\|Z_{t-1}\|_S \\
&+ 2\delta_{CF}(\|Y_{t-1}\|_S + \|Z_{t-1}\|_S)\}\{\|Y_{t-1}\|_S + \|Z_{t-1}\|_S\},
\end{aligned}
$$

which has every term bounded, so that

$$
\limsup_n n^{-1} \sum_{t=1}^n E\left(L_t^0\right) < \infty,
$$

and then $q_t(\omega, \theta)$ is Lipschitz-L_1 a.s. in Θ.

Appendix C: Proof of Lemma 4.1

C.1   *Proof of Lemma 4.1(i)*

We can write

$$
\begin{aligned}
|\nabla_\theta q_t(\omega, \theta)| &= |2DV \times (y_t - \beta_1^* y_{t-1} - \cdots - \beta_p^* y_{t-p} - \delta_1^* z_{t-1} - \cdots \\
&\quad - \delta_q^* z_{t-q} - f(y_{t-1}, \ldots, y_{t-p}, z_{t-1}, \ldots, z_{t-q}, \gamma^*))| \\
&\le |2DV| \times \|Y_t - BY_{t-1} - DZ_{t-1} - F(Y_{t-1}, Z_{t-1}, \gamma)\|_2 \\
&\le |2DV'| \times \|Y_t - BY_{t-1} - DZ_{t-1} - F(Y_{t-1}, Z_{t-1}, \gamma)\|_2 \\
&\le \delta_{DF}|2DV''| \times \|Y_t - BY_{t-1} - DZ_{t-1} - F(Y_{t-1}, Z_{t-1}, \gamma)\|_S
\end{aligned}
$$

for some constant δ_{DF}, where

$$
\begin{aligned}
DV' = [y_{t-1}, &\ldots, y_{t-p}, z_{t-1}, \ldots, z_{t-q}, \\
&\sqrt{\delta_L}(\|Y_{t-1}\|_2 + \|Z_{t-1}\|_2), \ldots, \sqrt{\delta_L}(\|Y_{t-1}\|_2 + \|Z_{t-1}\|_2)]', \\
DV'' = [y_{t-1}, &\ldots, y_{t-p}, z_{t-1}, \ldots, z_{t-q}, \\
&\sqrt{\delta_L}(\|Y_{t-1}\|_S + \|Z_{t-1}\|_S), \ldots, \sqrt{\delta_L}(\|Y_{t-1}\|_S + \|Z_{t-1}\|_S)]'.
\end{aligned}
$$

Let us study each element of the vector $|2DV''| \times \|Y_t - BY_{t-1} - DZ_{t-1} - F(Y_{t-1}, Z_{t-1}, \gamma)\|_S$. For the first type of elements we have

$$
\begin{aligned}
E^{1/g}|y_{t-1}\|Y_t &- BY_{t-1} - DZ_{t-1} - F(Y_{t-1}, Z_{t-1}, \gamma)\|_S|^g \\
&= E^{1/g}|y_{t-1}|^g\|Y_t - BY_{t-1} - DZ_{t-1} - F(Y_{t-1}, Z_{t-1}, \gamma)\|_S^g \\
&\le E^{1/2g}|y_{t-1}|^{2g} E^{1/2g}\|Y_t - BY_{t-1} - DZ_{t-1} - F(Y_{t-1}, Z_{t-1}, \gamma)\|_S^{2g} \\
&\le \Delta_Y^{2g}\Delta_{Y-Y}^{2g}.
\end{aligned}
$$

Since $g > 2$, recalling Lemma 3.4(iii), we have that $\|Y_t - BY_{t-1} - DZ_{t-1} - F(Y_{t-1}, Z_{t-1}, \gamma)\|_{S2g}$ is bounded by Δ_D^{2g} if $\|Z_t\|_{S4g} \le \Delta_Z^{4g}$ and $\|Y_t\|_{S4g} \le \Delta_Y^{4g}$. Then we take Assumption LF with $f = 8 + \varepsilon$. Analogously, elements of the type $|z_{t-i}\|Y_t - BY_{t-1} - DZ_{t-1} - F(Y_{t-1}, Z_{t-1}, \gamma)\|_S|$ are bounded by $\Delta_Z^{2g}\Delta_{Z-Z}^{2g}$.

For the other type of elements we have

$$E|\sqrt{L}(\|Y_{t-1}\|_S + \|Z_{t-1}\|_S)\|Y_t - BY_{t-1} - DZ_{t-1} - F(Y_{t-1}, Z_{t-1}, \gamma)\|_S|^g,$$

and if g is an integer we obtain summations as in Lemma 3.4(iv) which are bounded.

C.2   *Proof of Lemma 4.1(ii)*

Let us analyze each element of $\nabla_\theta^2 q_t(\omega, \theta)$. For types A, B, and D we recall the definition of the Lr norm and we have $E^{1/r}|y_{t-i}z_{t-j}|^r = \|y_{t-i}z_{t-j}\|_{Sr} \le \|y_{t-i}\|_{S2r}\|z_{t-j}\|_{S2r}$, and since r is $2 + \varepsilon$, we take $\delta_K \Delta_Y^5 \Delta_Z^5$ for some constant δ_K, because $\|y_t\|_{S2r} = E^{1/2r}\|y_t\|_S^{2r} \le E^{1/2r}\|Y_t\|_S^{2r} \times \delta_{(B\nabla F)'}$, and $\|Y_t\|_S \le \delta_{B\nabla F}\|Y_t\|_\infty$.

For type C recall Assumption LN(ii) and $\|\nabla_\gamma f(Y, Z, \gamma)\|_S^2 \le L(\|Y\|_S + \|Z\|_S)^2$. Then for some constant $\delta_{(B\nabla F)'}$

$$\left\| y_{t-i}\frac{\partial}{\partial \gamma_j} f(Y_{t-1}, Z_{t-1}, \gamma) \right\|_{Sr} \le \|y_{t-i}\|_{S2r}\left\| \frac{\partial}{\partial \gamma_j} f(Y_{t-1}, Z_{t-1}, \gamma) \right\|_{S2r}$$

$$\le \delta_{(B\nabla F)'}\|y_{t-i}\|_{S2r}\|\nabla_\gamma f(Y_{t-1}, Z_{t-1}, \gamma)\|_{S2r}$$

$$\le \delta_{(B\nabla F)'}\|y_{t-i}\|_{S2r}\sqrt{\delta_L}(\|Y_{t-1}\|_{S2r} + \|Z_{t-1}\|_{S2r}).$$

For type F recall Assumption LN'(i) and $\|\nabla_\gamma(\partial/\partial\gamma_j)f(Y, Z, \gamma)\|_S^2 \le L'(\|Y\|_S + \|Z\|_S)^2$; then

$$\left\| \frac{\partial^2}{\partial\gamma_i\partial\gamma_j}f(Y_{t-1}, Z_{t-1}, \gamma)\|Y_t - BY_{t-1} - DZ_{t-1} - F(Y_{t-1}, Z_{t-1}, \gamma)\|_S \right.$$

$$\left. + \frac{\partial}{\partial\gamma_i}f(Y_{t-1}, Z_{t-1}, \gamma)\frac{\partial}{\partial\gamma_i}f(Y_{t-1}, Z_{t-1}, \gamma) \right\|_{Sr}$$

$$\le \left\| \left\{ \frac{\partial^2}{\partial\gamma_i\partial\gamma_j}f(Y_{t-1}, Z_{t-1}, \gamma)\|Y_t - BY_{t-1} - DZ_{t-1} \right. \right.$$

$$\left. \left. - F(Y_{t-1}, Z_{t-1}, \gamma)\|_S \right\} \right\|_{S2r} \times \left\| \left\{ \frac{\partial}{\partial\gamma_i}f(Y_{t-1}, Z_{t-1}, \gamma) \right. \right.$$

$$\left. \left. \times \frac{\partial}{\partial\gamma_i}f(Y_{t-1}, Z_{t-1}, \gamma) \right\} \right\|_{S2r}$$

$$\le \left\| \frac{\partial^2}{\partial\gamma_i\partial\gamma_j}f(Y_{t-1}, Z_{t-1}, \gamma) \right\|_{S4r}\|Y_t - BY_{t-1} - DZ_{t-1}$$

$$- F(Y_{t-1}, Z_{t-1}, \gamma)\|_{S4r}\left\| \frac{\partial}{\partial\gamma_i}f(Y_{t-1}, Z_{t-1}, \gamma) \right\|_{S4r}$$

$$\times \left\| \frac{\partial}{\partial\gamma_i}f(Y_{t-1}, Z_{t-1}, \gamma) \right\|_{S4r}$$

$$\leq \sqrt{\delta_{L'}}(\|Y_{t-1}\|_{S4r} + \|Z_{t-1}\|_{S4r})\|Y_t - BY_{t-1} - DZ_{t-1}$$

$$- F(Y_{t-1}, Z_{t-1}, \gamma)\|_{S4r} \times \sqrt{\delta_L}(\|Y_{t-1}\|_S + \|Z_{t-1}\|_S)$$

$$\times \sqrt{\delta_L}(\|Y_{t-1}\|_S + \|Z_{t-1}\|_S),$$

and since $r = 2 + \varepsilon$, the terms are bounded by Assumption LF.

C.3   Proof of Lemma 4.1(iii)

Let us analyze each element of $\nabla_\theta q_t(\omega, \theta)$. Recall Lemma 3.4(i) and that $\{Y_t\}$ is NED on $\{(Z_t, V_t)\}$; therefore $\{y_t\}$ is NED on $\{(Z_t, V_t)\}$. We have from Lemma 3.4(iv) that $\{q_t(\omega, \theta)\}$ is NED on $\{(Z_t, V_t)\}$. Let us see that $q_t(\omega, \theta)^{1/2} \equiv \|Y_t - BY_{t-1} - DZ_{t-1} - F(Y_{t-1}, Z_{t-1}, \gamma)\|_2$ is NED. Let us take $b(X_t, \theta) = q_t(\omega, \theta)^{1/2}$, where X_t is defined in Lemma 3.4(iv), and analogously for $b(\tilde{X}_t, \theta)$. Now following the steps of Lemma 3.4(iv) we obtain

$$|b(X_t, \theta) - b(\tilde{X}_t, \theta)| = |\|Y_t - BY_{t-1} - DZ_{t-1} - F(Y_{t-1}, Z_{t-1}, \gamma)\|_2$$

$$- \|\tilde{Y}_t - B\tilde{Y}_{t-1} - DZ_{t-1} - F(\tilde{Y}_{t-1}, Z_{t-1}, \gamma)\|_2|$$

$$\leq |\|Y_t - \tilde{Y}_t\|_2 + \|B\|_2\|Y_t - \tilde{Y}_t\|_2$$

$$+ \|F(Y_{t-1}, Z_{t-1}, \gamma) - F(\tilde{Y}_{t-1}, Z_{t-1}, \gamma)\|_2|$$

$$\leq 1 \times (\|Y_t - \tilde{Y}_t\|_2 + (\delta_{CB} + \delta_{KY})\|Y_{t-1} - \tilde{Y}_{t-1}\|_2$$

$$\leq 1 \times (\|Y_t - \tilde{Y}_t\|_S + (\delta_{CB} + \delta_{KY})\|Y_{t-1} - \tilde{Y}_{t-1}\|_S$$

$$\equiv B(X_t, \tilde{X}_t, \theta) \times d(X_t, \tilde{X}_t)$$

(note that again the last inequality should be $\|X\| \leq \delta_{K(B\nabla F)}\|X\|_S$, but we drop the multiplicative constant); now, as in Lemma 3.4(iv), we obtain that $q_t(\omega, \theta)^{1/2}$ is NED. From Corollary 4.3 of Gallant and White (1988) we know that sums and products of NED sequences are NED; then we obtain that $\{y_{t-i}\|Y_t - BY_{t-1} - DZ_{t-1} - F(Y_{t-1}, Z_{t-1}, \gamma)\|_S\}$ is NED.

For the case $\{z_{t-i}\|Y_t - BY_{t-1} - DZ_{t-1} - F(Y_{t-1}, Z_{t-1}, \gamma)\|_S\}$ the proof is simpler because y_{t-i} is replaced by z_{t-i}.

For the case $\{(\partial/\partial\gamma_g)f(Y_{t-1}, Z_{t-1}, \gamma)\|Y_t - BY_{t-1} - DZ_{t-1} - F(Y_{t-1}, Z_{t-1}, \gamma)\|_S\}$ we see that $\{(\partial/\partial\gamma_g)f(Y_{t-1}, Z_{t-1}, \gamma)\}$ is NED on $\{(Z_t, V_t)\}$. With the same definitions as in Lemma 3.4(i) and for I_{t-m}^t the σ-algebra generated by $(v_t, v_{t-1}, z_{t-1}, \ldots, v_{t-m}, z_{t-m})$, and using Assumption CT'(i), we have from Young's theorem (see Sydsaeter and Hammond 1995) the equality of the partial derivatives under regularity conditions and then

$$
\left| \frac{\partial}{\partial \gamma_i} f(Y_{t-1}, Z_{t-1}, \gamma) - E\left[\frac{\partial}{\partial \gamma_i} f(Y_{t-1}, Z_{t-1}, \gamma) | I_{t-m}^t \right] \right|
$$

$$
\leq \left| \frac{\partial}{\partial \gamma_i} f(Y_{t-1}, Z_{t-1}, \gamma) - \frac{\partial}{\partial \gamma_i} f(\tilde{Y}_{t-1}, Z_{t-1}, \gamma) \right|
$$

$$
\leq \left| \frac{\partial}{\partial y_{t-1}} \frac{\partial}{\partial \gamma_i} f(\ddot{Y}_{t-1}, Z_{t-1}, \gamma)(y_{t-1} - \tilde{y}_{t-1}) + \cdots \right.
$$

$$
\left. + \frac{\partial}{\partial y_{t-p}} \frac{\partial}{\partial \gamma_i} f(\ddot{Y}_{t-1}, Z_{t-1}, \gamma)(y_{t-p} - \tilde{y}_{t-p}) \right|
$$

$$
= \left| \frac{\partial}{\partial \gamma_i} \frac{\partial}{\partial y_{t-1}} f(\ddot{Y}_{t-1}, Z_{t-1}, \gamma)(y_{t-1} - \tilde{y}_{t-1}) + \cdots \right.
$$

$$
\left. + \frac{\partial}{\partial \gamma_i} \frac{\partial}{\partial y_{t-p}} f(\ddot{Y}_{t-1}, Z_{t-1}, \gamma)(y_{t-p} - \tilde{y}_{t-p}) \right|
$$

$$
\leq \left\| \frac{\partial}{\partial \gamma_i} \nabla_Y f(\ddot{Y}_{t-1}, Z_{t-1}, \gamma) \right\|_2 \| Y_{t-1} - \tilde{Y}_{t-1} \|_2
$$

$$
\leq \delta_{KG} \| Y_{t-1} - \tilde{Y}_{t-1} \|_2
$$

$$
\leq \delta_{KG} \delta_{B\nabla F} \| Y_{t-1} - \tilde{Y}_{t-1} \|_S,
$$

and since $E\|Y_{t-1} - \tilde{Y}_{t-1}\|_S \leq \Delta_{Y-\tilde{Y}}^{(1)}$ as we saw in Theorem 3.4(i), then $(\partial/\partial\gamma_i) f(Y_{t-1}, Z_{t-1}, \gamma)$ is NED on $\{(Z_t, V_t)\}$. Since we know that products of NED sequences are NED, then $\{(\partial/\partial\gamma_i) f(Y_{t-1}, Z_{t-1}, \gamma) \|Y_t - BY_{t-1} - DZ_{t-1} - F(Y_{t-1}, Z_{t-1}, \gamma)\|_S\}$ is NED on $\{(Z_t, V_t)\}$.

C.4   *Proof of Lemma 4.1(iv)*

Let us analyze each element of $\nabla_\theta^2 q_t(\omega, \theta)$. For the cases A, B, and D we have that $y_{t-i} y_{t-j}$, $y_{t-i} z_{t-j}$, and $z_{t-i} z_{t-j}$ are NED. For the cases C and E we use the fact that $(\partial/\partial\gamma_j) f(Y_{t-1}, Z_{t-1}, \gamma)$ is NED, as we saw in Lemma 4.1(iii), and the result follows. For the case F we use Assumption CT′(ii) and we repeat the same steps as in Lemma 4.1(iii), but now the regularity conditions are applied to the derivative $(\partial/\partial\gamma_k) f(\cdot)$ in order to prove that $(\partial^2/\partial\gamma_i\partial\gamma_k) f(Y_{t-1}, Z_{t-1}, \gamma)$ is NED on $\{(Z_t, V_t)\}$, and the result follows.

C.5   *Proof of Lemma 4.1(v)*

Let us analyze each element of $\nabla_\theta q_t(\omega, \theta)$.

In the case of $y_{t-i} \|Y_t - BY_{t-1} - DZ_{t-1} - F(Y_{t-1}, Z_{t-1}, \gamma)\|_2$ we recall the steps of Lemma 3.1(v) and we write

$$|y_{t-i}\|Y_t - B^1 Y_{t-1} - D^1 Z_{t-1} - F(Y_{t-1}, Z_{t-1}, \gamma^1)\|_2$$
$$- y_{t-i}\|Y_t - B^2 Y_{t-1} - D^2 Z_{t-1} - F(Y_{t-1}, Z_{t-1}, \gamma^2)\|_2|$$
$$= |y_{t-1}| \times |\,\|Y_t - B^1 Y_{t-1} - D^1 Z_{t-1} - F(Y_{t-1}, Z_{t-1}, \gamma^1)\|_2$$
$$- \|Y_t - B^2 Y_{t-1} - D^2 Z_{t-1} - F(Y_{t-1}, Z_{t-1}, \gamma^2)\|_2|$$
$$= |y_{t-1}| \times |\,\|B^1 - B^2\|_2\|Y_{t-1}\|_2 + \|D^1 - D^2\|_2\|Z_{t-1}\|_2$$
$$\times \|F(Y_{t-1}, Z_{t-1}, \gamma^2) - F(Y_{t-1}, Z_{t-1}, \gamma^2)\|_2|$$
$$\le |y_{t-1}| \times (\|Y_{t-1}\|_2 + \|Z_{t-1}\|_2)$$
$$\times (\|B^2 - B^1\|_2 + \|D^2 - D^1\|_2 + \sqrt{\delta_L}\|\gamma^2 - \gamma^1\|_2)$$
$$\le |y_{t-1}| \times (\|Y_{t-1}\|_S + \|Z_{t-1}\|_S)$$
$$\times (\|B^2 - B^1\| + \|D^2 - D^1\| + \sqrt{\delta_L}\|\gamma^2 - \gamma^1\|)$$
$$\equiv |y_{t-1}| L_t^0 a^0(\rho(\theta^2, \theta^1)),$$

where $a^0(\cdot)$ is the identity function and $L_t^0 \equiv \|Y_{t-1}\|_S + \|Z_{t-1}\|_S$. Recalling Lemma 3.1(v), we conclude that

$$\limsup_n n^{-1} \sum_{t=1}^\infty E\left(|y_{t-i}| L_t^0\right) < \infty.$$

Analogous reasoning can be applied to the case $z_{t-i}\|Y_t - BY_{t-1} - DZ_{t-1} - F(Y_{t-1}, Z_{t-1}, \gamma)\|_2$. In the case $(\partial/\partial\gamma_i) f(Y_{t-1}, Z_{t-1}, \gamma)\|Y_t - BY_{t-1} - DZ_{t-1} - F(Y_{t-1}, Z_{t-1}, \gamma)\|_2$ we recall that $|AB - CD| = |AB - AD + AD - CD| \le |A||B - D| + |D||A - C|$, and we take

$$A = \frac{\partial}{\partial\gamma_i} f(Y_{t-1}, Z_{t-1}, \gamma^1),$$
$$C = \frac{\partial}{\partial\gamma_i} f(Y_{t-1}, Z_{t-1}, \gamma^2),$$
$$B = \|Y_t - B^1 Y_{t-1} - D^1 Z_{t-1} - F(Y_{t-1}, Z_{t-1}, \gamma^1)\|_2,$$
$$D = \|Y_t - B^2 Y_{t-1} - D^2 Z_{t-1} - F(Y_{t-1}, Z_{t-1}, \gamma^2)\|_2.$$

Then we have

$$\left|\frac{\partial}{\partial\gamma_i} f(Y_{t-1}, Z_{t-1}, \gamma^1)\|Y_t - B^1 Y_{t-1} - D^1 Z_{t-1} - F(Y_{t-1}, Z_{t-1}, \gamma^1)\|_2\right.$$
$$\left. - \frac{\partial}{\partial\gamma_i} f(Y_{t-1}, Z_{t-1}, \gamma^2)\|Y_t - B^2 Y_{t-1} - D^2 Z_{t-1} - F(Y_{t-1}, Z_{t-1}, \gamma^2)\|_2\right|$$

$$= \left| \frac{\partial}{\partial \gamma_i} f(Y_{t-1}, Z_{t-1}, \gamma^1) \right|$$

$$\times | \|Y_t - B^1 Y_{t-1} - D^1 Z_{t-1} - F(Y_{t-1}, Z_{t-1}, \gamma^1)\|_2$$

$$- \|Y_t - B^2 Y_{t-1} - D^2 Z_{t-1} - F(Y_{t-1}, Z_{t-1}, \gamma^2)\|_2 |$$

$$+ \|Y_t - B^2 Y_{t-1} - D^2 Z_{t-1} - F(Y_{t-1}, Z_{t-1}, \gamma^2)\|_2$$

$$\times \left| \frac{\partial}{\partial \gamma_i} f(Y_{t-1}, Z_{t-1}, \gamma^1) - \frac{\partial}{\partial \gamma_i} f(Y_{t-1}, Z_{t-1}, \gamma^2) \right|.$$

Now from Assumption LN we have $|(\partial/\partial \gamma_i) f(Y_{t-1}, Z_{t-1}, \gamma^1)| \leq \sqrt{\delta_L}(\|Y_{t-1}\|_2 + \|Z_{t-1}\|_2)$, and as in the first case we have

$$\|Y_t - B^1 Y_{t-1} - D^1 Z_{t-1} - F(Y_{t-1}, Z_{t-1}, \gamma^1)\|_2$$

$$- \|Y_t - B^2 Y_{t-1} - D^2 Z_{t-1} - F(Y_{t-1}, Z_{t-1}, \gamma^2)\|_2 |$$

$$\leq (\|Y_{t-1}\|_2 + \|Z_{t-1}\|_2) \times (\|B^2 - B^1\|_2 + \|D^2 - D^1\|_2$$

$$+ \sqrt{\delta_L}\|\gamma^2 - \gamma^1\|_2).$$

Recall that for the inner product $X \cdot Y = \sum_{i=1}^{n} x_i y_i$ that defines the norm $\|X\|_2 = (\sum_{i=1}^{n} x_i^2)^{1/2}$ we have $X \cdot Y \leq \|X\|_2 \|Y\|_2$; now, using Assumption LN', the following inequality results:

$$\left| \frac{\partial}{\partial \gamma_i} f(Y_{t-1}, Z_{t-1}, \gamma^1) - \frac{\partial}{\partial \gamma_i} f(Y_{t-1}, Z_{t-1}, \gamma^2) \right|$$

$$= \left| \nabla_\gamma \frac{\partial}{\partial \gamma_i} f(Y_{t-1}, Z_{t-1}, \ddot{\gamma}) \cdot (\gamma^2 - \gamma^1) \right|$$

$$\leq \left\| \nabla_\gamma \frac{\partial}{\partial \gamma_i} f(Y_{t-1}, Z_{t-1}, \ddot{\gamma}) \right\|_2 \|\gamma^2 - \gamma^1\|_2$$

$$\leq \sqrt{\delta_L}(\|Y_{t-1}\|_2 + \|Z_{t-1}\|_2)\|\gamma^2 - \gamma^1\|_2.$$

Then (assuming that $\sqrt{\delta_L} > \frac{1}{2}$ for the second inequality), we obtain

$$\left| \frac{\partial}{\partial \gamma_i} f(Y_{t-1}, Z_{t-1}, \gamma^1) \|Y_t - B^1 Y_{t-1} - D^1 Z_{t-1} - F(Y_{t-1}, Z_{t-1}, \gamma^1)\|_2 \right.$$

$$\left. - \frac{\partial}{\partial \gamma_i} f(Y_{t-1}, Z_{t-1}, \gamma^2) \|Y_t - B^2 Y_{t-1} - D^2 Z_{t-1} - F(Y_{t-1}, Z_{t-1}, \gamma^2)\|_2 \right|$$

$$\leq \sqrt{\delta_L}(\|Y_{t-1}\|_2 + \|Z_{t-1}\|_2)(\|Y_{t-1}\|_2 + \|Z_{t-1}\|_2)$$

$$\times (\|B^2 - B^1\|_2 + \|D^2 - D^1\|_2 + 2\sqrt{\delta_L}\|\gamma^2 - \gamma^1\|_2)$$

$$+ \|Y_t - B^2 Y_{t-1} - D^2 Z_{t-1} - F(Y_{t-1}, Z_{t-1}, \gamma^2)\|_2$$
$$\times \sqrt{\delta_L}(\|Y_{t-1}\|_2 + \|Z_{t-1}\|_2)\|\gamma^2 - \gamma^1\|_2$$
$$\leq \sqrt{\delta_L}(\|Y_{t-1}\|_2 + \|Z_{t-1}\|_2)(\|Y_{t-1}\|_2 + \|Z_{t-1}\|_2)$$
$$\times (\|B^2 - B^1\|_2 + \|D^2 - D^1\|_2 + 2\sqrt{\delta_L}\|\gamma^2 - \gamma^1\|_2)$$
$$+ \|Y_t - B^2 Y_{t-1} - D^2 Z_{t-1} - F(Y_{t-1}, Z_{t-1}, \gamma^2)\|_2$$
$$\times \sqrt{\delta_L}(\|Y_{t-1}\|_2 + \|Z_{t-1}\|_2)(\|B^2 - B^1\|_2 + \|D^2 - D^1\|_2$$
$$+ 2\sqrt{\delta_L}\|\gamma^2 - \gamma^1\|_2)$$
$$\leq \sqrt{\delta_L}(\|Y_{t-1}\|_2 + \|Z_{t-1}\|_2)[(\|Y_{t-1}\|_2 + \|Z_{t-1}\|_2)$$
$$+ \|Y_t\|_2 + \delta_{CB}\|Y_{t-1}\|_2 + \delta_{CD}\|Z_{t-1}\|_2 + \delta_{CF}(\|Y_{t-1}\|_2 + \|Z_{t-1}\|_2)]$$
$$\times (\|B^2 - B^1\|_2 + \|D^2 - D^1\|_2 + 2\sqrt{\delta_L}\|\gamma^2 - \gamma^1\|_2)$$
$$\leq \sqrt{\delta_L}(\|Y_{t-1}\|_S + \|Z_{t-1}\|_S)$$
$$\times [(\|Y_{t-1}\|_S + \|Z_{t-1}\|_S) + \|Y_t\|_S + \delta_{CB}\|Y_{t-1}\|_S + \delta_{CD}\|Z_{t-1}\|_S$$
$$+ \delta_{CF}(\|Y_{t-1}\|_S + \|Z_{t-1}\|_S)](\|B^2 - B^1\|_2$$
$$+ \|D^2 - D^1\|_2 + 2\sqrt{\delta_L}\|\gamma^2 - \gamma^1\|_2)$$
$$\equiv L_t^{01} a^0(\rho(\theta^2, \theta^1)),$$

and with analogous reasoning to that in the first case, $\limsup_n n^{-1}\sum_{t=1}^n E(L_t^{01}) < \infty$, and $\rho(\cdot)$ is a distance, which concludes the proof.

C.6   *Proof of Lemma 4.1(vi)*

Let us analyze each element of $\nabla_\theta^2 q_t(\omega, \theta)$. For case A there are no parameters and there is nothing to prove. Analogously for cases B and D.

In case C we have Assumption LN' and then

$$\left|2y_{t-i}\frac{\partial}{\partial\gamma_j}f(Y_{t-1}, Z_{t-1}, \gamma^2) - 2y_{t-i}\frac{\partial}{\partial\gamma_j}f(Y_{t-1}, Z_{t-1}, \gamma^1)\right|$$
$$= |2y_{t-i}|\left|\frac{\partial}{\partial\gamma_j}f(Y_{t-1}, Z_{t-1}, \gamma^2) - \frac{\partial}{\partial\gamma_j}f(Y_{t-1}, Z_{t-1}, \gamma^1)\right|$$
$$= |2y_{t-i}|\left|\nabla_\gamma\left(\frac{\partial}{\partial\gamma_j}f(Y_{t-1}, Z_{t-1}, \ddot{\gamma})\right) \cdot (\gamma^2 - \gamma^1)\right|$$
$$\leq |2y_{t-i}|\left\|\nabla_\gamma\left(\frac{\partial}{\partial\gamma_j}f(Y_{t-1}, Z_{t-1}, \ddot{\gamma})\right)\right\|_2 \|\gamma^2 - \gamma^1\|_2$$

$$\le |2y_{t-i}| \left\| \nabla_\gamma \left(\frac{\partial}{\partial \gamma_j} f(Y_{t-1}, Z_{t-1}, \ddot{\gamma}) \right) \right\|_2$$

$$\times (\|B^2 - B^1\|_2 + \|D^2 - D^1\|_2 + \|\gamma^2 - \gamma^1\|_2)$$

$$\le |2y_{t-i}| \sqrt{\delta_{L'}} (\|Y_{t-1}\|_2 + \|Z_{t-1}\|_2)$$

$$\times (\|B^2 - B^1\|_2 + \|D^2 - D^1\|_2 + \|\gamma^2 - \gamma^1\|_2)$$

$$\le |2y_{t-i}| \sqrt{\delta_{L'}} (\|Y_{t-1}\|_S + \|Z_{t-1}\|_S)$$

$$\times (\|B^2 - B^1\|_2 + \|D^2 - D^1\|_2 + \|\gamma^2 - \gamma^1\|_2)$$

$$\equiv L_t^{02} a^0 (\rho(\theta^2, \theta^1)),$$

and again $\limsup_n n^{-1} \sum_{t=1}^n E(L_t^{02}) < \infty$ as in Lemma 4.1(v). Case E is analogous to case C.

For case F we have

$$2 \left| \left\{ -\frac{\partial^2}{\partial \gamma_j \partial \gamma_i} f(Y_{t-1}, Z_{t-1}, \gamma^2) \|Y_t - B^2 Y_{t-1} - D^2 Z_{t-1} - F(Y_{t-1}, Z_{t-1}, \gamma^2)\|_2 \right. \right.$$

$$+ \left[\frac{\partial}{\partial \gamma_i} f(Y_{t-1}, Z_{t-1}, \gamma^2) \right] \left[\frac{\partial}{\partial \gamma_j} f(Y_{t-1}, Z_{t-1}, \gamma^2) \right]$$

$$+ \frac{\partial^2}{\partial \gamma_j \partial \gamma_i} f(Y_{t-1}, Z_{t-1}, \gamma^1) \|Y_t - B^1 Y_{t-1} - D^1 Z_{t-1}$$

$$- F(Y_{t-1}, Z_{t-1}, \gamma^1)\|_2 - \left[\frac{\partial}{\partial \gamma_i} f(Y_{t-1}, Z_{t-1}, \gamma^1) \right]$$

$$\left. \times \left[\frac{\partial}{\partial \gamma_j} f(Y_{t-1}, Z_{t-1}, \gamma^1) \right] \right\} \right|$$

$$= 2 \left| \left\{ \left[\frac{\partial}{\partial \gamma_i} f(Y_{t-1}, Z_{t-1}, \gamma^2) \frac{\partial}{\partial \gamma_j} f(Y_{t-1}, Z_{t-1}, \gamma^2) \right] \right. \right.$$

$$- \left[\frac{\partial}{\partial \gamma_i} f(Y_{t-1}, Z_{t-1}, \gamma^1) \frac{\partial}{\partial \gamma_j} f(Y_{t-1}, Z_{t-1}, \gamma^1) \right]$$

$$- \left[\frac{\partial^2}{\partial \gamma_j \partial \gamma_i} f(Y_{t-1}, Z_{t-1}, \gamma^1) \|Y_t - B^1 Y_{t-1} - D^1 Z_{t-1} \right.$$

$$\left. - F(Y_{t-1}, Z_{t-1}, \gamma^1)\|_2 \right] + \left[\frac{\partial^2}{\partial \gamma_j \partial \gamma_i} f(Y_{t-1}, Z_{t-1}, \gamma^1) \right.$$

$$\left. \left. \left. \times \|Y_t - B^1 Y_{t-1} - D^1 Z_{t-1} - F(Y_{t-1}, Z_{t-1}, \gamma^1)\|_2 \right] \right\} \right|$$

$$\equiv 2|AB - CD - EF - GH|$$

$$= 2|[AB - AD + AD - CD] - [EF - EH + EH - GH]|$$

$$= 2\{|A||B - D| + |D||A - C| + |E||F - H| + |H||E - G|\}$$

$$= 2\left\{\left[\left|\left|\frac{\partial}{\partial \gamma_i} f(Y_{t-1}, Z_{t-1}, \gamma^2)\right|\right| \frac{\partial}{\partial \gamma_j} f(Y_{t-1}, Z_{t-1}, \gamma^2)\right.\right.$$

$$\left.- \frac{\partial}{\partial \gamma_j} f(Y_{t-1}, Z_{t-1}, \gamma^1)\right| + \left|\frac{\partial}{\partial \gamma_j} f(Y_{t-1}, Z_{t-1}, \gamma^1)\right|$$

$$\times \left|\frac{\partial}{\partial \gamma_i} f(Y_{t-1}, Z_{t-1}, \gamma^2) - \frac{\partial}{\partial \gamma_i} f(Y_{t-1}, Z_{t-1}, \gamma^1)\right|$$

$$+ \left|\frac{\partial^2}{\partial \gamma_j \partial \gamma_i} f(Y_{t-1}, Z_{t-1}, \gamma^2)\right|$$

$$\times \|Y_t - B^2 Y_{t-1} - D^2 Z_{t-1} - F(Y_{t-1}, Z_{t-1}, \gamma^2)\|$$

$$- \|Y_t - B^1 Y_{t-1} - D^1 Z_{t-1} - F(Y_{t-1}, Z_{t-1}, \gamma^1)\|_2| + |\|Y_t - B^1 Y_{t-1}$$

$$- D^1 Z_{t-1} - F(Y_{t-1}, Z_{t-1}, \gamma^1)\|_2|$$

$$\times \left.\left.\left|\frac{\partial^2}{\partial \gamma_j \partial \gamma_i} f(Y_{t-1}, Z_{t-1}, \gamma^2) - \frac{\partial^2}{\partial \gamma_j \partial \gamma_i} f(Y_{t-1}, Z_{t-1}, \gamma^1)\right|\right]\right\},$$

and we recall the steps of Lemma 3.4(i), and then

$$\cdots \leq 2 \{(\sqrt{\delta_L}(\|Y_{t-1}\|_2 + \|Z_{t-1}\|_2)(\sqrt{\delta_L}(\|Y_{t-1}\|_2 + \|Z_{t-1}\|_2)\|\gamma^2 - \gamma^1\|_2)$$

$$+ (\sqrt{\delta_L}(\|Y_{t-1}\|_2 + \|Z_{t-1}\|_2)(\sqrt{\delta_L}(\|Y_{t-1}\|_2 + \|Z_{t-1}\|_2)\|\gamma^2 - \gamma^1\|_2)$$

$$+ \sqrt{\delta_{L'}}(\|Y_{t-1}\|_2 + \|Z_{t-1}\|_2)(\|Y_{t-1}\|_2 + \|Z_{t-1}\|_2)$$

$$\times (\|B^2 - B^1\|_2 + \|D^2 - D^1\|_2 + 2\sqrt{\delta_L}\|\gamma^2 - \gamma^1\|_2)$$

$$+ (\|Y_t\|_2 + \delta_{CB}\|Y_{t-1}\|_2 + \delta_{CD}\|Z_{t-1}\|_2 + \delta_{CF}(\|Y_{t-1}\|_2 + \|Z_{t-1}\|_2))$$

$$\times (\sqrt{\delta_{L''}}(\|Y_{t-1}\|_2 + \|Z_{t-1}\|_2)\|\gamma^2 - \gamma^1\|_2)\}.$$

Now with Assumption LN'(ii) (and assuming $\sqrt{\delta_L} > 1$) we obtain

$$\cdots \leq 2\{2(\sqrt{\delta_L}(\|Y_{t-1}\|_2 + \|Z_{t-1}\|_2)^2(\|B^2 - B^1\|_2 + \|D^2 - D^1\|_2$$

$$+ 2\sqrt{\delta_L}\|\gamma^2 - \gamma^1\|_2) + [\sqrt{\delta_{L'}}(\|Y_{t-1}\|_2 + \|Z_{t-1}\|_2)^2$$

$$+ \sqrt{\delta_{L''}}(\|Y_{t-1}\|_2 + \|Z_{t-1}\|_2) \times (\|Y_t\|_2 + \delta_{CB}\|Y_{t-1}\|_2$$

$$+ \delta_{CD}\|Z_{t-1}\|_2 + \delta_{CF}(\|Y_{t-1}\|_2 + \|Z_{t-1}\|_2))]$$

$$\times [\|B^2 - B^1\|_2 + \|D^2 - D^1\|_2 + \sqrt{\delta_L}\|\gamma^2 - \gamma^1\|_2]\}$$

$$\equiv (4\sqrt{\delta_L}(\|Y_{t-1}\|_2 + \|Z_{t-1}\|_2)^2 + [2\sqrt{\delta_{L'}}(\|Y_{t-1}\|_2 + \|Z_{t-1}\|_2)^2$$

$$+ 2\sqrt{\delta_{L''}}(\|Y_{t-1}\|_2 + \|Z_{t-1}\|_2)(\|Y_t\|_2 + \delta_{CB}\|Y_{t-1}\|_2$$

$$+ \delta_{CD}\|Z_{t-1}\|_2 + \delta_{CF}(\|Y_{t-1}\|_2 + \|Z_{t-1}\|_2))])$$

$$\times (\|B^2 - B^1\|_2 + \|D^2 - D^1\|_2 + \sqrt{\delta_L}\|\gamma^2 - \gamma^1\|_2)$$

$$\equiv (4\sqrt{\delta_L}(\|Y_{t-1}\|_S + \|Z_{t-1}\|_S)^2 + [2\sqrt{\delta_{L'}}(\|Y_{t-1}\|_S + \|Z_{t-1}\|_S)^2$$
$$+ 2\sqrt{\delta_{L''}}(\|Y_{t-1}\|_S + \|Z_{t-1}\|_S)(\|Y_t\|_S + \delta_{CB}\|Y_{t-1}\|_S + \delta_{CD}\|Z_{t-1}\|_S$$
$$+ \delta_{CF}(\|Y_{t-1}\|_S + \|Z_{t-1}\|_S))])(\|B^2 - B^1\|_2$$
$$+ \|D^2 - D^1\|_2 + \sqrt{\delta_L}\|\gamma^2 - \gamma^1\|_2)$$
$$\equiv [L_t^{03}][a^0(\rho(\theta^2, \theta^1))],$$

and again $\lim \sup_n n^{-1} \sum_{t=1}^n E(L_t^{02}) < \infty$ as in Lemma 4.1(v), which concludes the proof.

REFERENCES

Bierens, H. J. (1981), *Robust Methods and Asymptotic Theory in Nonlinear Econometrics,* Lecture Notes in Economics and Mathematical Systems, Vol. 192. Berlin: Springer-Verlag.

Burgess, S., A. Escribano, and G. A. Pfann (1993), "Asymmetric and Time-Varying Error-Correction: An Application to Labour Demand in the U.K.", Working Paper 93–31, Universidad Carlos III de Madrid.

(eds.) (1997), "Asymmetries and Non-Linearities in Dynamic Economic Models", *Journal of Econometrics,* Vol. 74.

Burguete, J., A. R. Gallant, and G. Souza (1982), "On the Unification of the Asymptotic Theory of Nonlinear Econometric Models", *Econometric Reviews* 1: 151–212.

Ciarlet, P. G. (1989), *Introduction to Numerical Linear Algebra and Optimisation.* Cambridge: Cambridge University Press.

Escribano, A. (1986), "Non-linear Error-Correction: The Case of Money Demand in the U.K. (1878–1970)", Ph.D. Dissertation, University of California, San Diego, Chapter IV.

(1996), "Nonlinear Error Correction: The Case of Money Demand in the U.K.", Working Paper 96-55, Universidad Carlos III de Madrid.

Escribano, A., and C. W. J. Granger (1998), "Investigating the Relationship between Gold and Silver Prices", *Journal of Forecasting* 17: 81–107.

Escribano, A., and O. Jordá (1994), "Testing Nonlinearity: Decision Rules for Selecting between Logistic and Exponential Star Models", Working Paper 94-47, Universidad Carlos III de Madrid.

(1999), "Improved Testing and Specification of Smooth Transition Regression Models", In Philip Rothman, ed., *Nonlinear Times Series Analysis of Economic and Financial Financial Data.* Kluwer Academic Publishers, pp. 289–319.

Escribano, A., and S. Mira (1997), "Nonlinear Error Correction Models", Working Paper 97-26, Universidad Carlos III de Madrid.

Escribano, A., and G. Pfann (1998), "Non-Linear Error-Correction, Asymmetric Adjustment and Cointegration", *Economic Modelling* 15: 197–216.

Gallant, A. R., and H. White (1988), *A Unified Theory of Estimation and Inference for Nonlinear Dynamic Models.* New York: Basil Blackwell.

Granger, C. W. J. (1995), "Modelling Nonlinear Relationships between Extended Memory Variables", *Econometrica* 63(2): 265–279.

Granger, C. W. J., and T. Teräsvirta (1993), *Modelling Nonlinear Economic Relationships.* New York: Oxford University Press.

Hannan, E. J. (1971), "Nonlinear Time Series Regression", *Journal of Applied Probability* 8: 767–780.

Jennrich, R. J. (1969), Asymptotic Properties of Non-linear Least Squares Estimators", *The Annals of Mathematical Statistics* 40(2): 633–643.

Klimko, L. A., and P. I. Nelson (1978), "On Conditional Least Squares Estimation for Stochastic Processes", *The Annals of Statistics* 6: 629–642.

Kwiatkowski, D., P. C. B. Phillips, P. Schmidt, and Y. Shin (1992), "Testing the Null Hypothesis of Stationarity against the Alternative of a Unit Root", *Journal of Econometrics* 54: 159–178.

Lo, A. W. (1991) "Long-Term Memory in Stock Market Prices", *Econometrica* 59: 1279–1313.

Meyn, S. P., and R. L. Tweedie (1992), "Stability of Markovian Processes I: Criteria for Discrete-Time Chains", *Advances in Applied Probability* 24: 542–574.

Mira, S. (1996), "Modelos Econométricos Dinámicos No Lineales con Tendencias Estocásticas", Ph.D. Dissertation, Universidad Carlos III de Madrid, Madrid.

Neftci, S. N. (1984), "Are Economic Time Series Asymmetric over the Business Cycles?", *Journal of Political Economy* 92: 307–328.

Newey, W. N., and D. McFadden (1994), "Large Sample Estimation and Hypothesis Testing", in R. F. Engle and D. L. McFadden (eds.), *Handbook of Econometrics*. Amsterdam: Elsevier Science, Vol. IV.

Potscher, B. M., and I. R. Prucha (1991a), "Basic Structure of the Asymptotic Theory in Dynamic Nonlinear Econometric Models, Part I: Consistency and Approximation Concepts", *Econometric Reviews* 10(2): 125–216.

(1991b), "Basic Structure of the Asymptotic Theory in Dynamic Nonlinear Econometric Models, Part II: Asymptotic Normality", *Econometric Reviews* 10(2): 253–325.

Priestley, M. B. (1980), "State-Dependent Models: A General Approach to Non-linear Time Series Analysis", *Journal of Time Series Analysis* 1: 47–71.

Robinson, P. M. (1972), "Non-linear Regression for Multiple Time Series", *Journal of Applied Probability* 9: 758–768.

Rothman, P. (1991), "Further Evidence on the Asymmetric Behavior of Unemployment Rates over the Business Cycle", *Journal of Macroeconomics* 13: 291–298.

Sydsaeter, K., and P. J. Hammond (1995), *Mathematics for Economic Analysis*. Englewood Cliffs, NJ: Prentice-Hall.

Teräsvirta, T. (1994), "Specification, Estimation, and Evaluation of Smooth Transition Autoregressive Models", *Journal of the American Statistical Association* 89(425): 208–218.

Teräsvirta, T., and H. M. Anderson (1992), "Characterizing Nonlinearities in Business Cycles Using Smooth Transition Autoregressive Models", *Journal of Applied Econometrics* 7: S119–S139.

Tjöstheim, D. (1986), "Some Doubly Stochastic Time Series Models", *Journal of Time Series Analysis* 7: 51–72.

Tong, H. (1978), "On a Threshold Model", in C. H. Chen (ed.), *Pattern Recognition and Signal Processing*. Amsterdam: Sijthoff and Noordhoff.

(1990), *A Dynamical System Approach*. Oxford: Oxford University Press.

White, H. (1982), "Maximum Likelihood Estimation of Misspecified Models", *Econometrica* 50: 1–25.

Wooldridge, J. M. (1994), "Estimation and Inference for Dependent Processes", in R. F. Engle and D. L. McFadden (eds.), Handbook of Econometrics. Amsterdam: Elsevier Science, Vol. IV.

CHAPTER 7

Asymptotic inference on nonlinear functions of the coefficients of infinite order cointegrated VAR processes

Pentti Saikkonen & Helmut Lütkepohl

1   Introduction

In practice most multiple time series analyses are based on vector autoregressive (VAR) models. In such analyses various nonlinear functions of the autoregressive coefficients are usually used for interpreting the empirical evidence summarized in such models. Examples include impulse responses and measures of persistence.

Traditionally, when the true data generation process (DGP) is assumed to be of finite order VAR type, an asymptotic theory for inference on these quantities is available (for a review of some results see Lütkepohl 1991). Unfortunately, finite order VAR models are hardly sufficiently general to truly represent the DGPs of the multiple time series of practical interest. Therefore, in a number of theoretical studies, it has been assumed that, although finite order VAR models are fitted to the data, the true DGP is from a larger class. The simplest possible generalization in this direction is to consider the class of infinite order VAR processes. In this framework the asymptotic theory is usually based on the assumption that the order of the process fitted to the data increases with the sample size. Thereby the approximation to the true DGP improves with growing sample size. The approach is thus nonparametric in nature, although parametric models are fitted.

For univariate time series this approach has been used by Berk (1974) and Bhansali (1978) among others. For multiple time series analysis it has been

Both authors gratefully acknowledge financial support by the DFG, Sonderforschungsbereich 373. This research was partly carried out while the first author was visiting the Institute of Statistics and Econometrics at the Humboldt University and partly while the second author was a Visiting Research Fellow at the Australian National University in Canberra. They thank Kirstin Hubrich, the participants of the (EC)[2] meeting in Aarhus, and a referee for comments.

adapted by Lewis and Reinsel (1985), Lütkepohl (1988, 1996), and Lütkepohl and Poskitt (1991, 1996) in the stationary case and by Saikkonen (1992), Saikkonen and Lütkepohl (1996), and Lütkepohl and Saikkonen (1997) for nonstationary, cointegrated processes. The latter paper also addresses the problem of asymptotic inference for nonlinear functions of the VAR coefficients such as impulse responses. However, in that paper the functions considered are assumed to depend on a fixed finite number of VAR coefficients only. This assumption is restrictive in practice, because some of the functions of interest depend on potentially infinitely many VAR coefficients when the DGP is an infinite order VAR process. For instance, this is true if the persistence of a shock to the system is considered and is measured as the limit of the response coefficients as time goes to infinity. Also, if interest centers on impulse responses, all future responses to an impulse may be of interest. In particular, it may be of interest whether there is a response at all in a certain variable when an impulse hits some other variable. To accommodate these cases we allow the nonlinear functions to depend on an increasing number of VAR coefficients when the sample size increases. It turns out that under this scenario somewhat unconventional assumptions are necessary for the nonlinear functions considered. If only nonlinear functions with fixed finite dimensional range and domain are considered, as in Lütkepohl and Saikkonen (1997), a standard mean value expansion can be used to obtain the asymptotic properties of the quantities of interest. Such a tool is difficult to use in the presently considered general case where the dimensions of the domain and range space of the functions of interest are allowed to increase with the sample size. Therefore alternative and less conventional assumptions will be devised.

This chapter is organized as follows. In the next section the model class is defined and the framework for asymptotic inference is laid out. The section summarizes results from Saikkonen (1992), Saikkonen and Lütkepohl (1996) (henceforth S&L), and Lütkepohl and Saikkonen (1997) (henceforth L&S). In Section 3 the main general results for nonlinear functions of the estimated VAR coefficients are stated and discussed. Applications are considered in Section 4. In particular, it will be shown that the main results are sufficiently general to apply for impulse responses. Conclusions are spelled out in Section 5, and most proofs are given in an Appendix.

The following notation will be used. The symbols vec and vech denote, respectively, the column stacking operator and the operator that stacks the columns of a symmetric matrix from the main diagonal downwards only. The $K \times K$ identity matrix is denoted by I_K, and for a symmetric positive semidefinite matrix A, $A^{1/2}$ denotes the symmetric positive semidefinite matrix satisfying $A^{1/2}A^{1/2} = A$. The notation $\Psi = [\Psi_1 : \Psi_2]$ means that the matrix Ψ is subdivided into matrices Ψ_1 consisting of the first columns and Ψ_2 consisting of the last columns. Similar notation is also used when a matrix is partitioned into more than two submatrices. The Euclidean norm is denoted by $\|\cdot\|$, that is, for a matrix A, $\|A\| = [\text{tr}(A'A)]^{1/2}$. The symbols $\lambda_{\max}(\cdot)$ and $\lambda_{\min}(\cdot)$ denote the largest and the smallest eigenvalue of a matrix, respectively, and the operator norm $\|\cdot\|_1$ is defined so that $\|A\|_1 = (\lambda_{\max}(A'A))^{1/2}$. The differencing operator is denoted by Δ, that is, for a time series variable or vector y_t, $\Delta y_t = y_t - y_{t-1}$. A variable or vector y_t is called $I(1)$ if it is nonstationary and Δy_t is stationary. A stationary variable is sometimes called $I(0)$. Variables or vectors y_{1t} and y_{2t} are cointegrated if they are $I(1)$ and a stationary linear combination exists. Moreover, $\xrightarrow{d}$ denotes convergence in distribution, and $o(\cdot)$, $O(\cdot)$, $o_p(\cdot)$, $O_p(\cdot)$ are the usual order symbols for boundedness and boundedness in probability. The sample size is denoted by T, and h is an autoregressive order or lag length. Associated with these two symbols is $N = T - h - 1$. Least squares is abbreviated as LS.

2   The basic framework

In the following the framework for asymptotic inference in infinite order cointegrated VAR processes developed by Saikkonen (1992) and S&L will be used. More details on the basic setup can be found in those articles. Let $y_t = (y_{1t}', y_{2t}')'$ be a process of the following type:

$$y_{1t} = C_1 y_{2t} + u_{1t}, \tag{2.1a}$$

$$\Delta y_{2t} = u_{2t}, \tag{2.1b}$$

$t = 1, 2, \ldots$, where y_{it} is $K_i \times 1$, $i = 1, 2$, C_1 is $K_1 \times K_2$, and

$$u_t = \begin{bmatrix} u_{1t} \\ u_{2t} \end{bmatrix}$$

is a strictly stationary process with $E(u_t) = 0$, positive definite covariance matrix $\Sigma_u = E(u_t u_t')$, and continuous spectral density matrix $f_{uu}(\lambda)$ which is nonsingular for $\lambda = 0$. In addition u_t is assumed to have an infinite order VAR representation. In the process (2.1) y_{2t} consists of integrated $I(1)$ variables, which are not cointegrated among themselves but are cointegrated with y_{1t}. In other words, the cointegration matrix is $C = [I_{K_1} : -C_1]$. It should be noted that every $I(1)$ process with cointegration rank K_1 can be arranged so that the cointegration matrix has the form assumed here. The initial vector y_0 may be any random vector.

The process (2.1) can be written in *triangular error correction model* (ECM) form

$$\Delta y_t = -\begin{bmatrix} I_{K_1} & -C_1 \\ 0 & 0 \end{bmatrix} y_{t-1} + v_t, \tag{2.2}$$

where

$$v_t = \begin{bmatrix} I_{K_1} & C_1 \\ 0 & I_{K_2} \end{bmatrix} u_t$$

also has an infinite order VAR representation, say

$$\sum_{j=0}^{\infty} G_j v_{t-j} = \varepsilon_t. \tag{2.3}$$

We make the following assumptions regarding this representation. The ε_t are i.i.d. white noise with continuous distribution such that $E(\varepsilon_t) = 0$, $E(\varepsilon_t \varepsilon_t') = \Sigma_\varepsilon$ is positive definite, and the sixth moments exist. Moreover, the $K \times K$ coefficient matrices G_j satisfy

$$\left| \sum_{j=0}^{\infty} G_j z^j \right| \neq 0 \quad \text{for } |z| \leq 1 \quad \text{and} \quad \sum_{j=1}^{\infty} j^a \|G_j\| < \infty \tag{2.4}$$

for $a \geq 1$. S&L show that under these assumptions y_t has a pure VAR representation

$$y_t = \sum_{j=1}^{h+1} A_j y_{t-j} + e_t, \qquad t = h+1, h+2, \ldots, \tag{2.5}$$

where $e_t = \varepsilon_t - \sum_{j=h+1}^{\infty} G_j v_{t-j}$ and the VAR coefficient matrices $A_1, \ldots, A_h$ do not depend on h, while A_{h+1} is a function of h.

For this process the asymptotic properties of the multivariate LS estimators of the VAR coefficients can be derived under the following assumption regarding the order h of the fitted process.

Assumption 1:

$$h = o(T^{1/3}) \quad \text{and} \quad \sum_{j=h+1}^{\infty} \|G_j\| = o(T^{-1/2}).$$

This assumption has become standard in the related literature. Its practical implications are discussed in more detail in S&L. For easy reference we state the following result from L&S.

Theorem 1. *Let L_h be a sequence of $K^2 h \times J$ matrices of full column rank. Then under the assumptions made in the foregoing,*

$$N^{1/2} \left(L_h' \left(H_h' \Gamma_{\mathrm{ECM}}^{-1} H_h \otimes \Sigma_\varepsilon \right) L_h \right)^{-1/2} L_h' \mathrm{vec}([\tilde{A}_1 : \ldots : \tilde{A}_h] - [A_1 : \ldots : A_h])$$

$$\xrightarrow{d} N(0, I_J), \tag{2.6}$$

where the $\tilde{A}_j$ are the LS estimators of the A_j,

$$
H_h = \begin{bmatrix}
C & 0 & \cdots & \cdots & 0 \\
I_K & -I_K & \ddots & & \vdots \\
0 & I_K & \ddots & \ddots & \vdots \\
\vdots & \ddots & \ddots & \ddots & 0 \\
\vdots & & \ddots & \ddots & -I_K \\
0 & \cdots & \cdots & 0 & I_K
\end{bmatrix}
\tag{2.7}
$$

is a $(K_1 + Kh) \times Kh$ matrix, and

$$
\Gamma_{\mathrm{ECM}} = E\left(\begin{bmatrix} u_{1,t-1} \\ \Delta y_{t-1} \\ \vdots \\ \Delta y_{t-h} \end{bmatrix} [u'_{1,t-1}, \ \Delta y'_{t-1}, \ \ldots, \ \Delta y'_{t-h}] \right).
\tag{2.8}
$$

Furthermore, denoting the LS residuals of (2.5) by $\tilde{e}_t$ and defining $\tilde{\Sigma}_e = N^{-1} \sum \tilde{e}_t \tilde{e}'_t$, we have

$$
N^{1/2} \operatorname{vech}(\tilde{\Sigma}_e - \Sigma_\varepsilon) \overset{d}{\to} N(0, \Omega),
\tag{2.9}
$$

where $\Omega = E[\operatorname{vech}(\varepsilon_t \varepsilon'_t - \Sigma_\varepsilon)(\operatorname{vech}(\varepsilon_t \varepsilon'_t - \Sigma_\varepsilon))']$. Moreover the quantities in (2.6) and (2.9) are asymptotically independent.

Notice that in the theorem the asymptotic distribution of the LS estimators of the first h VAR coefficient matrices only is given, while a VAR($h + 1$) model is fitted to the data, where h goes to infinity with the sample size. In other words, the last estimated coefficient matrix is ignored for inference purposes. In this procedure the number of cointegration relations K_1 is not assumed known, and it is not taken into account in the estimation of the coefficient matrices A_j. In the finite order VAR case overfitting the order and ignoring the last coefficient matrix has been used by Toda and Yamamoto (1995) and Dolado and Lütkepohl (1996) in order to construct Wald tests with asymptotic χ^2 distributions for linear restrictions on the coefficients. It has also been pointed out by these authors that this procedure may involve a loss in efficiency relative to the situation where the cointegration rank is known (see Toda and Phillips (1993, 1994) and the discussion in S&L, Section 4). On the other hand, in practice the cointegration rank will rarely be known a priori, and hence such an assumption will effectively result in a pretesting procedure with uncertain and potentially inferior statistical properties. Hence, the approach assumed in Theorem 1 may be reasonable in many practical situations.

If assumptions regarding the cointegration matrix C can be justified, some quantities involving C are also sometimes of interest. This matrix can then be estimated via the autoregressive ECM form of the process:

$$\Delta y_t = \Psi y_{t-1} + \sum_{j=1}^{h} \Pi_j \Delta y_{t-j} + e_t. \tag{2.10}$$

An estimator of Ψ is obtained from this form by applying multivariate LS. Denoting the estimator of Ψ by $\tilde{\Psi} = [\tilde{\Psi}_1 : \tilde{\Psi}_2]$, where $\tilde{\Psi}_i$ is $K \times K_i$, an estimator of C_1 is

$$\tilde{C}_1 = -\left(\tilde{\Psi}_1' \tilde{\Sigma}_e^{-1} \tilde{\Psi}_1\right)^{-1} \tilde{\Psi}_1' \tilde{\Sigma}_e^{-1} \tilde{\Psi}_2, \tag{2.11}$$

where $\tilde{\Sigma}_e$ is defined in Theorem 1. The LS residuals from (2.10) are, of course, identical to those from (2.5), since (2.10) is just a reparameterization. The estimator $\tilde{C}_1$ is superconsistent, that is, $\tilde{C}_1 = C_1 + O_p(T^{-1})$. Therefore the matrix $C = [I_{K_1} : -C_1]$ may be treated as known in deriving the asymptotic properties of the quantities of interest (see Saikkonen 1992). This result is the basis for some of the following derivations. Therefore (2.11) may be replaced by other superconsistent estimators of the cointegration matrix (see, e.g., Stock 1987, Johansen 1991, Phillips 1995, Park and Phillips 1988, 1989).

3   Main results

We assume that the estimated quantities of interest are obtained from the estimated coefficients $\tilde{A}_j$, $j = 1, \ldots, h$, and $\tilde{\Sigma}_e$ by a function

$$F_h : W \subset I\!R^{K^2 h + K(K+1)/2} \rightarrow I\!R^{ah+b}$$

where a and b are fixed constants. Thus, the dimension of the domain as well as the range space of the function of interest may increase with the VAR order h and hence with the sample size. For instance, defining $\alpha_h = [\text{vec}(A_1 : \cdots : A_h)', \text{vech}(\Sigma_\varepsilon)']'$, $F_h(\alpha_h)$ may represent a sequence of responses of variable i to an impulse in variable j for up to h future periods. The corresponding estimators are obtained from the LS estimators $\tilde{\alpha}_h$, that is, they are obtained as $F_h(\tilde{\alpha}_h)$.

Obviously, in order to derive asymptotic properties of these estimators it is necessary that $F_h(\cdot)$ satisfies suitable conditions. We use the following notation in stating them.

$$J = \begin{bmatrix} 0 & 0 \\ 0 & I_{K_2} \end{bmatrix} \quad (K \times K)$$

$$\underline{C} = \begin{bmatrix} I_{K_1} & -C_1 \\ 0 & I_{K_2} \end{bmatrix} \qquad (K \times K)$$

$$P_2 = \begin{bmatrix} \underline{C} & -J & 0 & \cdots & 0 \\ 0 & \underline{C} & -J & \ddots & \vdots \\ \vdots & \ddots & \ddots & \ddots & 0 \\ \vdots & & \ddots & \ddots & -J \\ 0 & \cdots & \cdots & \ddots & \underline{C} \end{bmatrix} \qquad (Kh \times Kh)$$

In addition to P_2 we also define the $(K_1 + Kh) \times Kh$ matrix $P_1 = [P_2' : 0]'$. The parameter vector α_h is written as $\alpha_h = [a_h' \; \sigma']'$, where, of course, $a_h = \text{vec}[A_1 : \cdots : A_h]$ and $\sigma = \text{vech}(\Sigma_\varepsilon)$. The notation $\tilde{\alpha}_h = [\tilde{a}_h' \; \tilde{\sigma}']'$ is defined similarly.

The matrix P_1 is a transformation matrix which is used to study the asymptotic properties of the LS estimators of the coefficient matrices $A_1, \ldots, A_h$ in the pure VAR representation (2.5). Specifically, note that the model (2.5) can be rewritten as

$$\Delta y_t = \Psi_0 y_{2,t-1} + \sum_{j=1}^{h} \Xi_j u_{t-j} + \Xi_{h+1,1} u_{1,t-h-1} + e_t, \qquad (3.1)$$

where $\Psi_0 = 0$ and the matrix $\Xi = [\Xi_1 : \cdots : \Xi_h : \Xi_{h+1,1}]$ is linearly related to the coefficient matrices $A = [A_1 : \cdots : A_h]$ of (2.5) as

$$A = \Xi P_1 + [I_K : 0 : \cdots : 0] \qquad (3.2)$$

(see equation (A.2) of S&L). Since the transformation matrix P_1 involves the unknown cointegration matrix C_1, recovering A from Ξ requires knowledge of this matrix. For our purposes this is no problem, however, since in the following we are interested in asymptotic properties only. In order to derive those, the LS estimator $\tilde{A}$ of A can be studied by using the infeasible LS estimator $\tilde{\Xi}$ of Ξ obtained from (3.1). Indeed, from the proof of Theorem 2 of S&L (see (A.7)) we have the result

$$\tilde{A} - A = (\tilde{\Xi} - \Xi)P_1 + Z, \qquad (3.3)$$

where $\|Z\| = O_p(N^{-1})$. This result was used extensively in S&L, and it will also be used in this chapter. Due to the previous important role of the transformation matrix P_1, it is perhaps no surprise that conditions required for the function $F_h(\cdot)$ will also involve this matrix. However, since $P_1 = [P_2' : 0]'$ and the role of the null matrix is unimportant, we can focus on P_2 instead of P_1. This is convenient because the matrix P_2 is nonsingular.

The function $F_h(\cdot)$ is assumed to satisfy the following conditions.

Assumption 2:

(i) For every h and all $\alpha_h^* = [\ a_h^{*\prime}\ \sigma^{*\prime}]'$ in some neighbourhood of the true parameter value α_h the function $F_h(\cdot)$ is continuously differentiable and has the representation

$$F_h(\alpha_h^*) = F_h(\alpha_h) + \frac{\partial F_h(\alpha_h)}{\partial \alpha_h'}(\alpha_h^* - \alpha_h)$$

$$+ R_h^{(a)}(\alpha_h^*, \alpha_h)(a_h^* - a_h) + R_h^{(\sigma)}(\alpha_h^*, \alpha_h)(\sigma^* - \sigma),$$

$$(3.4)$$

where $R_h^{(a)}(\alpha_h^*, \alpha_h)$ and $R_h^{(\sigma)}(\alpha_h^*, \alpha_h)$ are matrix functions with appropriate orders.

(ii) If

$$\Delta_h = \frac{\partial F_h(\alpha_h)}{\partial \alpha_h'} \begin{bmatrix} H_h' \Gamma_{\text{ECM}}^{-1} H_h \otimes \Sigma_\varepsilon & 0 \\ 0 & \Omega \end{bmatrix} \frac{\partial F_h(\alpha_h)'}{\partial \alpha_h}$$

and L_h is a sequence of $(ah + b) \times J$ matrices of full column rank (cf. Theorem 1), then the matrix functions $R_h^{(a)}(\alpha_h^*, \alpha_h)$ and $R_h^{(\sigma)}(\alpha_h^*, \alpha_h)$ satisfy the following conditions: For every $\delta > 0$ there exists a finite constant c_δ such that

$$\left\| (L_h' \Delta_h L_h)^{-1/2} L_h' R_h^{(a)}(\alpha_h^*, \alpha_h)(P_2' \otimes I_K) \right\|_1 \leq c_\delta h / T^{1/2}$$

$$(3.5)$$

and

$$\left\| (L_h' \Delta_h L_h)^{-1/2} L_h' R_h^{(\sigma)}(\alpha_h^*, \alpha_h) \right\|_1 \leq c_\delta h^{3/2} / T^{1/2} \qquad (3.6)$$

as $\|(P_2'^{-1} \otimes I_K)(a_h^* - a_h)\| \leq \delta h^{1/2}/T^{1/2}$ and $\|\sigma^* - \sigma\| \leq \delta/T^{1/2}$.

Since this assumption is a bit unconventional, several remarks may be in order. Basically Assumption 2 ensures that the function $F_h(\cdot)$ is sufficiently smooth so that a linear approximation and Theorem 1 can be used. If one considers a parameter vector with a fixed and finite dimension, this follows in the conventional way from a standard mean value expansion (cf. L&S, Theorem 3). In the more general context of this paper the situation is more difficult and requires more specific assumptions. Note, for instance, that we do not assume that (3.4) is based on a mean value expansion, although it may result from such an expansion. If a standard mean value expansion is used, the interpretation of the matrices $R_h^{(a)}(\alpha_h^*, \alpha_h)$ and $R_h^{(\sigma)}(\alpha_h^*, \alpha_h)$ is straightforward. Then $R_h^{(a)}(\alpha_h^*, \alpha_h)$, for example, becomes $\partial F_h(\bar{\alpha}_h)/\partial a_h' - \partial F_h(\alpha_h)/\partial a_h'$, where each row of $\partial F_h(\bar{\alpha}_h)/\partial a_h'$ is evaluated at a potentially different intermediate point. Since we let the number of rows in $\partial F_h(\alpha_h)/\partial a_h'$ increase with the sample size, this feature can actually make the use of mean value expansions difficult. The discussion below illustrates some of the difficulties involved. Another reason for not necessarily assuming the mean value expansion is that in the applications we have in mind (see Section 4) it is possible to consider alternative and, in fact, slightly more convenient expansions.

Let us now consider how the linear approximation obtained from (3.4) may be used. Since we wish to apply Theorem 1, we have to consider the difference $F_h(\tilde{\alpha}_h) - F_h(\alpha_h)$ premultiplied by a suitable standardization matrix, which is obviously $(L_h' \Delta_h L_h)^{-1/2} L_h'$ with Δ_h and L_h as defined in Assumption 2. This means that we have to obtain the results

$$\left\| N^{1/2} (L_h' \Delta_h L_h)^{-1/2} L_h' R_h^{(a)}(\tilde{\alpha}_h, \alpha_h)(\tilde{a}_h - a_h) \right\| = o_p(1) \tag{3.7}$$

and

$$\left\| N^{1/2} (L_h' \Delta_h L_h)^{-1/2} L_h' R_h^{(\sigma)}(\tilde{\alpha}_h, \alpha_h)(\tilde{\sigma} - \sigma) \right\| = o_p(1). \tag{3.8}$$

In (3.7) we can use the result in (3.3) and write

$$R_h^{(a)}(\tilde{\alpha}_h, \alpha_h)(\tilde{a}_h - a_h) = R_h^{(a)}(\tilde{\alpha}_h, \alpha_h)[(P_2' \otimes I_K)(\tilde{\xi}_h - \xi_h) + z],$$

where $\tilde{\xi}_h = \text{vec}([\tilde{\Xi}_1 : \cdots : \tilde{\Xi}_h])$, and similarly for ξ_h and z. An important point here is that in applications involving, for instance, impulse responses, postmultiplying the matrix $R_h^{(a)}(\tilde{\alpha}_h, \alpha_h)$ by $P_2' \otimes I_K$ reduces its operator norm by $O_p(h^{-1})$. This is the reason why the matrix $P_2' \otimes I_K$ appears in the latter part of Assumption 2. Moreover, on the basis of the above discussion we can bound the left hand side of (3.7) by

$$\left\| N^{1/2} (L_h' \Delta_h L_h)^{-1/2} L_h' R_h^{(a)}(\tilde{\alpha}_h, \alpha_h)(P_2' \otimes I_K) \right\|_1$$
$$\times \left(\| \tilde{\xi}_h - \xi_h \| + \left\| (P_2'^{-1} \otimes I_K)z \right\| \right)$$
$$= O_p(h^{1/2}) \left\| (L_h' \Delta_h L_h)^{-1/2} L_h' R_h^{(a)}(\tilde{\alpha}_h, \alpha_h)(P_2' \otimes I_K) \right\|_1, \tag{3.9}$$

where the equality makes use of the result $\| \tilde{\xi}_h - \xi_h \| = O_p(h^{1/2}/T^{1/2})$ (see Lemma A.1 in the Appendix) and $\| (P_2'^{-1} \otimes I_K)z \| = O_p(h/N)$. Since we also have $\| (P_2'^{-1} \otimes I_K)(\tilde{a}_h - a_h) \| = O_p(h^{1/2}/T^{1/2})$ (again by Lemma A.1), the conditions of Assumption 2(ii) involving (3.5) readily show that the last norm in (3.9) is $O_p(h/T^{1/2})$ (see Lemma A.2 in the Appendix). Thus, it follows that the left hand side of (3.7) is $O_p(h^{3/2}/T^{1/2})$, or $o_p(1)$ when Assumption 1 holds.

Notice that when the value of h is fixed and finite and a conventional mean value expansion is used, the continuity of the derivative matrix $\partial F_h(\alpha_h)/\partial \alpha_h$ implies that $\|R_h^{(a)}(\tilde{\alpha}_h, \alpha_h)(P_2' \otimes I_K)\|_1 = o_p(1)$, and, if $\|(L_h'\Delta_h L_h)^{-1/2}L_h'\|_1 = O(1)$, the right hand side of (3.9) is $o_p(1)$ and (3.7) follows. However, when h is allowed to increase with the sample size this argument clearly breaks down, as equation (3.9) shows. This happens even if the dimension of the range space of the function $F_h(\cdot)$ is fixed and finite, because under our assumptions the dimension of the domain still increases. To deal with this problem an appropriate rate of convergence is needed for the norm on the right hand side of (3.9), and this is obtained by the assumption involving inequality (3.5).

The above discussion concerning equation (3.7) applies with obvious modifications to equation (3.8) as well. In this case the situation is simpler, however, and some problems encountered with equation (3.7) disappear. The reason is that the dimension of the parameter vector σ is fixed and finite and no analog of the matrix $P_2' \otimes I_K$ is needed. Obtaining $R_h^{(\sigma)}(\alpha_h^*, \alpha_h)$ from a mean value expansion may also be quite feasible here. Since the estimator $\tilde{\sigma}$ is consistent of order $O_p(T^{1/2})$, the right hand side of (3.6) is larger than that of (3.5), where the estimation accuracy of the associated parameter vector is only of order $O_p(h^{1/2}/T^{1/2})$.

At this point it may be worth noting that inequalities (3.5) and (3.6) are only assumed to hold locally in suitably shrinking balls centered at the true parameter values a_h and σ. The rates at which these balls shrink are directly related to the orders of consistency of the estimators $\tilde{a}_h$ and $\tilde{\sigma}$. This is of course natural, because these consistency results essentially mean that we do not need to worry about what happens outside such shrinking balls. Often inequalities (3.5) and (3.6) can be verified by showing that the left hand sides can be bounded by $c_1 h^{1/2}\|(P_2'^{-1} \otimes I_K)(a_h^* - a_h)\|$ and $c_2 h^{3/2}\|\sigma^* - \sigma\|$, respectively. If this can be done, equalities (3.7) and (3.8) immediately follow from the consistency properties of the estimators $\tilde{a}_h$ and $\tilde{\sigma}$.

One might consider the possibility of using the norm inequality $\|AB\|_1 \leq \|A\|_1\|B\|_1$ to bound the left hand sides of (3.5) and (3.6). In the case of (3.5) this means using the inequality

$$\left\|(L_h'\Delta_h L_h)^{-1/2}L_h'R_h^{(a)}(\alpha_h^*, \alpha_h)(P_2' \otimes I_K)\right\|_1$$
$$\leq \left\|(L_h'\Delta_h L_h)^{-1/2}L_h'\right\|_1\left\|R_h^{(a)}(\alpha_h^*, \alpha_h)(P_2' \otimes I_K)\right\|_1. \tag{3.10}$$

Applying Lemma A.1 of S&L, one can bound the first norm on the right hand side by $1/\lambda_{\min}(\Delta_h)$. In some applications, including impulse responses, it may be possible to show that $\lambda_{\min}(\Delta_h) \geq c > 0$, so that the problem is reduced to finding a suitable upper bound for the latter norm on the right hand side of (3.10). The main reason why we have not used this approach in the formulation of Assumption 2(ii) is simply that it cannot be used directly for impulse responses (which are considered in the next section), because the latter norm on the right hand side of (3.10) is too large in this case. However, the left hand side still has the desired upper bound, because the two matrices on the right hand side are related in a suitable way. In the case of impulse responses this can be expressed by noting that the idea in (3.10) can be applied after premultiplying the matrix $R_h^{(a)}(\alpha_h^*, \alpha_h)(P_2' \otimes I_K)$ by a suitable nonsingular matrix, W say, and postmultiplying the matrix $(L_h' \Delta_h L_h)^{-1/2} L_h'$ by W^{-1}. We have not used this idea in formulating Assumption 2(ii), because it does not simplify the interpretation of the assumption. Of course, such a device may still be used in proofs if it simplifies the arguments.

The use of the operator norm on the left hand sides of (3.5) and (3.6) is essential because, when the dimensions of the involved matrices increase, the euclidean norm is often too large to be useful. This is particularly so in the nonstationary case, where matrices with norms increasing with h have to be considered. For instance, in the case of impulse responses, $\|\partial F_h(\alpha_h)/\partial \alpha_h'\|_1$ may not be bounded when the process is nonstationary, while the corresponding quantity is bounded in the stationary case. This example also explains why the use of a linear approximation is typically more difficult to justify in the nonstationary case than in the stationary case. For this reason it may also be worth pointing out that our results can be specialized to the stationary case. Then $K = K_1$ and $y_t = u_t = v_t$. The cointegrating matrix C and the transformation matrix $P_2' \otimes I_K$ are replaced by appropriate identity matrices. In the stationary case there is also no need to ignore the estimator $\tilde{A}_{h+1}$ – or, equivalently, the order of the VAR representation (2.5) may be reduced from $h + 1$ to h.

Now we can state the following theorem, which is proven in the Appendix.

Theorem 2. *Let L_h be a sequence of $(ah + b) \times J$ matrices such that $L_h'(\partial F_h/\partial \alpha_h')$ has full row rank. Here J is a fixed number. Then, if the assumptions of Theorem 1 and Assumption 2 hold,*

$$N^{1/2}(L_h' \Delta_h L_h)^{-1/2} L_h'[F_h(\tilde{\alpha}_h) - F_h(\alpha_h)] \xrightarrow{d} N(0, I_J).$$

Occasionally it is of interest to construct tests for infinite sets of restrictions. As mentioned earlier, it may be of interest to check whether an impulse in one variable has no effect at all (at any lead time) on some other variable. In that case one would like to test whether an infinite sequence of impulse responses is zero. For stationary VAR processes such tests have been considered by Lütkepohl and Poskitt (1996) and Lütkepohl (1996). In the present framework this type of hypothesis of interest can be tested by letting the number of restrictions grow with the VAR order h and hence with the sample size. The following theorem, which is also proven in the Appendix, provides the basis for testing hypotheses of the form $H_0 : L_h' F_h(\alpha) = l_h$, where l_h is a given $J \times 1$ vector whose dimension may increase with the sample size. In particular $J = h$ is permitted. The corresponding Wald statistic is

$$\lambda_{\text{Wald}} = N[F_h(\tilde{\alpha}_h) - F_h(\alpha_h)]' L_h(L_h' \tilde{\Delta}_h L_h)^{-1} L_h'[F_h(\tilde{\alpha}_h) - F_h(\alpha_h)], \quad (3.11)$$

where

$$\tilde{\Delta}_h = \frac{\partial F_h(\tilde{\alpha}_h)}{\partial \alpha_h'} \begin{bmatrix} \tilde{\Gamma}_y^h \otimes \tilde{\Sigma}_e & 0 \\ 0 & \tilde{\Omega} \end{bmatrix} \frac{\partial F_h(\tilde{\alpha}_h)'}{\partial \alpha_h} \quad (3.12)$$

with $\tilde{\Omega}$ being a consistent estimator of Ω, and $\tilde{\Gamma}_y^h$ the upper left hand $Kh \times Kh$ submatrix of the inverse of

$$\hat{\Gamma}_y = \frac{1}{N} \sum_{t=h+2}^{T} \begin{bmatrix} y_{t-1} \\ \vdots \\ y_{t-h-1} \end{bmatrix} [y_{t-1}', \dots, y_{t-h-1}'].$$

The proof requires, however, that we make the following additional assumption for $F_h(\cdot)$.

Assumption 3: For every $\delta > 0$ there exists a finite constant c_δ such that

$$\left\| (L_h' \Delta_h L_h)^{-1/2} L_h' \left[\frac{\partial F_h(\alpha_h^*)}{\partial a_h'} - \frac{\partial F_h(\alpha_h)}{\partial a_h'} \right] (P_2' \otimes I_K) \right\|_1 \leq \frac{c_\delta h}{T^{1/2}} \quad (3.13)$$

and

$$\left\| (L_h' \Delta_h L_h)^{-1/2} L_h' \left[\frac{\partial F_h(\alpha_h^*)}{\partial \sigma'} - \frac{\partial F_h(\alpha_h)}{\partial \sigma'} \right] \right\|_1 \leq \frac{c_\delta h}{T^{1/2}} \quad (3.14)$$

as $\|(P_2'^{-1} \otimes I_K)(a_h^* - a_h)\| \leq \delta h^{1/2} / T^{1/2}$ and $\|\sigma^* - \sigma\| \leq \delta / T^{1/2}$.

Assumption 3 is used to show that the derivative matrix $\partial F_h(\tilde{\alpha}_h)/\partial \alpha_h'$ in the definition of the Wald statistic λ_{Wald} can be replaced by its theoretical counterpart. This means that we have to show, for instance, that

$$\left\| (L_h' \Delta_h L_h)^{-1/2} L_h' \left[\frac{\partial F_h(\tilde{\alpha}_h)}{\partial a_h'} - \frac{\partial F_h(\alpha_h)}{\partial a_h'} \right] (P_2' \otimes I_K) \right\|_1 = o_p(1) \quad (3.15)$$

(see the proof of Lemma A.3(iii)). Thus, the idea is to give sufficient conditions for results similar to (3.7) and (3.8) for which Assumption 2(ii) was formulated. This explains the similarity of the two assumptions and the discussion given for Assumption 2(ii) above applies with appropriate modifications also here. It may be noted, however, that in (3.15) the transformation matrix $P_2' \otimes I_K$ is obtained from the matrix $H_h' \Gamma_{\text{ECM}}^{-1} H_h$ which appears in the definition of Δ_h. For this reason it is not possible to consider alternative transformation matrices here. A major difference between Assumptions 2 and 3 is that the upper bounds in inequalities (3.6) and (3.14), which are used for similar purposes, are different.

Now we can set up a Wald test for an infinite number of restrictions on the quantities of interest obtained through $F_h(\cdot)$.

Theorem 3. *Suppose that in addition to the assumptions of Theorem 2 and Assumption 3 the following conditions are satisfied:*

(i)   *(2.4) holds with $a = 2$;*
(ii) *the ε_t have finite eighth moments;*
(iii) $h = o(T^{1/3})$ *and* $\sum_{j=h+1}^{\infty} \|G_j\| = o((TJ)^{-1/2})$;
(iv) $J \to \infty$ *at the same rate as h when $T \to \infty$, i.e., $J = O(h)$;*
(v) $\tilde{\Omega}$ *is an estimator of Ω satisfying* $\tilde{\Omega} = \Omega + O_p(h/T^{1/2})$.

Then $(\lambda_{\text{Wald}} - J)/\sqrt{2J} \overset{d}{\to} N(0, 1)$, *where λ_{Wald} is defined in (3.11).*

The theorem may be interpreted as justification for using λ_{Wald} in conjunction with the critical values of a $\chi^2(J)$ distribution even if the number of restrictions, J, goes to infinity with the sample size. In fact the proof proceeds by showing that λ_{Wald} is essentially a χ^2 statistic with increasing number of degrees of freedom and by using that $(\chi^2(J) - J)/\sqrt{2J} \overset{d}{\to} N(0, 1)$. Of course, if J is large (e.g., $J \geq 100$), the statistic $(\chi^2(J) - J)/\sqrt{2J}$ and thus $(\lambda_{\text{Wald}} - J)/\sqrt{2J}$ may be used with critical values from a standard normal distribution. Simulation results reported in Lütkepohl (1996) for impulse response functions of stationary processes indicate that in some cases the asymptotic distribution is not a very reliable guide for small sample performance. A χ^2 approximation to the distribution of λ_{Wald} clearly works better than the standard normal approximation to the distribution of $(\lambda_{\text{Wald}} - J)/\sqrt{2J}$. In any case, for a reliable inference, rather large samples seem to be required, especially if large lags are important to capture the dynamic structure of the process.

Note that the additional assumptions (i)–(v) in Theorem 3 are fairly general. For instance, (i) is just a technical condition which is satisfied if v_t in (2.3) is a finite order vector autoregressive moving average process. Also, since finite order sixth moments have been assumed for the ε_t from the outset, requiring eighth moments now may not be problematic from a practical point of view. In (iii) the lower bound for the VAR order h is tightened relative to the one in Assumption 1. More precisely, $\sum_{j=h+1}^{\infty} \|G_j\|$ now has to be $o((TJ)^{-1/2})$, whereas in Assumption 1 $o(T^{-1/2})$ is sufficient. Since this is just an asymptotic condition, it is of little or no practical relevance. Of course, this does not mean that the choice of h is irrelevant for the small sample properties of the test statistic. In (iv) the number of restrictions, J, is required to approach infinity at the same rate as h. A simpler (although more restrictive) version of this condition would be that $J = h$. We prefer the more general version of the condition given in the theorem because the proofs do not simplify significantly by assuming $J = h$.

A plausible estimator of Ω is

$$\tilde{\Omega} = N^{-1} \sum_{t=h+2}^{T} \text{vech}(\tilde{e}_t \tilde{e}_t' - \tilde{\Sigma}_e)[\text{vech}(\tilde{e}_t \tilde{e}_t' - \tilde{\Sigma}_e)]'. \tag{3.16}$$

L&S show that for this estimator $\tilde{\Omega} = \Omega + o_p(1)$ (see their Theorem 2). In general it seems to be unknown whether a better result as required in Theorem 3(v) can be obtained for this estimator in the present context. However, if the white noise process ε_t is Gaussian, $\Omega = 2D_K^+(\Sigma_\varepsilon \otimes \Sigma_\varepsilon)D_K^{+'}$, where $D_K^+ = (D_K' D_K)^{-1} D_K'$ is the Moore–Penrose generalized inverse of the $K^2 \times K(K+1)/2$ duplication matrix D_K (see, e.g., Magnus and Neudecker 1988, pp. 48–49, for the definition of this matrix). It follows from Theorem 1 that the estimator

$$\tilde{\Omega} = 2D_K^+(\tilde{\Sigma}_e \otimes \tilde{\Sigma}_e)D_K^{+'}$$

under the present assumptions satisfies condition (v) of Theorem 3. Hence, for this case a suitable estimator of Ω is available.

Theorem 3 covers the situation where effectively infinitely many restrictions have to be tested and hence $J \to \infty$. It is also possible that the number of restrictions, J, is fixed and finite and still the number of matrices A_i involved increases with h or the sample size. For instance, this situation occurs when the total effect of an impulse (the sum of the individual effects for different lags) is of interest. In our framework in this case the dimension of the domain of the function $F_h(\cdot)$ increases although the range space may have a fixed constant dimension. For that situation a better result than Theorem 3 is possible. The following result follows from the proof of the latter theorem.

Theorem 3'. *Suppose that J is a fixed finite number and the assumptions of Theorem 2 and Assumption 3 hold. Furthermore, $\tilde{\Omega}$ is an estimator of Ω satisfying $\tilde{\Omega} = \Omega + O_p(h/T^{1/2})$. Then λ_{Wald} has an asymptotic $\chi^2(J)$ distribution.*

Note that this result differs from Corollary 1 of L&S because in that corollary a function $F_i(\cdot)$ is considered with domain of fixed finite dimension.

It is perhaps worth pointing out that all the foregoing results remain valid if an intercept term is included in the process (2.1), as long as that term can be absorbed into the cointegration relation. Hence our results remain valid if (2.1a) is replaced by

$$y_{1t} = \nu + C_1 y_{2t} + u_{1t},$$

where ν denotes a fixed $K_1 \times 1$ vector. Such an assumption implies that the variables of the system do not have deterministic trends, which is a reasonable assumption for many economic time series. S&L discuss the necessary adjustments in the computation of the estimators and related quantities in more detail.

4 Applications

As mentioned earlier, Theorems 2 and 3 are useful because the assumptions used are satisfied by relevant functions $F_h(\cdot)$. In this section we will consider some such functions. An example of some importance are the forecast error impulse responses (see, e.g., Lütkepohl 1991, Chapter 2). Although these impulse responses have been criticized and as a result their use in applied work is limited, they are of special interest here because a number of other quantities that have been used in practice can easily be obtained from them.

For stationary processes forecast error impulse responses are just the coefficients of the Wold moving average representation. For cointegrated processes they are determined from the VAR coefficients in the same recursive way as

$$\Phi_0 = I_K,$$

$$\Phi_j = \sum_{k=1}^{j} \Phi_{j-k} A_k, \qquad j = 1, 2, \dots. \tag{4.1}$$

The mnth element $\phi_{mn,j}$ of Φ_j is sometimes interpreted as the response of variable m to an impulse in variable n that has occurred j periods ago. In the stationary case the Φ_j are absolutely summable and therefore they are relatively easy to handle. In the nonstationary case they are more difficult to deal with because they are generally not absolutely summable. Of course, this is no problem if just a fixed finite number of them is of interest. However, in some situations interest centers on an infinite sequence, say $\phi_{mn,j}, j = 1, 2, \dots$. If all these impulse responses are, for instance, zero, then variable m does not react at all to an impulse in variable n, and hence the latter variable is in this sense noncausal for variable m. Theorem 3 lends itself to testing such a hypothesis provided it can be established that the corresponding function

$$F_h(A_1, \dots, A_h, \Sigma_\varepsilon) = \text{vec}[\Phi_1 : \cdots : \Phi_h] \tag{4.2}$$

satisfies Assumptions 2 and 3. The next theorem states that this is true. The proof is also given in the Appendix.

Theorem 4. *The function $F_h(\cdot)$ in (4.2) satisfies Assumptions 2 and 3.*

A similar result may be obtained for orthogonal impulse responses, which are often considered in applied work. Formally they are defined as $\Theta_j = \Phi_j P$, where P is a Choleski component of Σ_ε, that is, P is lower triangular with positive diagonal such that $P P' = \Sigma_\varepsilon$ (see once again Lütkepohl 1991, Chapter 2).

Defining

$$F_h(A_1, \ldots, A_h, \Sigma_\varepsilon) = \text{vec}[\Theta_1 : \cdots : \Theta_h], \qquad (4.3)$$

we get an analogous result to Theorem 4.

Theorem 5. *The function $F_h(\cdot)$ in (4.3) satisfies Assumptions 2 and 3.*

Again the proof is given in the Appendix, and again this result can be used to draw on Theorems 2 and 3 for inference on orthogonal impulse responses. Without giving detailed proofs, we also note that similar results hold for quantities such as $C\Phi_j$ and $C\Theta_j$, where C is the cointegration matrix as before. These quantities are sometimes interpreted as responses to disturbances of the cointegration relations and have also been used in applied work for interpreting cointegrated systems (see e.g., Mellander, Vredin, and Warne 1992).

5   Conclusions

The coefficients of VAR processes are often difficult to interpret directly. Therefore a number of quantities derived from these coefficients have been used in emprical studies in order to highlight the relationships between the variables described by a VAR system. Examples are impulse responses. Many of the quantities of interest are nonlinear functions of the VAR coefficients. Therefore we have developed a general theory for inference on nonlinear functions of the coefficients of cointegrated VAR processes.

Our results extend previous work in a number of respects. First, in contrast to most work on cointegrated processes, we allow the true VAR order to be infinite. The order of the processes fitted to the data is assumed to be finite but increasing with the sample size. Second, the domains as well as the ranges of the nonlinear functions considered are permitted to increase with the sample size. Thereby we may, for instance, obtain the joint asymptotic distribution of an increasing number of impulse responses when the sample size increases and construct tests of the null hypothesis that some variable does not react at all to an impulse in some other variable.

The price for this generality is that the assumptions look relatively complicated. Therefore a detailed discussion of our assumptions is also provided so that their relevance for theoretical and applied work can be understood more easily. In addition to the "usual" assumptions for the data generation process, we also need conditions on the nonlinear functions considered.

Appendix: Proofs

We shall first prove two lemmas which can be used in the proofs of Theorems 2 and 3. The first lemma is related to quantities discussed at the beginning of Section 3 (see in particular Equation (3.3)).

Lemma A.1.

(i) $\|P_2\|_1 = O(1)$ and $\|P_2^{-1}\|_1 = O(h)$.

(ii) $(P_2'^{-1} \otimes I_K)(\tilde{a}_h - a_h) = \tilde{\xi}_h - \xi_h + z$, where $\|z\| = O_p(N^{-1})$ and $\|\tilde{\xi}_h - \xi_h\| = O_p(h^{1/2}/T^{1/2})$.

Proof: Noting that $P_1 = D_h^{-1} H_h = [P_2' : 0]'$, the first result in (i) is a straightforward consequence of the definition of P_2, and the second one is justified in the proof of Theorem 2 of S&L. The first equality in (ii) follows from (i) and the result given in equation (3.3), while the second one is obtained from (A.12) of S&L. □

Lemma A.2.

(i) If the function $F_h(\cdot)$ satisfies Assumption 2, then

$$\left\| (L_h' \Delta_h L_h)^{-1/2} L_h' R_h^{(a)}(\tilde{\alpha}_h, \alpha_h)(P_2' \otimes I_K) \right\|_1 = O_p(h/T^{1/2})$$

and

$$\left\| (L_h' \Delta_h L_h)^{-1/2} L_h' R_h^{(\sigma)}(\tilde{\alpha}_h, \alpha_h) \right\|_1 = O_p(h^{3/2}/T^{1/2}).$$

(ii) If Assumption 3 holds, then

$$\left\| (L_h' \Delta_h L_h)^{-1/2} L_h' \left[\frac{\partial F_h(\tilde{\alpha}_h)}{\partial a_h'} - \frac{\partial F_h(\alpha_h)}{\partial a_h'} \right] (P_2' \otimes I_K) \right\|_1$$
$$= O_p(h/T^{1/2})$$

and

$$\left\| (L_h' \Delta_h L_h)^{-1/2} L_h' \left[\frac{\partial F_h(\tilde{\alpha}_h)}{\partial \sigma'} - \frac{\partial F_h(\alpha_h)}{\partial \sigma'} \right] \right\|_1 = O_p(h/T^{1/2}).$$

Proof: Consider the first result in part (i), and, for brevity, let B_h be the matrix whose norm is taken, so that we have to show that $\|B_h\|_1 = O_p(h/T^{1/2})$. Let

$\epsilon > 0$, and use Lemma A.1(ii) to find $\delta > 0$ such that

$$\Pr\{\|(P_2'^{-1} \otimes I_K)(\tilde{a}_h - a_h)\| \leq \delta h^{1/2}/T^{1/2}\} > 1 - \epsilon.$$

For this choice of δ we can now use Assumption 2(ii) and choose c_δ such that $\Pr\{\|B_h\| \leq c_\delta h/T^{1/2}\}$ is at least as large as the l.h.s. of the above inequality. This proves the desired result. The proof of the second result in part (i) uses the result $\tilde{\sigma} = \sigma + O_p(T^{-1/2})$ obtained from Theorem 1, but is otherwise similar. The proof of part (ii) may be obtained by the same argument.   □

A.1   *Proof of Theorem 2*

Using Asumption 2 and Lemmas A.1 and A.2, we can now readily show that Theorem 2 follows from Theorem 1. First replace α_h^* in Assumption 2(i) by $\tilde{\alpha}_h$ and observe that we only need to justify equations (3.7) and (3.8). Using the norm inequality $\|AB\| \leq \|A\|_1\|B\|$, it is immediately seen that (3.7) follows from Lemma A.1(ii) and the first result in Lemma A.2(ii). Since $\|\tilde{\sigma} - \sigma\| = O_p(T^{-1/2})$, one can obtain (3.8) similarly from the last result in Lemma A.2(i).

A.2   *Proof of Theorem 3*

We shall use the notation from Section 3 (particularly from the definition of the test statistic λ_{Wald}) and define

$$\tilde{\gamma}_h = N^{1/2}[F_h(\tilde{\alpha}_h) - F_h(\alpha_h)]$$

and

$$\underline{\tilde{\Delta}}_h = L_h'\tilde{\Delta}_h L_h.$$

Then $\lambda_{\text{Wald}} = \tilde{\gamma}_h' L_h \underline{\tilde{\Delta}}_h^{-1} L_h' \tilde{\gamma}_h$. Next we shall introduce the matrix

$$\hat{\Delta}_h = \frac{\partial F_h(\tilde{\alpha}_h)}{\partial \alpha_h'} \begin{bmatrix} P_1'\hat{\Gamma}_u^{-1}P_1 \otimes \tilde{\Sigma}_e & 0 \\ 0 & \tilde{\Omega} \end{bmatrix} \frac{\partial F_h(\tilde{\alpha}_h)'}{\partial \alpha_h},$$

where

$$\hat{\Gamma}_u = N^{-1} \sum_{t=h+2}^{T} U_t U_t', \qquad U_t = [u_{t-1}' \cdots u_{t-h}', u_{1,t-h-1}']'.$$

Analogously to $\underline{\tilde{\Delta}}_h$, we finally define $\underline{\Delta}_h = L_h'\Delta_h L_h$ and $\underline{\hat{\Delta}}_h = L_h'\hat{\Delta}_h L_h$, so that we can write

$$\frac{\lambda_{\text{Wald}} - J}{\sqrt{2J}} = \frac{\tilde{\gamma}_h' L_h \underline{\tilde{\Delta}}_h^{-1} L_h' \tilde{\gamma}_h - J}{\sqrt{2J}} + \frac{r_{1h}}{\sqrt{2J}} + \frac{r_{2h}}{\sqrt{2J}},$$

where

$$r_{1h} = \tilde{\gamma}_h' L_h (\underline{\tilde{\Delta}}_h^{-1} - \underline{\hat{\Delta}}_h^{-1}) L_h' \tilde{\gamma}_h$$

and

$$r_{2h} = \tilde{\gamma}_h' L_h \big(\hat{\underline{\Delta}}_h^{-1} - \underline{\Delta}_h^{-1} \big) L_h' \tilde{\gamma}_h.$$

We wish to show that

$$r_{ih}/\sqrt{2J} = o_p(1), \qquad i = 1, 2. \tag{A.1}$$

The following intermediate results are proven first.

Lemma A.3. *Under the conditions of Theorem 3, we have*

(i) $\| \Delta_h^{1/2} \hat{\Delta}_h^{-1/2} \|_1 = O_p(1),$

(ii) $\| \hat{\underline{\Delta}}_h^{1/2} \tilde{\underline{\Delta}}_h^{-1} \hat{\underline{\Delta}}_h^{1/2} - I_J \|_1 = O_p(h^2/T),$

(iii) $\| \Delta_h^{1/2} \hat{\Delta}_h^{-1} \Delta_h^{1/2} - I_J \|_1 = O_p(h/T^{1/2}).$

Proof: First note that

$$\big\| \Delta_h^{1/2} \hat{\Delta}_h^{-1/2} \big\|_1^2 = \big\| \Delta_h^{1/2} \hat{\Delta}_h^{-1} \Delta_h^{1/2} \big\|_1$$
$$\leq \big\| \Delta_h^{1/2} \hat{\Delta}_h^{-1} \Delta_h^{1/2} - I_J \big\|_1 + 1.$$

This shows that the first result follows from the third one, so that it suffices to prove (ii) and (iii). Consider the former, and notice that by Lemma A.2 of S&L it suffices to prove the same result for the corresponding inverse, that is, to prove

$$\big\| \hat{\underline{\Delta}}_h^{-1/2} (\tilde{\underline{\Delta}}_h - \hat{\underline{\Delta}}_h) \hat{\underline{\Delta}}_h^{-1/2} \big\|_1 = O_p(h^2/T). \tag{A.2}$$

In what follows it will be convenient to simplify the notation and denote $\tilde{F}_{h,a} = \partial F_h(\tilde{\alpha}_h)'/\partial a_h$ and $F_{h,a} = \partial F_h(\alpha_h)'/\partial a_h$, and similarly for $\tilde{F}_{h,\sigma}$ and $F_{h,\sigma}$. From the definitions of $\tilde{\underline{\Delta}}_h$ and $\hat{\underline{\Delta}}_h$ it then follows that

$$\tilde{\underline{\Delta}}_h - \hat{\underline{\Delta}}_h = L_h' \tilde{F}_{h,a}' \big[\big(\tilde{\Gamma}_y^h - P_1' \hat{\Gamma}_u^{-1} P_1 \big) \otimes \tilde{\Sigma}_e \big] \tilde{F}_{h,a} L_h.$$

Noting that $P_1 = D_h^{-1} H_h$ and using Lemma A.5 of S&L we can now apply an expansion of the difference $\tilde{\Gamma}_y^h - P_1' \hat{\Gamma}_u^{-1} P_1$. Specifically, if

$$\tilde{L}_h' = L_h' \tilde{F}_{h,a}' \quad \text{and} \quad \tilde{L}_h^{*'} = \tilde{L}_h' P_1',$$

we can write

$$\tilde{\underline{\Delta}}_h - \hat{\underline{\Delta}}_h = \tilde{L}_h^{*'}(A_1 \otimes \tilde{\Sigma}_e)\tilde{L}_h^* + \tilde{L}_h'(A_2 \otimes \tilde{\Sigma}_e)\tilde{L}_h^* + \tilde{L}_h^{*'}(A_2' \otimes \tilde{\Sigma}_e)\tilde{L}_h$$
$$+ \tilde{L}_h'(A_3 \otimes \tilde{\Sigma}_e)\tilde{L}_h$$
$$\stackrel{\text{def}}{=} B_{1h} + B_{2h} + B_{2h}' + B_{3h},$$

where $\|A_1\|_1 = O_p(h/T)$, $\|A_2\|_1 = O_p(h^{1/2}/T)$, and $\|A_3\|_1 = O_p(T^{-1})$. Hence, (A.2) is implied by

$$\left\|\hat{\underline{\Delta}}_h^{-1/2} B_{ih}\hat{\underline{\Delta}}_h^{-1/2}\right\|_1 = O_p(h^2/T), \qquad i = 1, 2, 3. \tag{A.3}$$

In order to show this, we need upper bounds for $\|\hat{\underline{\Delta}}_h^{-1/2}\tilde{L}_h'\|_1$ and $\|\hat{\underline{\Delta}}_h^{-1/2}\tilde{L}_h^{*\prime}\|_1$. For the former we have

$$\left\|\hat{\underline{\Delta}}_h^{-1/2}\tilde{L}_h'\right\|_1^2 = \lambda_{\max}\left(\tilde{L}_h\hat{\underline{\Delta}}_h^{-1}\tilde{L}_h'\right)$$

$$= \lambda_{\max}\left(\tilde{L}_h\left[L_h'\left(\tilde{F}_{h,a}'P_1'\hat{\Gamma}_u^{-1}P_1\tilde{F}_{h,a} + \tilde{F}_{h,\sigma}'\tilde{\Omega}\tilde{F}_{h,\sigma}\right)L_h\right]^{-1}\tilde{L}_h'\right),$$

where the equalities follow from definitions. The largest eigenvalue of the inverse in the last expression is bounded by $\lambda_{\max}((L_h'\tilde{F}_{h,a}'P_1'\hat{\Gamma}_u^{-1}P_1\tilde{F}_{h,a}L_h)^{-1})$. Hence,

$$\left\|\hat{\underline{\Delta}}_h^{-1/2}\tilde{L}_h'\right\|_1^2 \leq \lambda_{\max}\left(\tilde{L}_h\left[\tilde{L}_h'P_1'\hat{\Gamma}_u^{-1}P_1\tilde{L}_h\right]^{-1}\tilde{L}_h'\right)$$

$$= \left\|\left(\tilde{L}_h'P_1'\hat{\Gamma}_u^{-1}P_1\tilde{L}_h\right)^{-1/2}\tilde{L}_h'\right\|_1^2$$

$$\leq 1/\lambda_{\min}\left(P_1'\hat{\Gamma}_u^{-1}P_1\right),$$

where the last relation is obtained from Lemma A.1 of S&L. Now, since $1/\lambda_{\min}(\hat{\Gamma}_u^{-1}) = O_p(1)$ (see Lewis and Reinsel 1985, p. 397) and $1/\lambda_{\min}(P_1'P_1) = \|P_2^{-1}\|_1^2 = O(h^2)$ (see Lemma A.1(i)), we can conclude that

$$\left\|\hat{\underline{\Delta}}_h^{-1/2}\tilde{L}_h'\right\|_1 = O_p(h). \tag{A.4}$$

Replacing $\tilde{L}_h'$ in the above derivation by $\tilde{L}_h^{*\prime} = \tilde{L}_h'P_1'$, it can similarly be shown that

$$\left\|\hat{\underline{\Delta}}_h^{-1/2}\tilde{L}_h^{*\prime}\right\|_1 = O_p(1). \tag{A.5}$$

Now we can readily justify (A.3). Since the proof is similar in each of the three cases, we shall only give details of the second one, which contains all the essential features. Using (A.4) and (A.5) in conjunction with the results $\|A_2\|_1 = O_p(h^{1/2}/T)$ and $\|\tilde{\Sigma}_e\|_1 = O_p(1)$ yields

$$\left\|\hat{\underline{\Delta}}_h^{-1/2} B_{2h}\hat{\underline{\Delta}}_h^{-1/2}\right\|_1 = \left\|\hat{\underline{\Delta}}_h^{-1/2}\tilde{L}_h'(A_2 \otimes \tilde{\Sigma}_e)\tilde{L}_h^*\hat{\underline{\Delta}}_h^{-1/2}\right\|_1$$

$$\leq \left\|\hat{\underline{\Delta}}_h^{-1/2}\tilde{L}_h'\right\|_1\|A_2 \otimes \tilde{\Sigma}_e\|_1\left\|\tilde{L}_h^*\hat{\underline{\Delta}}_h^{-1/2}\right\|_1$$

$$= O_p(h)O_p(h^{1/2}/T)O_p(1)$$

$$= O_p(h^{3/2}/T).$$

In the same way one finds that the l.h.s. of (A.3) is of order $O_p(h/T)$ for $i = 1$ and of order $O_p(h^2/T)$ for $i = 3$. Hence, we have proven (ii).

Now consider the proof of (iii). As above, we may prove the corresponding result for the inverse instead of (iii). Thus, we need to consider the norm

$$
\left\| \Delta_h^{-1/2} (\hat{\underline{\Delta}}_h - \Delta_h) \Delta_h^{-1/2} \right\|_1
$$
$$
\leq \left\| \Delta_h^{-1/2} (\hat{\underline{\Delta}}_h - \overline{\underline{\Delta}}_h) \Delta_h^{-1/2} \right\|_1 + \left\| \Delta_h^{-1/2} (\overline{\underline{\Delta}}_h - \Delta_h) \Delta_h^{-1/2} \right\|_1,
$$
(A.6)

where $\overline{\underline{\Delta}}_h$ is defined with replacing $\tilde{\alpha}_h$ in the definition of $\hat{\underline{\Delta}}_h$ with α_h, that is, $\overline{\underline{\Delta}}_h = L'_h \overline{\Delta}_h L_h$ with

$$
\overline{\Delta}_h = \frac{\partial F_h(\alpha_h)}{\partial \alpha'_h} \begin{bmatrix} P'_1 \hat{\Gamma}_u^{-1} P_1 \otimes \tilde{\Sigma}_e & 0 \\ 0 & \tilde{\Omega} \end{bmatrix} \frac{\partial F_h(\alpha_h)'}{\partial \alpha_h}.
$$

We have to show that the two norms on the r.h.s. of (A.6) are $O_p(h/T^{1/2})$. As for the first one, denote

$$
F'_{h,\alpha} = \partial F_h(\alpha_h)/\partial \alpha'_h,
$$
$$
\hat{Q}_u = \begin{bmatrix} \hat{\Gamma}_u^{-1} \otimes \tilde{\Sigma}_e & 0 \\ 0 & \tilde{\Omega} \end{bmatrix} \quad \text{and} \quad S_1 = \begin{bmatrix} P_1 \otimes I_K & 0 \\ 0 & I_{K(K+1)/2} \end{bmatrix}.
$$

Then

$$
\hat{\underline{\Delta}}_h = L'_h \tilde{F}'_{h,\alpha} S'_1 \hat{Q}_u S_1 \tilde{F}_{h,\alpha} L_h
$$

and

$$
\overline{\underline{\Delta}}_h = L'_h F'_{h,\alpha} S'_1 \hat{Q}_u S_1 F_{h,\alpha} L_h.
$$

Thus, we may write the matrix in the first norm on the r.h.s. of (A.6) as a sum of three matrices and then bound its norm by a sum of three norms. To show that each of these is $O_p(h/T^{1/2})$, it suffices to consider one of them, which is

$$
\left\| \Delta_h^{-1/2} L'_h (\tilde{F}'_{h,\alpha} - F'_{h,\alpha}) S'_1 \hat{Q}_u S_1 F_{h,\alpha} L_h \Delta_h^{-1/2} \right\|_1
$$
$$
\leq \left\| \Delta_h^{-1/2} L'_h (\tilde{F}'_{h,\alpha} - F'_{h,\alpha}) S'_1 \right\|_1 \left\| \hat{Q}_u \right\|_1 \left\| S_1 F_{h,\alpha} L_h \Delta_h^{-1/2} \right\|_1
$$
$$
\overset{\text{def}}{=} \| B_{4h} \|_1 \| \hat{Q}_u \|_1 \| B_{5h} \|_1.
$$

We have to bound the three last norms. It was noticed above that $1/\lambda_{\min}(\hat{\Gamma}_u^{-1}) = O_p(1)$, so that we clearly have $\| \hat{Q}_u \|_1 = O_p(1)$. Now consider

$$
\| B_{4h} \|_1^2 \leq \| (L'_h \Delta_h L_h)^{-1/2} L'_h (\tilde{F}'_{h,a} - F'_{h,a})(P'_2 \otimes I_K) \|_1^2
$$
$$
+ \| (L'_h \Delta_h L_h)^{-1/2} L'_h (\tilde{F}'_{h,\sigma} - F'_{h,\sigma}) \|_1^2
$$
$$
= O_p(h^2/T).
$$

Here the first relation follows from definitions and the simple inequality $\| [A : B] \|_1^2 \leq \| A \|_1^2 + \| B \|_1^2$, and the second one from Lemma A.2(ii).

As for the third norm $\|B_{5h}\|_1$, denote

$$Q_u = \begin{bmatrix} \Gamma_u^{-1} \otimes \Sigma_\varepsilon & 0 \\ 0 & \Omega \end{bmatrix},$$

and recall from the proof of Theorem 2 of S&L that $H_h'\Gamma_{\text{ECM}}^{-1}H_h = P_1'\Gamma_u^{-1}P_1$. Thus, we can write

$$\|B_{5h}\|_1 = \left\|\left(L_h'\Delta_h^{-1/2}L_h\right)^{-1/2}L_h'F_{h,\alpha}'S_1'\right\|_1,$$

where $\Delta_h = F_{h,\alpha}'S_1'Q_uS_1F_{h,\alpha}$. From Lemma A.1 of S&L it therefore follows that $\|B_{5h}\|_1 \le [\lambda_{\min}(Q_u)]^{-1} = O(1)$, since $1/\lambda_{\min}(\Gamma_u^{-1}) = O(1)$. Thus, we can conclude that the first norm on the r.h.s. of (A.6) is $O_p(h/T^{1/2})$.

To complete the proof, we still have to study the latter norm on the r.h.s. of (A.6). Using the notation introduced above, we can write

$$\left\|\Delta_h^{-1/2}(\bar{\Delta}_h - \Delta_h)\Delta_h^{-1/2}\right\|_1 = \left\|\Delta_h^{-1/2}L_h'F_{h,\alpha}'S_1'(\hat{Q}_u - Q_u)S_1F_{h,\alpha}L_h\Delta_h^{-1/2}\right\|_1$$

$$\le \|B_{5h}\|_1^2\|\hat{Q}_u - Q_u\|_1.$$

Above we noticed that $\|B_{5h}\|_1 = O(1)$. Moreover, $\|\hat{Q}_u - Q_u\|_1 = O_p(h/T^{1/2})$ holds as a straightforward consequence of the definitions and the facts $\|\hat{\Gamma}_u^{-1} - \Gamma_u\|_1 = O_p(h/T^{1/2})$ (Saikkonen 1991, Lemma A.4), $\tilde{\Sigma}_e - \Sigma_\varepsilon = O_p(T^{-1/2})$ (Theorem 1), and $\tilde{\Omega} - \Omega = O_p(h/T^{1/2})$ (Assumption (v) of Theorem 3). Thus, we have shown that (A.6) is of order $O_p(h/T^{1/2})$ and thereby completed the proof of Lemma A.3. □

Now we can prove (A.1). First note that

$$|r_{1h}| = \left|\tilde{\gamma}_h'L_h\Delta_h^{-1/2}\Delta_h^{1/2}(\tilde{\Delta}_h^{-1} - \hat{\Delta}_h^{-1})\Delta_h^{1/2}\Delta_h^{-1/2}L_h'\tilde{\gamma}_h\right|$$

$$\le \left\|\Delta_h^{1/2}(\tilde{\Delta}_h^{-1} - \hat{\Delta}_h^{-1})\Delta_h^{1/2}\right\|_1\left\|\Delta_h^{-1/2}L_h'\tilde{\gamma}_h\right\|^2.$$

Moreover,

$$\left\|\Delta_h^{1/2}(\tilde{\Delta}_h^{-1} - \hat{\Delta}_h^{-1})\Delta_h^{1/2}\right\|_1 = \left\|\Delta_h^{1/2}\hat{\Delta}_h^{-1/2}(\hat{\Delta}_h^{1/2}\tilde{\Delta}_h^{-1}\hat{\Delta}_h^{1/2} - I_J)\hat{\Delta}_h^{-1/2}\Delta_h^{1/2}\right\|_1$$

$$\le \left\|\Delta_h^{1/2}\hat{\Delta}_h^{-1/2}\right\|_1^2\left\|\hat{\Delta}_h^{1/2}\tilde{\Delta}_h^{-1}\hat{\Delta}_h^{1/2} - I_J\right\|_1.$$

Hence,

$$|r_{1h}| \le \left\|\Delta_h^{1/2}\hat{\Delta}_h^{-1/2}\right\|_1^2\left\|\hat{\Delta}_h^{1/2}\tilde{\Delta}_h^{-1}\hat{\Delta}_h^{1/2} - I_J\right\|_1\left\|\Delta_h^{-1/2}L_h'\tilde{\gamma}_h\right\|^2. \quad (A.7)$$

For r_{2h} we have

$$|r_{2h}| = \left|\tilde{\gamma}_h'L_h\Delta_h^{-1/2}(\Delta_h^{1/2}\hat{\Delta}_h^{-1}\Delta_h^{1/2} - I_J)\Delta_h^{-1/2}L_h'\tilde{\gamma}_h\right|$$

$$\le \left\|\Delta_h^{1/2}\hat{\Delta}_h^{-1}\Delta_h^{1/2} - I_J\right\|_1\left\|\Delta_h^{-1/2}L_h'\tilde{\gamma}_h\right\|^2. \quad (A.8)$$

From Lemma A.3 one thus obtains

$$|r_{1h}| \leq O_p(h^2/T)\left\|\Delta_h^{-1/2} L_h' \tilde{\gamma}_h\right\|^2$$

and

$$|r_{2h}| \leq O_p(h/T^{1/2})\left\|\Delta_h^{-1/2} L_h' \tilde{\gamma}_h\right\|^2.$$

Thus, since $J = O(h)$ and $h = o(T^{1/3})$, (A.1) follows if

$$\left\|\Delta_h^{-1/2} L_h' \tilde{\gamma}_h\right\| = O_p(J^{1/2}). \tag{A.9}$$

Using the notation $F_{h,\alpha}' = \partial F_h(\alpha_h)/\partial \alpha_h'$ and the definition of $\tilde{\gamma}_h$, we can write

$$\Delta_h^{-1/2} L_h' \tilde{\gamma}_h = N^{1/2} \Delta_h^{-1/2} L_h' [F_h(\tilde{\alpha}_h) - F_h(\alpha_h)]$$
$$= N^{1/2} \Delta_h^{-1/2} L_h' F_{h,\alpha}'(\tilde{\alpha}_h - \alpha_h) + R_{T,h}, \tag{A.10}$$

where the latter equality is obtained as in the proof of Theorem 2 and $\|R_{T,h}\| = o_p(1)$. We have to show that the norm of the first vector in the last expression in (A.10) is $O_p(J^{1/2})$. Define

$$S_2' = \begin{bmatrix} P_2' \otimes I_K & 0 \\ 0 & I_{K(K+1)/2} \end{bmatrix},$$

and note that

$$\left\|N^{1/2}\Delta_h^{-1/2} L_h' F_{h,\alpha}'(\tilde{\alpha}_h - \alpha_h)\right\| \leq N^{1/2}\left\|\Delta_h^{-1/2} L_h' F_{h,\alpha}' S_2'\right\|_1 \left\|S_2'^{-1}(\tilde{\alpha}_h - \alpha_h)\right\|. \tag{A.11}$$

Here we first have

$$\left\|S_2'^{-1}(\tilde{\alpha}_h - \alpha_h)\right\|^2 = \left\|\left(P_2'^{-1} \otimes I_K\right)(\tilde{a}_h - a_h)\right\|^2 + \|\tilde{\sigma} - \sigma\|^2$$
$$= O_p(h/T),$$

where the latter equality follows from Lemma A.1(i) and Theorem 1. Using the definitions of the matrices S_1 and B_{5h}, from the proof of Lemma A.3 we also have

$$\left\|\Delta_h^{-1/2} L_h' F_{h,\alpha}' S_2'\right\|_1 = \left\|\Delta_h^{-1/2} L_h' F_{h,\alpha}' S_1'\right\|_1$$
$$= \|B_{5h}\|_1$$
$$= O_p(1),$$

where the last equality was justified in the proof of Lemma A.3. Combining the above results shows that the r.h.s. of (A.11) is $O_p(h^{1/2})$, so that, since $J = O(h)$, we have established (A.9), and thereby (A.1) as well.

To complete the proof of Theorem 3, denote the first term in the last expression of (A.10) or the one in (A.11) by $\tilde{\delta}_h$. Then, since we have established (A.1),

$$
\frac{\lambda_{\text{Wald}} - J}{\sqrt{2J}} = \frac{(\tilde{\delta}_h + R_{T,h})'(\tilde{\delta}_h + R_{T,h}) - J}{\sqrt{2J}} + o_p(1)
$$

$$
= \frac{\tilde{\delta}_h'\tilde{\delta}_h - J}{\sqrt{2J}} + \frac{2\tilde{\delta}_h' R_{T,h}}{\sqrt{2J}} + \frac{\|R_{T,h}\|^2}{\sqrt{2J}} + o_p(1).
$$

It was shown above that $\|\tilde{\delta}_h\|$ is $O_p(J^{1/2})$. Hence, since it has also been shown that $\|R_{T,h}\| = o_p(1)$, we can conclude that

$$
\frac{\lambda_{\text{Wald}} - J}{\sqrt{2J}} = \frac{\tilde{\delta}_h'\tilde{\delta}_h - J}{\sqrt{2J}} + o_p(1) \xrightarrow{d} N(0, 1),
$$

where the second relation follows from the definition of $\tilde{\delta}_h$ and Theorem 7 of S&L with L_h' redefined as $L_h' \partial F_h(\alpha_h)/\partial \alpha_h'$. This completes the proof of Theorem 3.

A.3   *Proof of Theorem 4*

Let $\phi_h = \text{vec}[\Phi_1 : \cdots : \Phi_h]$, and define $\tilde{\phi}_h$ similarly by using the estimators $\tilde{\Phi}_1, \ldots, \tilde{\Phi}_h$. Furthermore, set $\phi_h^* = \text{vec}[\Phi_1^* : \cdots : \Phi_h^*]$, where $\Phi_1^*, \ldots, \Phi_h^*$ are impulse responses corresponding to autoregressive coefficient matrices $A_1^*, \ldots, A_h^*$, for which we define $a_h^* = \text{vec}[A_1^* : \cdots : A_h^*]$. From (A.4) of Lütkepohl (1988) one obtains

$$
\phi_h^* - \phi_h = \nabla_h(\phi_h)(a_h^* - a_h) + (\nabla_h(\phi_h^*) - \nabla_h(\phi_h))(a_h^* - a_h), \quad \text{(A.12)}
$$

where

$$
\nabla_h(\phi_h) = \begin{bmatrix} \sum_{j=0}^{0} J_h A_h'^{0-j} \otimes \Phi_j \\ \vdots \\ \sum_{j=0}^{h-1} J_h A_h'^{h-1-j} \otimes \Phi_j \end{bmatrix} \quad (hK^2 \times hK^2)
$$

with

$$
A_h = \begin{bmatrix} A_1 & A_2 & \ldots & A_{h-1} & A_h \\ I_K & 0 & \ldots & 0 & 0 \\ 0 & I_K & \ldots & 0 & 0 \\ \vdots & \vdots & \ddots & \vdots & \vdots \\ 0 & 0 & \ldots & I_K & 0 \end{bmatrix} \quad (Kh \times Kh)
$$

and

$$
J_h = [I_K : 0 : \cdots : 0] \quad (K \times Kh).
$$

In (A.12) $\nabla_h(\phi_h^*)$ is defined in the same way as $\nabla_h(\phi_h)$ except that the matrices Φ_j are replaced by Φ_j^*. The matrix A_h thus appears in both $\nabla_h(\phi_h)$ and $\nabla_h(\phi_h^*)$. Since $\nabla_h(\phi)$ is the derivative matrix of the transformation $a_h \to \phi_h$ evaluated at the true parameter vector a_h (see e.g., Lütkepohl 1991, p. 99), equation (A.12) gives the representation required in Assumption 2 with $R_T(\alpha_h^*, \alpha_h) = \nabla_h(\phi_h^*) - \nabla_h(\phi_h)$. Hence, we need to check that the condition imposed on $R_T(\alpha_h^*, \alpha_h)$ is satisfied in the present context.

Since $J_h A_h^{\prime j} = [\Phi_j' : \cdots : \Phi_1' : I_K : 0 : \cdots : 0]$ (cf. Lütkepohl 1988, equation (A.8)), it is easy to see that

$$
\nabla_h(\phi_h) = \begin{bmatrix} I_K \otimes I_K & 0 & \ldots & 0 \\ I_K \otimes \Phi_1 & \ddots & \ddots & \vdots \\ \vdots & & \ddots & 0 \\ I_K \otimes \Phi_{h-1} & \ldots & \ldots & I_K \otimes I_K \end{bmatrix}
$$

$$
\times \begin{bmatrix} I_K \otimes I_K & 0 & \ldots & 0 \\ \Phi_1' \otimes I_K & \ddots & \ddots & \vdots \\ \vdots & & \ddots & 0 \\ \Phi_{h-1}' \otimes I_K & \ldots & \ldots & I_K \otimes I_K \end{bmatrix}. \tag{A.13}
$$

Defining the $h \times h$ matrix

$$
\mathcal{L} = \begin{bmatrix} 0 & 0 & \ldots & 0 & 0 \\ 1 & 0 & \ldots & 0 & 0 \\ 0 & 1 & \ddots & & \vdots \\ \vdots & \vdots & \ddots & \ddots & \vdots \\ 0 & 0 & \ldots & 1 & 0 \end{bmatrix},
$$

the above equation can be written as

$$
\nabla_h(\phi_h) = \left(\sum_{j=0}^{h-1} \mathcal{L}^j \otimes I_K \otimes \Phi_j \right) \left(\sum_{j=0}^{h-1} \mathcal{L}^j \otimes \Phi_j' \otimes I_K \right)
$$

$$
\stackrel{\text{def}}{=} \Phi_1(\mathcal{L})\Phi_2(\mathcal{L}),
$$

where, of course, $\mathcal{L}^0 = I_h$. Furthermore, if we replace Φ_j in the definition of $\Phi_1(\mathcal{L})$ by Φ_j^* and denote the resulting matrix by $\Phi_1^*(\mathcal{L})$, we have

$$
\nabla_h(\phi_h^*) = \Phi_1^*(\mathcal{L})\Phi_2(\mathcal{L}).
$$

Hence, we can write

$$\nabla_h(\phi_h^*) - \nabla_h(\phi_h) = (\Phi_1^*(\mathcal{L}) - \Phi_1(\mathcal{L}))\Phi_2(\mathcal{L}). \tag{A.14}$$

In order to verify the remaining part of Assumption 2, we first prove a series of intermediate results.

Lemma A.4. *The impulse responses* $\Phi_1, \ldots, \Phi_h$ *can be expressed as*

$$\Phi_j = \Lambda_j + \Gamma \sum_{i=0}^{j-1} \Lambda_i, \qquad j = 1, \ldots, h,$$

where $\Lambda_0 = I_K$,

$$\Gamma = \begin{bmatrix} C_1 \\ I_{K_2} \end{bmatrix} \begin{bmatrix} 0 & I_{K_2} \end{bmatrix},$$

and

$$\sum_{j=1}^{h} \|\Lambda_j\| = O(1).$$

Proof: From (3.2) we first obtain the equations

$$A_1 = \Xi_1 \underline{C} + I_K,$$
$$A_j = \Xi_j \underline{C} - \Xi_{j-1} J, \qquad j = 2, \ldots, h$$

(see also S&L, proof of Theorem 2). Now set

$$\Xi_1^{(1)} = \Xi_1 + \begin{bmatrix} I_{K_1} & 0 \\ 0 & 0 \end{bmatrix},$$

$$\Xi_j^{(1)} = \Xi_j, \qquad j = 2, \ldots, h.$$

Then, since $\underline{C}^{-1} J = \Gamma$, it is readily seen that $A_1, \ldots, A_h$ can also be expressed as

$$A_1 = \Xi_1^{(1)} \underline{C} + \Gamma,$$
$$A_j = \Xi_j^{(1)} \underline{C} - \Xi_{j-1}^{(1)} \underline{C} \Gamma, \qquad j = 2, \ldots, h.$$

Comparing these representations of the A_i with those below (2.8) of S&L shows that $\Xi_j^{(1)} \underline{C} = -G_j$ (see (2.3)). Thus we have

$$A_1 = -G_1 + \Gamma,$$
$$A_j = -G_j + G_{j-1} \Gamma, \qquad j = 2, \ldots, h.$$

Let $\mathcal{L}_{h+1}$ be an $(h+1) \times (h+1)$ analog of the matrix $\mathcal{L}$ defined earlier. Then the previous equations can be written in matrix form as

$$[I_K : -A_1 : \cdots : -A_h]$$

$$= [I_K : G_1 : \cdots : G_h] \begin{bmatrix} I_K & -\Gamma & 0 & \cdots & \cdots & 0 \\ 0 & \ddots & \ddots & \ddots & & \vdots \\ \vdots & \ddots & \ddots & \ddots & \ddots & \vdots \\ & & \ddots & \ddots & \ddots & 0 \\ \vdots & & & \ddots & \ddots & -\Gamma \\ 0 & \cdots & 0 & \cdots & 0 & I_K \end{bmatrix},$$

$$\stackrel{\text{def}}{=} [I_K : G_1 : \cdots : G_h] \, \Gamma(\mathcal{L}'_{h+1}),$$

where $\Gamma(\mathcal{L}'_{h+1}) = \mathcal{L}^0_{h+1} \otimes I_K - \mathcal{L}'_{h+1} \otimes \Gamma$. We also define

$$\Phi(\mathcal{L}'_{h+1}) = \sum_{j=0}^{h} \mathcal{L}'^{j}_{h+1} \otimes \Phi_j$$

and

$$A(\mathcal{L}'_{h+1}) = \mathcal{L}^0_{h+1} \otimes I_K - \sum_{j=1}^{h} \mathcal{L}'^{j}_{h+1} \otimes A_j.$$

Then it follows from the definitions that

$$\Phi(\mathcal{L}'_{h+1}) = A(\mathcal{L}'_{h+1})^{-1}. \tag{A.16}$$

If we now further set

$$G(\mathcal{L}'_{h+1}) = \sum_{j=0}^{h} \mathcal{L}'^{j}_{h+1} \otimes G_j$$

we can conclude from (A.15) that $A(\mathcal{L}'_{h+1}) = G(\mathcal{L}'_{h+1})\Gamma(\mathcal{L}'_{h+1})$, so that, by (A.16),

$$\Phi(\mathcal{L}'_{h+1}) = \Gamma(\mathcal{L}'_{h+1})^{-1}G(\mathcal{L}'_{h+1})^{-1}. \tag{A.17}$$

It is not difficult to check that $\Gamma^i = \Gamma$, $i \geq 1$, and

$$\Gamma(\mathcal{L}'_{h+1})^{-1} = \begin{bmatrix} I_K & \Gamma & \cdots & \Gamma \\ 0 & \ddots & \ddots & \vdots \\ \vdots & \ddots & \ddots & \Gamma \\ 0 & \cdots & 0 & I_K \end{bmatrix}.$$

Now, define $\Lambda_1, \ldots, \Lambda_h$ by

$$\sum_{j=0}^{h} \mathcal{L}_{h+1}^{\prime j} \otimes \Lambda_j = \Lambda(\mathcal{L}_{h+1}^{\prime}) = G(\mathcal{L}_{h+1}^{\prime})^{-1},$$

where $\Lambda_0 = I_K$. Then, observing that the first K rows of the matrix $\Phi(\mathcal{L}_{h+1}^{\prime})$ are given by $[I_K : \Phi_1 : \cdots : \Phi_h]$, the required representation of Φ_j follows from (A.17). Since $\Lambda_1, \ldots, \Lambda_h$ can also be obtained from

$$\sum_{j=0}^{h} \Lambda_j z^j = \left(\sum_{j=0}^{h} G_j z^j \right)^{-1},$$

the required absolute summability condition follows from Hannan and Deistler (1988, Theorem 7.4.2).   □

For the next lemma we define the matrix

$$P_3 = \mathcal{L}^0 \otimes I_K \otimes \underline{C} - \mathcal{L} \otimes I_K \otimes J$$

and note that in place of the matrix Δ_h in Assumptions 2 and 3 we now have

$$\Delta_h(\phi) = \Phi_1(\mathcal{L})\Phi_2(\mathcal{L})(P_1^{\prime} \otimes I_K)(\Gamma_u^{-1} \otimes \Sigma_{\varepsilon})(P_1 \otimes I_K)\Phi_2(\mathcal{L})^{\prime}\Phi_1(\mathcal{L})^{\prime}.$$

Lemma A.5.

(i) $\|\Phi_2(\mathcal{L})(P_2^{\prime} \otimes I_K)\|_1 = O(1),$
(ii) $\|P_3\Phi_1(\mathcal{L})\|_1 = O(1),$
(iii) $\|(L_h^{\prime}\Delta_h(\phi)L_h)^{-1/2}L_h^{\prime}P_3^{-1}\|_1 = O(1).$

Proof: From the definition of P_2 we get

$$P_2^{\prime} = \mathcal{L}^0 \otimes \underline{C}^{\prime} - \mathcal{L} \otimes J,$$

where $J = J^{\prime}$ is used. Hence,

$$\Phi_2(\mathcal{L})(P_2^{\prime} \otimes I_K) = \sum_{j=0}^{h-1} \mathcal{L}^j \otimes \Phi_j^{\prime}\underline{C}^{\prime} \otimes I_K - \sum_{j=0}^{h-1} \mathcal{L}^{j+1} \otimes \Phi_j^{\prime}J \otimes I_K$$

$$\stackrel{\text{def}}{=} \sum_{j=0}^{h-1} \mathcal{L}^j \otimes X_j^{\prime} \otimes I_K,$$

where $\mathcal{L}^h = 0$ has been used and X_j is to be determined. Equating powers of $\mathcal{L}$, we find from above that

$$X_0^{\prime} = \Phi_0^{\prime}\underline{C}^{\prime} = \underline{C}^{\prime},$$

$$X_1' = \Phi_1'\underline{C}' - \Phi_0 J = \Phi_1'\underline{C}' - J,$$

$$X_j' = \Phi_j'\underline{C}' - \Phi_{j-1}'J, \qquad j = 2, \ldots, h-1.$$

From the definitions it can be seen directly that $\Gamma'J = J$ and $\Gamma'\underline{C}' = J$. From the above equations and Lemma A.4 it thus follows that

$$X_1' = \Lambda_1'\underline{C}',$$

$$X_j' = \Lambda_j'\underline{C}' + \sum_{i=0}^{j-1}\Lambda_i'J - \Lambda_{j-1}'J - \sum_{i=0}^{j-2}\Lambda_i'J$$

$$= \Lambda_j'\underline{C}', \qquad j = 2, \ldots, h-1.$$

Hence, since we clearly have $\|\mathcal{L}^j\|_1 = 1$ for $j < h$, we can conclude that

$$\|\Phi_2(\mathcal{L})(P_2' \otimes I_K)\|_1 = \left\|\sum_{j=0}^{h-1}\mathcal{L}^j \otimes \Lambda_j'\underline{C}' \otimes I_K\right\|_1$$

$$\le \sum_{j=0}^{h-1}\|\mathcal{L}^j \otimes \Lambda_j'\underline{C}' \otimes I_K\|_1$$

$$\le \text{const} \cdot \sum_{j=0}^{h-1}\|\Lambda_j\|.$$

The last quantity is $O(1)$ by Lemma A.4, which proves (i).

To prove (ii), observe that

$$P_3\Phi_1(\mathcal{L}) = \sum_{j=0}^{h-1}\mathcal{L}^j \otimes I_K \otimes \underline{C}\Phi_j - \sum_{j=0}^{h-1}\mathcal{L}^{j+1} \otimes I_K \otimes J\Phi_j$$

$$= \sum_{j=0}^{h-1}\mathcal{L}^j \otimes I_K \otimes \underline{C}\Lambda_j,$$

where the latter equality follows in the same way as its analog in the case (i), and in the same way as there we can complete the proof of (ii).

As for (iii), write $P_3\Phi_1(\mathcal{L}) = B_1$, $\Phi_2(\mathcal{L})(P_1' \otimes I_K) = B_2$, and $\Phi_2(\mathcal{L})(P_2' \otimes I_K) = B_3$, so that, by the definition of $\Delta_h(\phi)$,

$$\left\|(L_h'\Delta_h(\phi)L_h)^{-1/2}L_h'P_3^{-1}\right\|_1$$

$$= \left\|\left(L_h'P_3^{-1}B_1B_2\left(\Gamma_u^{-1} \otimes \Sigma_\varepsilon\right)B_2'B_1'P_3'^{-1}L_h\right)^{-1/2}L_h'P_3^{-1}\right\|_1$$

$$\le 1\big/\left[\lambda_{\min}\left(B_1B_2\left(\Gamma_u^{-1} \otimes \Sigma_\varepsilon\right)B_2'B_1'\right)\right]^{1/2}$$

$$\leq \text{const} \cdot [\lambda_{\min}(B_1 B_2 B_2' B_1')]^{-1/2}$$

$$= \text{const} \cdot [\lambda_{\min}(B_1 B_3 B_3' B_1')]^{-1/2}$$

$$= \text{const} \cdot \left\| B_3^{-1} B_1^{-1} \right\|_1$$

$$\leq \text{const} \cdot \left\| B_3^{-1} \right\|_1 \left\| B_1^{-1} \right\|_1.$$

Here the second relation follows from Lemma A.1 of S&L, and the third one from the well-known result $\lambda_{\min}(\Gamma_u^{-1} \otimes \Sigma_\varepsilon) \geq c > 0$. The last two relations are immediate consequences of the properties of eigenvalues and the operator norm. The desired result follows from the above, since $\|B_i^{-1}\|_1 = O(1)$, $i = 1, 2$.

To see this, consider the case $i = 1$. For convenience, denote $R_0 = I_K$ and $R_j = -G_j$, $j = 1, \ldots, h$. Then

$$B_1^{-1} = \Phi_1(\mathcal{L})^{-1} P_3^{-1}$$

$$= \left(\sum_{j=0}^{h-1} \mathcal{L}^j \otimes I_K \otimes \underline{C} \Lambda_j \right)^{-1}$$

$$= \sum_{j=0}^{h-1} \mathcal{L}^j \otimes I_K \otimes R_j \underline{C}^{-1}, \tag{A.18}$$

where the third equality follows from the definition of the sequence Λ_j given in the proof of Lemma A.4. Hence, since the sequence R_j is absolutely summable (see (2.4)), an argument similar to that already used in (i) shows that

$$\left\| B_1^{-1} \right\|_1 \leq \text{const} \cdot \sum_{j=0}^{h-1} \|R_j\| = O(1).$$

$\|B_2^{-1}\|_1 = O(1)$ follows analogously.   □

To see that Assumption 2 holds in the case of impulse responses, recall that we now have $R_h^{(a)}(\alpha_h^*, \alpha_h) = \nabla_h(\phi_h^*) - \nabla_h(\phi_h)$ (see (A.14)) and $R_h^{(\sigma)}(\alpha_h^*, \alpha_h) = 0$. Also, as before, set $A = [A_1 : \cdots : A_h]$ and let $A^* = [A_1^* : \cdots : A_h^*]$ be an alternative parameter value which corresponds to ϕ_h^* and for which we can assume that $\|(A^* - A)P_2^{-1}\| \leq \delta h^{1/2}/T^{1/2}$ (see Assumption 2(ii)). We need to obtain a suitable upper bound for

$$\|(L_h' \Delta_h(\phi)L_h)^{-1/2} L_h'[\nabla_h(\phi_h^*) - \nabla_h(\phi_h)](P_2' \otimes I_K)\|_1$$

$$\leq \left\| (L_h' \Delta_h(\phi)L_h)^{-1/2} L_h' P_3^{-1} \right\|_1 \|P_3(\nabla_h(\phi_h^*) - \nabla_h(\phi_h))(P_2' \otimes I_K)\|_1$$

$$\leq \text{const} \cdot \|P_3[\Phi_1^*(\mathcal{L}) - \Phi_1(\mathcal{L})]\Phi_2(\mathcal{L})(P_2' \otimes I_K)\|_1$$

$$\leq \text{const} \cdot \|P_3(\Phi_1^*(\mathcal{L}) - \Phi_1(\mathcal{L}))\|_1. \tag{A.19}$$

Here the first inequality follows from the norm inequality $\|AB\|_1 \leq \|A\|_1\|B\|_1$, the second one from (A.14), and Lemma A.5(iii), and the third from Lemma A.5(i) and, again, the above norm inequality. Thus, we need to bound

$$\|P_3(\Phi_1^*(\mathcal{L}) - \Phi_1(\mathcal{L}))\|_1 = \left\|P_3\Phi_1(\mathcal{L})[\Phi_1^*(\mathcal{L})^{-1} - \Phi_1(\mathcal{L})^{-1}]P_3^{-1}P_3\Phi_1^*(\mathcal{L})\right\|_1$$
$$\leq \mathrm{const} \cdot \left\|(\Phi_1^*(\mathcal{L})^{-1} - \Phi_1(\mathcal{L})^{-1})P_3^{-1}\right\|_1 \|P_3\Phi_1^*(\mathcal{L})\|_1,$$
$$(A.20)$$

where the latter relation makes use of Lemma A.5(ii). Next recall that in (A.18), we obtained a representation for $\Phi_1(\mathcal{L})^{-1}P_3^{-1}$ in terms of the matrices $R_j = -G_j$ ($j = 1, \ldots, h$). Let R_j^* be an analog of R_j ($j = 1, \ldots, h$) obtained from the parameter values $A_1^*, \ldots, A_h^*$. Then the relation between R_j and Ξ_j given in the proof of Lemma A.4 shows that

$$R_j\underline{C}^{-1} - R_j^*\underline{C}^{-1} = \Xi_j - \Xi_j^*,$$

where Ξ_j^* ($j = 1, \ldots, h$) is defined by

$$[\Xi_1^* : \cdots : \Xi_h^*]P_1 = [A_1^* : \cdots : A_h^*] - [I_K : 0 : \cdots : 0]$$

(cf. (3.2)). Hence, it follows from (A.18) and the above that

$$\left\|(\Phi_1(\mathcal{L})^{-1} - \Phi_1^*(\mathcal{L})^{-1})P_3^{-1}\right\|_1 = \left\|\sum_{j=0}^{h-1}\mathcal{L}^j \otimes I_K \otimes (R_j - R_j^*)\underline{C}^{-1}\right\|_1$$
$$\leq \sum_{j=0}^{h-1}\|\mathcal{L}^j\|_1\|I_K\|_1\|\Xi_j - \Xi_j^*\|$$
$$= \sum_{j=0}^{h-1}\|\Xi_j - \Xi_j^*\|$$
$$\leq h^{1/2}\left(\sum_{j=0}^{h-1}\|\Xi_j - \Xi_j^*\|^2\right)^{1/2}$$
$$= h^{1/2}\left\|(A - A^*)P_2^{-1}\right\|$$
$$\leq \delta h/T^{1/2}. \qquad (A.21)$$

Here the second inequality follows from the Cauchy–Schwarz inequality. From (A.19), (A.20), and (A.21) we can now conclude that Assumption 2 holds provided $\|P_3\Phi_1^*(\mathcal{L})\|_1 = O(1)$. This last requirement is, however, an immediate consequence of Lemma A.5(ii), (A.21), and Lemma A.2 of S&L.

In order to show that Assumption 3 holds, recall that we now have $\partial F_h(\alpha_h)/\partial a_h' = \Phi_1(\mathcal{L})\Phi_2(\mathcal{L})$ and $\partial F_h(\alpha_h)/\partial \sigma' = 0$. Thus, assuming that $\|(A^* - A)P_2\| \leq \delta h^{1/2}/T^{1/2}$, we have to find a suitable upper bound for

$$\|(L_h'\Delta_h(\phi)L_h)^{-1/2}L_h'[\Phi_1^*(\mathcal{L})\Phi_2^*(\mathcal{L}) - \Phi_1(\mathcal{L})\Phi_2(\mathcal{L})](P_2' \otimes I_K)\|_1$$
$$\leq \text{const} \cdot \|P_3[\Phi_1^*(\mathcal{L})\Phi_2^*(\mathcal{L}) - \Phi_1(\mathcal{L})\Phi_2(\mathcal{L})](P_2' \otimes I_K)\|_1,$$

where the inequality can be justified in the same way as the first two inequalities in (A.19). The last norm can be bounded by

$$\|P_3[\Phi_1^*(\mathcal{L}) - \Phi_1(\mathcal{L})]\Phi_2(\mathcal{L})(P_2' \otimes I_K)\|_1$$
$$+ \|P_3\Phi_1^*(\mathcal{L})[\Phi_2^*(\mathcal{L}) - \Phi_2(\mathcal{L})](P_2' \otimes I_K)\|_1$$
$$\leq \text{const} \cdot \|P_3(\Phi_1^*(\mathcal{L}) - \Phi_1(\mathcal{L}))\|_1$$
$$+ \|P_3\Phi_1^*(\mathcal{L})\|_1\|[\Phi_2^*(\mathcal{L}) - \Phi_2(\mathcal{L})](P_2' \otimes I_K)\|_1$$
$$\leq \text{const} \cdot h/T^{1/2} + \text{const} \cdot \|(\Phi_2^*(\mathcal{L}) - \Phi(\mathcal{L}))(P_2' \otimes I_K)\|_1,$$

where we have made use first of Lemma A.5(i) one and then of the arguments used to justify Assumption 2 above (see (A.20) and the derivations following it). The last norm is seen to be $O(h/T^{1/2})$ by following the arguments used in (A.20) and (A.21). Details are entirely similar and therefore omitted. Hence, we have established the validity of Assumption 3 as well and thereby completed the proof of Theorem 4.

A.3   *Proof of Theorem 5*

Let $F_h(A_1, \ldots, A_h, \Sigma_\varepsilon) = \text{vec}(\Theta_1 : \cdots : \Theta_h) \overset{\text{def}}{=} \theta_h$, where $\Theta_i = \Phi_i P$ and $\Sigma_\varepsilon = PP'$. Furthermore let A^* and A_i^* be as in the proof of Theorem 4, and σ^* such that $\|\sigma^* - \sigma\| \leq \delta/T^{1/2}$. The notation θ_h^*, P^*, Φ_i^*, etc., is defined in an obvious way. Write

$$\Theta_i^* - \Theta_i = (\Phi_i^* - \Phi_i)P^* + \Phi_i(P^* - P),$$

so that

$$\theta_h^* - \theta_h = \begin{bmatrix} \text{vec}(\Theta_1^* - \Theta_1) \\ \vdots \\ \text{vec}(\Theta_h^* - \Theta_h) \end{bmatrix}$$

$$= (I_h \otimes P^{*'} \otimes I_K)(\phi_h^* - \phi_h) + \begin{bmatrix} I_K \otimes \Phi_1 \\ \vdots \\ I_K \otimes \Phi_h \end{bmatrix} D_K(p^* - p)$$

$$\overset{\text{def}}{=} D(p^*)(\phi_h^* - \phi_h) + B(\phi_h)(p^* - p), \tag{A.22}$$

where $p = \mathrm{vech}(P)$, $p^* = \mathrm{vech}(P^*)$, and D_K is the $K^2 \times K(K+1)/2$ dupli-cation matrix as before. Notice that $\|B(\phi_h)\|_1 \le \|B(\phi_h)\| = O(h^{1/2})$ and that $\|D(p^*)\|_1 = \|P^*\|_1 = O(1)$, and similarly of course $\|D(p)\|_1 = O(1)$.

Let

$$p^* - p = H(\bar{\sigma})(\sigma^* - \sigma)$$

be a conventional mean value expansion of the function $\sigma \to p$ so that, if we write $p = p(\sigma)$, then $H(\sigma) = \partial p(\sigma)/\partial \sigma'$, and the notation $H(\bar{\sigma})$ should be interpreted as explained earlier, that is, a different intermediate value between σ^* and σ may be required in each row of $H(\cdot)$. Using (A.12) and (A.22), we can now write

$$\theta_h^* - \theta_h = D(p)\nabla_h(\phi_h)(a_h^* - a_h) + B(\phi_h)H(\sigma)(\sigma^* - \sigma)$$
$$+ R_h^{(a)}(\alpha_h^*, \alpha_h)(a_h^* - a_h) + R_h^{(\sigma)}(\alpha_h^*, \alpha_h)(\sigma^* - \sigma),$$

where

$$R_h^{(a)}(\alpha_h^*, \alpha_h) = D(p^*)(\nabla_h(\phi_h^*) - \nabla_h(\phi_h)) + (D(p^*) - D(p))\nabla_h(\phi_h) \quad \text{(A.23)}$$

and

$$R_h^{(\sigma)}(\alpha_h^*, \alpha_h) = B(\phi_h)(H(\bar{\sigma}) - H(\sigma)). \quad \text{(A.24)}$$

This is the representation required in Assumption 2(i). (Of course, $R_h^{(a)}(\alpha_h^*, \alpha_h)$ and $R_h^{(\sigma)}(\alpha_h^*, \alpha_h)$ can be expressed in terms of α_h^* and α_h, but this is not necessary and, in fact, not convenient. See also the discussion at the beginning of Section A.3.) It is straightforward to check that in place of the matrix Δ_h in Assumption 2 we now have

$$\Delta_h(\theta) = D(p)\Delta_h(\phi)D(p)' + B(\phi_h)H(\sigma)\Omega H(\sigma)'B(\phi_h)',$$

where $\Delta_h(\phi)$ was defined in the case of ordinary impulse responses (see above Lemma A.5, and recall that $\nabla_h(\phi_h) = \Phi_1(\mathcal{L})\Phi_2(\mathcal{L})$).

To show that Assumption 2(ii) holds, it suffices to consider the two terms on the r.h.s. of (A.23) separately. First we have to obtain a suitable upper bound for

$$\|(L_h'\Delta_h(\theta)L_h)^{-1/2}L_h'D(p^*)[\nabla_h(\phi_h^*) - \nabla_h(\phi_h)](P_2' \otimes I_K)\|_1$$

$$\le \left\|(L_h'\Delta_h(\theta)L_h)^{-1/2}L_h'D(p^*)P_3^{-1}\right\|_1 \|P_3[\nabla_h(\phi_h^*) - \nabla_h(\phi_h)](P_2' \otimes I_K)\|_1$$

$$\le \frac{\delta h}{T^{1/2}}\left\|(L_h'\Delta_h(\theta)L_h)^{-1/2}L_h'D(p^*)P_3^{-1}\right\|_1,$$

where the latter inequality is obtained from (A.19)–(A.21). Thus, we need to show that the last norm above is bounded. To see this, recall that $\|D(p^*)\|_1 =$

$O(1)$, and observe that $D(p^*)P_3^{-1} = P_3^{-1}D(p^*)$. Hence, we have

$$\left\|(L_h'\Delta_h(\theta)L_h)^{-1/2}L_h'D(p^*)P_3^{-1}\right\|_1$$

$$\leq \text{const} \cdot \left\|\left(L_h'P_3^{-1}P_3\Delta_h(\theta)P_3'P_3'^{-1}L_h\right)^{-1/2}L_h'P_3^{-1}\right\|_1$$

$$\leq \text{const} \cdot [\lambda_{\min}(P_3\Delta_h(\theta)P_3')]^{-1/2}, \tag{A.25}$$

where the latter inequality follows from Lemma A.1 of S&L. By the definition of $\Delta_h(\theta)$,

$$\lambda_{\min}(P_3\Delta_h(\theta)P_3') \geq \lambda_{\min}(P_3 D(p)\Delta_h(\phi)D(p)'P_3')$$

$$= \lambda_{\min}(D(p)P_3\Delta_h(\phi)P_3'D(p)')$$

$$\geq \lambda_{\min}(P_3\Delta_h(\phi)P_3')\lambda_{\min}(D(p)D(p)')$$

$$= \lambda_{\min}\left(B_1 B_2\left(\Gamma_u^{-1} \otimes \Sigma_\varepsilon\right)B_2'B_1'\right)\lambda_{\min}(D(p)D(p)').$$

Here the second relation is based on the fact that the matrices $D(p)$ and P_3 commute, while the last expression only expresses the preceeding one by using the notation used in the proof of Lemma A.5(iii). There it was shown that the reciprocal of the first eigenvalue on the r.h.s. is bounded, so that the desired result follows.

Next consider the second term on the r.h.s. of (A.23). We have

$$\|(L_h'\Delta_h(\theta)L_h)^{-1/2}L_h'[D(p^*) - D(p)]\nabla_h(\phi_h)(P_2' \otimes I_K)\|_1$$

$$= \left\|(L_h'\Delta_h(\theta)L_h)^{-1/2}L_h'P_3^{-1}[D(p^*) - D(p)]P_3\nabla_h(\phi_h)(P_2' \otimes I_K)\right\|_1$$

$$\leq \left\|(L_h'\Delta_h(\theta)L_h)^{-1/2}L_h'P_3^{-1}\right\|_1\|D(p^*) - D(p)\|_1$$

$$\|P_3\Phi_1(\mathcal{L})\|_1\|\Phi_2(\mathcal{L})(P_2' \otimes I_K)\|_1$$

$$\leq \text{const} \cdot \|D(p^*) - D(p)\|$$

$$\leq \text{const} \cdot \|p^* - p\|$$

$$\leq \text{const} \cdot \|\sigma^* - \sigma\|$$

$$= O(T^{-1/2}).$$

Here the first relation follows because the matrix $D(p)$ commutes with P_3 and P_3^{-1}, while the second one makes use of the norm inequality and the definition $\nabla_h(\phi_h) = \Phi_1(\mathcal{L})\Phi_2(\mathcal{L})$. The third relation is based on Lemma A.5(i) and (ii) and the upper bound obtained for (A.25). The fourth relation is an immediate consequence of the definitions of $D(p)$ and p, and the fifth one is justified by the mean value expansion of the transformation $\sigma \to p$ and continuity of the partial derivatives. Hence, we can conclude that the first part of Assumption 2(ii) (inequality (3.5)) has been verified.

To verify inequality (3.6), we have to consider the r.h.s. of (A.24) and bound

$$\|(L_h'\Delta_h(\theta)L_h)^{-1/2}L_h'B(\phi_h)[H(\bar{\sigma}) - H(\sigma)]\|_1$$

$$\leq \left\|(L_h'\Delta_h(\theta)L_h)^{-1/2}L_h'P_3^{-1}\right\|_1\|P_3\|_1\|B(\phi_h)\|_1\|H(\bar{\sigma}) - H(\sigma)\|_1$$

$$\leq \text{const} \cdot h^{1/2} \| H(\bar{\sigma}) - H(\sigma) \|_1$$

$$\leq \text{const} \cdot h^{1/2} \| \sigma^* - \sigma \|$$

$$= O(h^{1/2}/T^{1/2}),$$

which is even of a smaller order of magnitude than required in (3.6). Above, the first inequality is obvious, while the second one is based on the boundedness result obtained for (A.25) and the facts $\| P_3 \|_1 = O(1)$ and $\| B(\phi_h) \|_1 = O(h^{1/2})$. The third inequality follows because the function $H(\sigma)$ is continuously differentiable. Thus we have verified Assumption 2 for orthogonalized impulse responses.

To verify Assumption 3, we first have to consider

$$\| (L_h' \Delta_h(\theta) L_h)^{-1/2} L_h' [D(p^*) \Phi_1^*(\mathcal{L}) \Phi_2^*(\mathcal{L}) - D(p) \Phi_1(\mathcal{L}) \Phi_2(\mathcal{L})] (P_2' \otimes I_K) \|_1$$

$$\leq \text{const} \cdot \| P_3 [D(p^*) \Phi_1^*(\mathcal{L}) \Phi_2^*(\mathcal{L}) - D(p) \Phi_1(\mathcal{L}) \Phi_2(\mathcal{L})] (P_2' \otimes I_K) \|_1,$$

where the inequality is again based on the upper bound obtained for (A.25). Since the matrices P_3 and $D(p^*)$ commute the last norm can be bounded by

$$\| (D(p^*) - D(p)) P_3 \Phi_1(\mathcal{L}) \Phi_2(\mathcal{L}) (P_2' \otimes I_K) \|_1$$

$$+ \| D(p^*) P_3 [\Phi_1^*(\mathcal{L}) \Phi_2^*(\mathcal{L}) - \Phi_1(\mathcal{L}) \Phi_2(\mathcal{L})] (P_2' \otimes I_K) \|_1$$

$$\leq \text{const} \cdot \| D(p^*) - D(p) \|$$

$$+ \text{const} \cdot \| P_3 [\Phi_1^*(\mathcal{L}) \Phi_2^*(\mathcal{L}) - \Phi_1(\mathcal{L}) \Phi_2(\mathcal{L})] (P_2' \otimes I_K) \|_1$$

$$= O(T^{-1/2}) + O(h/T^{1/2}),$$

as required by inequality (3.13). Here the inequality makes use of Lemma A.5(i) and (ii), and the equality is based on the arguments used to justify Assumption 2(ii) above and Assumption 3 in the case of ordinary impulse responses. Hence, we have verified the first part of Assumption 3.

The verification of the second one requires that we bound

$$\| (L_h' \Delta_h(\theta) L_h)^{-1/2} L_h' [B(\phi_h^*) H(\sigma^*) - B(\phi_h) H(\sigma)] \|_1$$

$$\leq \text{const} \cdot \| P_3 [B(\phi_h^*) H(\sigma^*) - B(\phi_h) H(\sigma)] \|_1,$$

where the inequality once again makes use of the upper bound obtained for (A.25). Hence, we need to bound the last norm, which is at most

$$\| P_3 [B(\phi_h^*) - B(\phi_h)] H(\sigma^*) \|_1 + \| P_3 B(\phi_h) [H(\sigma^*) - H(\sigma)] \|_1$$

$$\leq \text{const} \cdot \| P_3 (B(\phi_h^*) - B(\phi_h)) \|_1 + \text{const} \cdot O(h^{1/2}) \| H(\sigma^*) - H(\sigma) \|_1$$

$$\leq \text{const} \cdot \| P_3 (B(\phi_h^*) - B(\phi_h)) \|_1 + O(h^{1/2}/T^{1/2}),$$

as $\| \sigma^* - \sigma \| = O(T^{-1/2})$. Here we have used the results $P_3 = O(1)$ and $\| B(\phi_h) \|_1 = O(h^{1/2})$ and also the fact that $H(\sigma)$ is continuosly differentiable. Thus, we need to bound the last operator norm above. From the definition of

P_3 and $B(\phi_h)$ it follows by calculations similar to those in the proof of Lemma A.5(i) that

$$
P_3 B(\phi_h) =
\begin{bmatrix}
I_K \otimes (\underline{C}\Lambda_1 + J) \\
I_K \otimes \underline{C}\Lambda_2 \\
\vdots \\
I_K \otimes \underline{C}\Lambda_h
\end{bmatrix}
D_K .
$$

This implies that

$$
\| P_3 (B(\phi_h^*) - B(\phi_h)) \|_1 \leq \text{const} \cdot \left\| \sum_{j=0}^{h} \mathcal{L}_{h+1}^j \otimes I_K \otimes \underline{C}\Lambda_j^* \right.
$$

$$
\left. - \sum_{j=0}^{h} \mathcal{L}_{h+1}^j \otimes I_K \otimes \underline{C}\Lambda_j \right\|_1
$$

where the $(h+1) \times (h+1)$ matrix $\mathcal{L}_{h+1}$ is defined as in the proof of Lemma A.4. The last norm is seen to be $O(h/T^{1/2})$, as $\|(P_2'^{-1} \otimes I_K)(a_h^* - a_h)\| = O(h^{1/2}/T^{1/2})$, by observing that the same result holds for the corresponding inverses (Lemma A.2 of S&L) which with h replaced by $h-1$ was obtained in (A.21). The applicability of Lemma A.2 of S&L in the present situation follows because $\|(\sum_{j=0}^{h} \mathcal{L}_{h+1}^j \otimes I_K \otimes \underline{C}\Lambda_j)^{-1}\|_1 = O(1)$ by arguments analogous to those following (A.18). Hence we have verified Assumption 3 as well, and the proof of Theorem 5 is thereby complete.

REFERENCES

Berk, K. N. (1974), "Consistent Autoregressive Spectral Estimates", *The Annals of Statistics* 2: 489–502.

Bhansali, R. J. (1978), "Linear Prediction by Autoregressive Model Fitting in the Time Domain", *The Annals of Statistics* 6: 224–31.

Dolado, J. J., and H. Lütkepohl (1996), "Making Wald Tests Work for Cointegrated Systems", *Econometric Reviews* 15: 369–86.

Hannan, E. J., and M. Deistler (1988), *The Statistical Theory of Linear Systems*. New York: Wiley.

Johansen, S. (1991), "Estimation and Hypothesis Testing of Cointegrated Vectors in Gaussian Vector Autoregressive Models", *Econometrica* 59: 1551–80.

Lewis, R., and G. C. Reinsel (1985), "Prediction of Multivariate Time Series by Autoregressive Model Fitting", *Journal of Multivariate Analysis* 16: 393–411.

Lütkepohl, H. (1988), "Asymptotic Distribution of the Moving Average Coefficients of an Estimated Vector Autoregressive Process", *Econometric Theory* 4: 77–85.

(1991), *Introduction to Multiple Times Series Analysis*. Berlin: Springer-Verlag.

(1996), "Testing for Nonzero Impulse Responses in Vector Autoregressive Processes", *Journal of Statistical Planning and Inference* 50: 1–20.

Lütkepohl, H., and D. S. Poskitt (1991), "Estimating Orthogonal Impulse Responses via Vector Autoregressive Models", *Econometric Theory* 7: 487–96.

(1996), "Testing for Causation Using Infinite Order Vector Autoregressive Processes", *Econometric Theory* 12: 61–87.

Lütkepohl, H., and P. Saikkonen (1997), "Impulse Response Analysis in Infinite Order Cointegrated Vector Autoregressive Processes", *Journal of Econometrics*, 81: 127–57.

Magnus, J. R., and H. Neudecker (1988), *Matrix Differential Calculus with Applications in Statistics and Econometrics*. Chichester: Wiley.

Mellander, E., A. Vredin, and A. Warne (1992), "Stochastic Trends and Economic Fluctuations in a Small Open Economy", *Journal of Applied Econometrics* 7: 369–94.

Park, J. Y., and P. C. B. Phillips (1988), "Statistical Inference in Regressions with Integrated Processes: Part 1," *Econometric Theory* 4: 468–97.

(1989), "Statistical Inference in Regressions with Integrated Processes: Part 2", *Econometric Theory* 5: 95–131.

Phillips, P. C. B. (1995), "Fully Modified Least Squares and Vector Autoregression", *Econometrica* 63: 1023–78.

Saikkonen, P. (1991), "Asymptotically Efficient Estimation of Cointegrating Regressions", *Econometric Theory* 7: 1–21.

(1992), "Estimating and Testing of Cointegrated Systems by an Autoregressive Approximation", *Econometric Theory* 8: 1–27.

Saikkonen, P., and H. Lütkepohl (1996), "Infinite Order Cointegrated Vector Autoregressive Processes: Estimation and Inference", *Econometric Theory* 12: 814–44.

Stock, J. H. (1987), "Asymptotic Properties of Least Squares Estimators of Cointegrating Vectors", *Econometrica* 55: 1035–56.

Toda, H. Y., and P. C. B. Phillips (1993), "Vector Autoregressions and Causality", *Econometrica* 61: 1367–93.

(1994), "Vector Autoregression and Causality: A Theoretical Overview and Simulation Study", *Econometric Reviews* 13: 259–85.

Toda, H. Y., and T. Yamamoto (1995), "Statistical Inference in Vector Autoregressions with Possibly Integrated Processes", *Journal of Econometrics* 66: 225–50.

CHAPTER 8

Nonlinear error-correction models for interest rates in the Netherlands

Dick van Dijk & Philip Hans Franses

1   Introduction

Many economic variables, while nonstationary individually, are linked by long-run equilibrium relationships. The concept of cointegration, introduced by Granger (1981) and Engle and Granger (1987), together with the corresponding error-correction models, allows these two characteristics to be modelled simultaneously. In the "standard" error-correction model, adjustment toward the long-run equilibrium is linear, i.e., it is always present and of the same strength under all circumstances. There are, however, economic situations for which the validity of this assumption might be questioned. Recently, several attempts have been made to construct empirical econometric models which allow for nonlinear adjustment; see Anderson (1997), Balke and Fomby (1997), Dwyer, Locke, and Yu (1996), Hansen and Kim (1996), and Kunst (1992, 1995). It appears that relevant forms of nonlinear error correction often concern some sort of asymmetry, i.e., distinction is to be made between adjustment of positive and negative or between adjustment of large and small deviations from equilibrium. Both types of asymmetry arise in a rather natural way when applying cointegration techniques to modelling prices of so-called equivalent assets in financial markets; see Yadav, Pope, and Paudyal (1994) and Anderson (1997) for elaborate discussions. Equivalent assets in a certain sense represent the same value; examples include stock and futures, and bonds of different maturity. Since they are traded in the same market, or in markets which are linked by arbitrage-related forces, the prices of equivalent assets should be such that investors are indifferent between holding either one of them. If prices deviate from

We would like to thank Heather Anderson, Herman Bierens, Robert Kunst, and an anonymous referee for useful comments and suggestions.

equilibrium, arbitrage opportunities are created which will result in the prices being driven back together again. However, market frictions can give rise to asymmetric adjustment of such deviations. Due to short-selling restrictions, for example, the response to negative deviations from equilibrium will be different from the response to positive deviations. Alternatively, transaction costs prevent adjustment of equilibrium errors as long as the benefits from adjustment, which equal the price difference, are smaller than those costs. These market frictions suggest that the degree of error correction is a function of the sign and/or size of the deviation from equilibrium. In practice, short-selling restrictions and transaction costs are not the same for all market participants. Because of this heterogeneity among traders, it might be expected that the adjustment behaviour does not suddenly become different at sharply determined boundaries, but rather that the aggregate force on the prices to return to equilibrium changes gradually.

The purpose of this chapter is twofold. First, we document that both types of asymmetric adjustment discussed above can be modelled by means of smooth-transition error-correction models (STECMs). Second, and more important, we aim to review the practical issues involved in the empirical specification of STECMs and to provide useful guidelines for practitioners. These practical issues concern (i) cointegration, i.e., whether we can establish, using linear methods, whether there is a linear long-run relation while there is nonlinear adjustment, (ii) nonlinearity, i.e., whether we can determine which form of nonlinear adjustment is appropriate, (iii) outliers, i.e., whether we can prevent our results from being due to only a few influential data points, and (iv) aggregation, i.e., what is the effect of aggregation on finding nonlinear adjustment. We address these issues using an example of a monthly bivariate interest rate series in the Netherlands, of which we also have weekly observed data. We have chosen to follow this particular sequence in order to clearly demonstrate the effects of outliers and aggregation. Our empirical results can be viewed as substantiating the arguments in Granger and Lee (1999).

The outline of our chapter is as follows. Section 2 introduces the general idea of smooth-transition error correction by discussing a simple model, which subsequently will be used in simulation experiments. That section also contains an outline of an empirical specification procedure for STECMs. Section 3 focuses on the first step in this specification procedure by presenting some Monte Carlo evidence on the performance of standard linearity-based tests for cointegration, when applied in the presence of nonlinear error correction. Section 4 reviews the results of Lagrange multiplier (LM) tests for nonlinear error correction when applied to the interest-rate series, and it also presents estimates of an STECM for these series. Section 5 deals with the issues of outliers and sampling frequency. Finally, Section 6 contains some recommendations for practitioners.

2   Smooth-transition error correction

The concept of smooth-transition error correction can conveniently be introduced by considering the following system for a bivariate time series $\{(y_t, x_t)'$, $t = 1, \ldots, T\}$:

$$y_t + \beta x_t = z_t, \qquad z_t = (\rho_1 + \rho_2 F(z_{t-d}))z_{t-1} + \varepsilon_t, \qquad (1)$$

$$y_t + \alpha x_t = w_t, \qquad w_t = w_{t-1} + \eta_t, \qquad (2)$$

where the so-called transition function $F(z_{t-d})$ is continuous and bounded between 0 and 1, $d \in \{1, 2, \ldots\}$, $\alpha \neq \beta$, and

$$\begin{pmatrix} \varepsilon_t \\ \eta_t \end{pmatrix} \sim \text{i.i.d.}(0, \Sigma), \qquad \Sigma = \begin{pmatrix} 1 & \theta\sigma \\ \theta\sigma & \sigma^2 \end{pmatrix}. \qquad (3)$$

The standard linear setup, which is used by, *inter alia*, Banerjee, Dolado, Hendry, and Smith (1986) and Engle and Granger (1987), is obtained by taking $F(z_{t-d})$ equal to zero and imposing the restriction $|\rho_1| < 1$. The series y_t and x_t then are cointegrated with cointegrating vector $(1, \beta)'$. Put differently, the series y_t and x_t are linked by the (long-run) relationship $y_t = -\beta x_t$, and z_t represents the deviation from this "equilibrium". In the general system (1)–(3), we assume that z_t follows a smooth-transition autoregressive (STAR) model; see Granger and Teräsvirta (1993) and Teräsvirta (1994) for detailed discussions of this class of nonlinear time series models. For y_t and x_t still to be cointegrated in this case, z_t has to be stationary. This implies that, depending on the specific form of the function $F(z_{t-d})$, certain restrictions have to be put on ρ_1 and ρ_2. For example, $|\rho_1| < 1$ and $|\rho_1 + \rho_2| < 1$ are sufficient conditions for z_t to be stationary for all possible choices of $F(z_{t-d})$ which are bounded between zero and one.

It is useful to rewrite the system (1)–(2) in error-correction format as

$$\Delta y_t = \frac{\alpha(\rho_1 + \rho_2 F(z_{t-d}) - 1)}{\alpha - \beta} z_{t-1} + \xi_{1t}, \qquad (4)$$

$$\Delta x_t = \frac{-(\rho_1 + \rho_2 F(z_{t-d}) - 1)}{\alpha - \beta} z_{t-1} + \xi_{2t}, \qquad (5)$$

where ξ_{it}, $i = 1, 2$, are linear combinations of ε_t and η_t. From (4)–(5), the meaning of the term smooth-transition error correction is obvious. For example, in the equation for Δy_t, the strength of error correction changes smoothly from $\alpha(\rho_1 - 1)/(\alpha - \beta)$ to $\alpha(\rho_1 + \rho_2 - 1)/(\alpha - \beta)$ as $F(z_{t-d})$ changes from 0 to 1.

The function $F(z_{t-d})$ can be used to obtain many different kinds of nonlinear error-correcting behaviour. As argued in the introduction, in empirical applications (in particular those involving financial variables), one might be especially interested in modelling asymmetric adjustment. Asymmetric effects of positive and negative deviations can be obtained by taking $F(z_{t-d})$ to be the logistic function,

$$F(z_{t-d}) \equiv F(z_{t-d}; \gamma, c) = (1 + \exp\{-\gamma(z_{t-d} - c)\})^{-1}, \qquad \gamma > 0.$$

(6)

In the resulting logistic STECM (LSTECM), the strength of reversion of z_t to its attractor zero changes monotonically from ρ_1 to $\rho_1 + \rho_2$ for increasing values of z_{t-d}. The logistic model therefore allows for different effects of positive and negative deviations (relative to the threshold c) from the equilibrium. The parameter γ determines the speed of the transition; the higher γ, the faster the change from ρ_1 to $\rho_1 + \rho_2$. If $\gamma \to 0$, the LSTECM becomes linear, while if $\gamma \to \infty$, the logistic function approaches a Heaviside function, taking the value 0 for $z_{t-d} < c$ and 1 for $z_{t-d} > c$.

The second type of asymmetry, which distinguishes between small and large equilibrium errors, is obtained when $F(z_{t-d})$ is taken to be the exponential function,

$$F(z_{t-d}; \gamma, c) = 1 - \exp\{-\gamma(z_{t-d} - c)^2\}, \qquad \gamma > 0,$$

(7)

which results in gradually changing strength of adjustment for larger (both positive and negative) deviations from equilibrium. In the resulting STECM, the strength of mean reversion changes from $\rho_1 + \rho_2$ to ρ_1 and back again with increasing z_{t-d}, and this change is symmetric around c. Although the exponential function (7) is commonly used in applications of STAR models (see, among others, Teräsvirta and Anderson 1992), a possible drawback of this choice for the transition function is that if either $\gamma \to 0$ or $\gamma \to \infty$, the model becomes linear. This can be avoided by using the *quadratic logistic* function,

$$F(z_{t-d}; \gamma, c_1, c_2) = (1 + -\exp\{-\gamma(z_{t-d} - c_1)(z_{t-d} - c_2)\})^{-1},$$

$$\gamma > 0, \quad c_1 \le c_2,$$

(8)

as proposed by Jansen and Teräsvirta (1996). In this case, if $\gamma \to 0$, the model becomes linear, while if $\gamma \to \infty$, the function $F(\cdot)$ is equal to 1 for $z_{t-d} < c_1$ and $z_{t-d} > c_2$, and equal to 0 in between. Hence, the quadratic LSTECM (QLSTECM) includes the three-regime threshold error-correction model (TECM) of Balke and Fomby (1997) as a limiting case. For our example to be discussed below, the specification in (8) seems the most relevant. Note that for finite γ, the minimum value taken by the function (8), which is attained for $z_{t-d} = (c_1 + c_2)/2$, is not equal to zero. This has to be kept in mind when interpreting estimates from models with this particular transition function.

It is fairly straightforward to extend the specification strategy for STAR models of Teräsvirta (1994) to the error-correction case considered here. Empirical specification of an STECM then involves the following steps: (i) testing for cointegration and estimating the cointegrating relationship, (ii) testing for nonlinearity of the adjustment process and investigating the type of nonlinearity, and (iii) estimating and evaluating the STECM. Each of these steps is addressed in turn in the following sections.

3   Testing for cointegration

In this section we address the first step involved in specifying an STECM, i.e., testing for cointegration and estimating the cointegrating relationship. Escribano and Mira (1996) show that the cointegrating vector(s) can still be estimated superconsistently in the presence of neglected nonlinearity in the adjustment process; see also Corradi, Swanson, and White (1995). In this section we aim to substantiate these theoretical results using Monte Carlo experiments. Additionally, we also examine the finite-sample properties of linearity-based tests for cointegration. These simulations complement and extend the Monte Carlo results in Pippenger and Goering (1993) and Balke and Fomby (1997). Both studies only consider the case of threshold error correction, and furthermore, Pippenger and Goering (1993) only consider situations in which the cointegrating relationship can be assumed to be known, something which may not always be possible in practice.

We only consider those cointegration tests that are most popular among practitioners: the residual-based test suggested by Engle and Granger (1987) and the likelihood-ratio test introduced by Johansen (1988). To save space, we discuss the tests only for (bivariate) cases where no deterministic regressors are included in the model.

The residual-based test for cointegration is performed via the two-step procedure of Engle and Granger (1987). That is, we first estimate the cointegrating regression

$$y_t = -\beta x_t + u_t \tag{9}$$

by ordinary least squares (OLS), and second test for the presence of a unit root in the regression residuals $\hat{u}_t = y_t + \hat{\beta} x_t$. The latter is done by using the augmented Dickey–Fuller (ADF) test of Dickey and Fuller (1979), which is the familiar t-ratio of ρ in the auxiliary AR(p) regression

$$\Delta \hat{u}_t = \rho \hat{u}_{t-1} + \sum_{i=1}^{p-1} \phi_i \, \Delta \hat{u}_{t-i} + \eta_t. \tag{10}$$

The ADF statistic requires the choice of an appropriate value for p in (10). The number of lagged differences included should be such that the residuals $\hat{\eta}_t$ obtained from (10) resemble white noise. In our Monte Carlo experiments, we follow Gregory (1994) in initially setting p fairly large (here equal to 6) and then reducing this number until the last lag included is significant at the 5% level, using normal critical values.

The likelihood-ratio tests developed by Johansen (1988) are derived from the vector ECM (VECM) for the 2×1 vector time series $\{X_t = (y_t, x_t)', \; t = 1 - p, \ldots, T\}$,

$$\Delta X_t = \Pi X_{t-1} + \sum_{i=1}^{p-1} \Gamma_i \Delta X_{t-i} + \varepsilon_t, \qquad t = 1, \ldots T, \tag{11}$$

where $\varepsilon_t \sim \text{NID}(0, \Sigma)$. If y_t and x_t are cointegrated, the matrix Π has rank 1 and can be decomposed as $\Pi = \alpha \beta'$ for 2×1 vectors α and β. Johansen (1988) advocates test for cointegration by testing the rank r of Π. This can be done by applying likelihood ratio (LR) tests to test the significance of the squared partial canonical correlations between ΔX_t and X_{t-1}, denoted $\hat{\lambda}_1$ and $\hat{\lambda}_2$, which can be obtained by solving a generalized eigenvalue problem. Ordering them so that $\hat{\lambda}_1 > \hat{\lambda}_2$, the trace statistics can be used to test $H_0 : r = r_0$ against the alternative hypothesis $H_1 : r \geq r_0 + 1$ for $r_0 = 0, 1$, and is given by

$$\text{LR}_{\text{trace}} = -T \sum_{i=r_0+1}^{2} \ln(1 - \hat{\lambda}_i). \tag{12}$$

The asymptotic distribution of the trace statistic is nonstandard and depends on the number of zero canonical correlations; see Johansen (1988, 1991). If the trace test points toward cointegration between y_t and x_t, an estimate of the cointegrating vector β is given by the eigenvector corresponding to the largest canonical correlation $\hat{\lambda}_1$. In our Monte Carlo experiments, the VAR order p in (11) is determined by minimizing the Schwarz criterion BIC.

3.1   Monte Carlo experiments

The tests for cointegration discussed above assume that the adjustment process driving the variables toward the equilibrium is linear. In this section we investigate the empirical rejection frequencies of the tests, as well as the corresponding estimates of the cointegrating vector, when the series of interest are characterized by smooth-transition error correction. For convenience, we call the rejection frequency of the cointegration tests its *power*.

Monte Carlo design

The generalized bivariate system (1)–(3) is used as DGP for the artificial time series y_t and x_t. Both types of asymmetric error correction which have been discussed before are investigated, by using (6) and (8) as transition functions with $d = 1$, i.e., z_{t-1} is used as transition variable. In the Monte Carlo experiments, we investigate the effects of the autoregressive parameters in the STAR model for z_t, ρ_1, and ρ_2, on the power of the tests for cointegration and the estimates of the cointegrating parameter β.

In case (6) is used as the transition function (which will be referred to as case I), the threshold c is fixed at zero, so that adjustment is different for positive and negative equilibrium errors, while the parameter γ, which determines the speed of the transition, is set equal to 0.5 and 5. Finally, ρ_1 and ρ_2 are chosen such that ρ_1 and $\rho_1 + \rho_2$, which are the effective autoregressive parameters for $F(z_{t-1}) = 0$ and $F(z_{t-1}) = 1$, respectively, vary between 0.4 and 1.

In case (8) is taken as the transition function (case II), γ is set equal to 0.1 and 1. For $\gamma = 1$, the transition is already almost instantaneous at the thresholds, so it is not very useful to consider larger values for this parameter. The thresholds c_1 and c_2 are varied between 0 and 6, with $c_1 = -c_2 \equiv c$. For larger value of c_1 and c_2, cointegration hinges upon only a very small fraction of the data. The function $F(z_{t-1})$ is rescaled by applying the transformation $F^*(z_{t-1}) = (F(z_{t-1}) - F(0))/(1 - F(0))$. The function $F^*(z_{t-1})$ attains a minimum value of 0 at $z_{t-1} = 0$ and approaches 1 for large negative and positive values of z_{t-1}. By using $F^*(z_{t-1})$ instead of $F(z_{t-1})$ in (1) it is assured that ρ_1 and $\rho_1 + \rho_2$ indicate the strength of mean reversion in z_t for $z_{t-1} = 0$ and for large absolute values of z_{t-1}, respectively, for both selected values of γ. In all experiments ρ_1 is fixed at 1, while ρ_2 is varied between -0.2 and -0.8. Hence, no error correction is present at $z_{t-1} = 0$ (and for $\gamma = 1$, almost no error correction occurs for $-c < z_{t-1} < c$), and adjustment is stronger for larger deviations from equilibrium.

The remaining parameters in the model are fixed at the following values for both cases I and II: $\beta = -1$, $\alpha = -2$, $\sigma^2 = 1$, $\theta = 0$. For each experiment we generate 2500 series of $T = 100$ or 250 observations. The starting values for both y_t and x_t are set equal to zero, and the first 100 observations are discarded. All calculations are performed using GAUSS.

Strictly speaking, power comparisons of the various tests are possible only when size-adjusted critical values are used. The power calculations presented in this section are made using asymptotic critical values, since this corresponds more closely to empirical practice. The asymptotic critical values for the ADF test are taken from Phillips and Ouliaris (1990), Table IIa, while Table 0 in Osterwald-Lenum (1992) provides critical values for the Johansen trace test.

Results of Monte Carlo experiments
The results for case I are set out in Tables 1 and 2. Table 1 shows the rejection frequencies of the ADF and trace tests. Note that the cells corresponding to $\rho_1 = 1.0$ and $\rho_1 + \rho_2 = 1.0$ contain the size of the tests, and the cells corresponding to $\rho_1 = \rho_1 + \rho_2$ contain the power of the tests in the case of linear error correction. The remaining cells contain estimates of the power in the case of asymmetric error correction.

Dick van Dijk & Philip Hans Franses

Table 1. *Size and power of cointegration tests, case I*[a]

T	γ	ρ_1	Test	$\rho_1 + \rho_2 =$	Rejection frequency (%)			
					1.0	0.8	0.6	0.4
100	0.5	1.0	ADF		6.6	19.2	39.8	62.8
			LR_{trace}		4.4	11.1	27.6	52.6
		0.8	ADF		17.4	80.3	93.4	95.9
			LR_{trace}		10.8	60.5	88.4	94.0
		0.6	ADF		39.8	93.8	96.2	97.2
			LR_{trace}		28.5	89.4	94.8	94.7
		0.4	ADF		62.7	95.6	96.9	98.0
			LR_{trace}		51.2	94.1	94.6	94.6
	5.0	1.0	ADF		5.7	10.2	10.3	10.6
			LR_{trace}		3.2	6.3	7.0	7.2
		0.8	ADF		10.1	79.4	90.0	89.8
			LR_{trace}		5.4	60.0	81.8	87.7
		0.6	ADF		10.2	90.0	96.3	97.0
			LR_{trace}		5.8	83.5	93.4	93.6
		0.4	ADF		10.0	90.5	97.0	98.0
			LR_{trace}		6.0	88.2	93.7	93.8
250	0.5	1.0	ADF		4.6	27.7	55.2	78.2
			LR_{trace}		4.6	21.1	48.4	74.0
		0.8	ADF		29.7	99.8	100.0	100.0
			LR_{trace}		22.2	94.4	94.4	94.6
		0.6	ADF		56.5	100.0	100.0	100.0
			LR_{trace}		49.2	94.4	94.6	94.5
		0.4	ADF		77.8	100.0	100.0	100.0
			LR_{trace}		74.1	94.4	94.5	94.5
	5.0	1.0	ADF		4.6	10.3	10.0	9.9
			LR_{trace}		4.6	7.2	7.5	7.0
		0.8	ADF		11.4	99.8	99.9	99.8
			LR_{trace}		7.7	94.4	94.5	94.5
		0.6	ADF		11.2	100.0	100.0	100.0
			LR_{trace}		7.8	94.4	94.6	94.6
		0.4	ADF		10.8	99.8	100.0	100.0
			LR_{trace}		7.7	94.5	94.5	94.5

[a] Rejection frequencies at 5% significance level using asymptotic critical values for series generated by (1)–(3) and (6) with $\beta = -1, \alpha = -2, \sigma^2 = 1$, and $\theta = 0$. The table is based on 2500 replications. Critical values are taken from Phillips and Ouliaris (1990, Table IIa) for the ADF test, and from Osterwald-Lenum (1992, Table 0) for the LR_{trace} tests.

Table 2. *Mean and standard deviation of $\hat{\beta} - \beta$, case I* [a]

				Mean (standard deviation)			
T	γ	ρ_1	Estimate	$\rho_1 + \rho_2 =$ 1.0	0.8	0.6	0.4
100	0.5	1.0	OLS	0.505(0.607)	0.135(0.380)	0.073(0.163)	0.050(0.118)
			VECM	0.553(1.593)	−0.021(0.805)	−0.012(1.179)	0.393(14.49)
		0.8	OLS	0.126(0.302)	0.049(0.084)	0.043(0.071)	0.037(0.062)
			VECM	0.023(0.754)	−0.078(0.332)	−0.026(0.185)	−0.035(0.868)
		0.6	OLS	0.070(0.177)	0.043(0.069)	0.036(0.056)	0.032(0.050)
			VECM	−0.060(0.434)	−0.213(8.906)	−0.010(0.090)	−0.019(0.358)
		0.4	OLS	0.048(0.120)	0.037(0.062)	0.031(0.050)	0.029(0.044)
			VECM	−0.060(0.289)	0.006(0.506)	−0.012(0.046)	−0.014(0.052)
250	0.5	1.0	OLS	0.519(0.475)	0.094(0.240)	0.058(0.133)	0.043(0.104)
			VECM	−0.294(6.776)	−0.065(0.754)	−0.028(0.653)	−0.417(17.51)
		0.8	OLS	0.099(0.228)	0.035(0.049)	0.026(0.038)	0.022(0.034)
			VECM	−0.028(0.380)	−0.013(0.074)	−0.017(0.040)	−0.020(0.030)
		0.6	OLS	0.055(0.135)	0.027(0.039)	0.021(0.029)	0.018(0.025)
			VECM	−0.072(1.386)	−0.018(0.035)	−0.021(0.026)	−0.022(0.018)
		0.4	OLS	0.040(0.086)	0.022(0.034)	0.018(0.025)	0.016(0.022)
			VECM	−0.029(0.329)	−0.020(0.031)	−0.022(0.017)	−0.022(0.014)

[a] Mean (standard deviation) of $\hat{\beta} - \beta$ for series generated by (1)–(3) and (6) with $\beta = -1$, $\alpha = -2$, $\sigma^2 = 1$, and $\theta = 0$. OLS and VECM refer to the estimates obtained from the cointegrating regression (9) and the vector error-correction model (11) respectively. The entries for the respective estimators are based on those replications for which the ADF and LR_{trace} statistic reject the null of no cointegration.

From Table 1, it is seen that the power of both tests is almost unaffected by the nonlinearity of the error-correction process. The only exception to this general observation is the case where either ρ_1 or $\rho_1 + \rho_2$ is equal to 1, i.e., in case there is no correction at all of either negative or positive errors, respectively.

Table 2 shows the means and standard deviations of the bias in the estimates of the cointegrating parameter β, obtained from the static regression (9) (OLS) and the Johansen procedure (VECM). Only the results for $\gamma = 0.5$ are shown; the results for $\gamma = 5.0$ are very similar. It should be noted that all entries are only based on those replications for which the respective test procedures detect cointegration at the 5% significance level.

It is seen that on average, the cointegrating parameter is over- and underestimated by the static regression and maximum-likelihood procedure, respectively. This, however, can simply be a consequence of the choice of the DGP. Other conclusions which emerge from Table 2 are that the mean of the bias from the static regression is larger, but the variance is smaller. Both the mean and the variance of the bias decrease as the strength of mean reversion of the equilibrium error becomes stronger, i.e., for increasing values of ρ_1 and $\rho_1 + \rho_2$.

The results for case II are shown in Tables 3 and 4.

It appears that in this case the simple ADF test is more powerful than the trace statistic, although the difference in power is not very large. Increasing values of c imply that the strength of error correction increases more slowly as z_{t-1} gets larger (in absolute value). It is seen that the power of the tests decreases accordingly. More negative values of ρ_2 imply that the strength of attraction of z_t to zero becomes larger for given values of z_{t-1}. It might be expected that in this case the power of the tests increases, which is confirmed by Table 3. Finally, increasing γ, while keeping c and ρ_2 fixed, has two opposing effects. On the one hand, for large values of γ the strength of error correction is virtually zero as long as $z_{t-1} \in (-c, c)$, while for small γ, error correction becomes active as soon as there is a deviation from equilibrium. This effect might be expected to decrease the power of the cointegration tests as γ increases. On the other hand, for larger values of γ, the transition to the maximum strength of attraction is much quicker, which might be expected to increase the power of the tests. The simulation results seem to suggest that the second effect dominates, since for a very large majority of combinations of c and ρ_2 the power of the tests is higher for $\gamma = 1.0$.

Table 4 displays the means and standard deviations of the bias in the estimates of the cointegrating parameter β for case II. From this table, roughly the same conclusions can be drawn as for case I. The main difference is that for $T = 100$ and $\gamma = 0.1$, the mean of the bias from the static regression now is smaller (in absolute value) than the mean bias from the VECM.

In general, we observe that the bias in estimating the cointegrating rank and the cointegrating vector is not larger for asymmetric and nonlinear adjustment than for linear adjustment. These findings serve to substantiate some of the theoretical results in Escribano and Mira (1996) and Corradi, Swanson, and White (1995).

3.2   *Data analysis*

In this subsection we examine the cointegration properties of our bivariate sample series. It is generally accepted that interest rates can be characterized as nonstationary processes or, to be more precise, processes which are integrated of order 1 ($I(1)$). Hall, Anderson, and Granger (1992) argue that many theories of the term structure of interest rates imply that n interest rates of different maturity are cointegrated with cointegrating rank $n - 1$, with the differences between the interest rates, or *spreads*, being the stationary linear combinations. If interest rates are such that the spread deviates from its equilibrium value, arbitrage opportunities are created, and these will drive the interest rates back toward

Table 3. *Power of cointegration tests, case II*[a]

					Rejection frequency (%)			
T	γ	ρ_2	Test	$c =$	0	2	4	6
100	0.1	−0.2	ADF		27.8	25.2	17.5	11.6
			LR_{trace}		13.9	13.0	9.6	6.2
		−0.4	ADF		59.2	52.3	31.0	14.0
			LR_{trace}		32.7	28.2	16.2	8.3
		−0.6	ADF		79.7	73.9	46.2	16.6
			LR_{trace}		57.3	49.4	24.3	9.7
		−0.8	ADF		88.0	84.4	60.1	19.3
			LR_{trace}		78.8	70.2	34.8	11.1
	1.0	−0.2	ADF		74.2	59.2	17.1	11.1
			LR_{trace}		52.1	33.3	10.1	8.1
		−0.4	ADF		95.2	89.4	27.7	13.6
			LR_{trace}		94.3	83.8	16.5	9.5
		−0.6	ADF		96.8	92.9	43.1	17.6
			LR_{trace}		94.7	94.3	26.6	11.3
		−0.8	ADF		97.5	95.1	63.5	21.8
			LR_{trace}		94.6	94.8	44.1	14.3
250	0.1	−0.2	ADF		93.3	91.3	78.2	32.6
			LR_{trace}		81.8	76.4	51.4	18.2
		−0.4	ADF		98.6	98.1	93.6	58.5
			LR_{trace}		94.2	94.2	88.8	33.4
		−0.6	ADF		99.6	99.5	97.1	74.4
			LR_{trace}		94.3	94.4	94.1	49.4
		−0.8	ADF		99.9	99.7	98.8	83.5
			LR_{trace}		94.4	94.3	94.3	65.2
	1.0	−0.2	ADF		99.8	98.9	78.3	23.4
			LR_{trace}		94.4	94.3	56.0	15.0
		−0.4	ADF		100.0	99.9	88.8	40.0
			LR_{trace}		94.5	94.4	89.8	25.4
		−0.6	ADF		100.0	100.0	95.7	65.1
			LR_{trace}		94.6	94.4	94.4	48.8
		−0.8	ADF		100.0	100.0	98.4	82.7
			LR_{trace}		94.5	94.5	94.5	73.8

[a] Rejection frequencies at 5% significance level using asymptotic critical values for series generated by (1)–(3) and (8) with $\beta = -1, \alpha = -2$, $\sigma^2 = 1, \theta = 0$, and $\rho_1 = 1$. The function $F(z_{t-1})$ is rescaled by applying the transformation $F^*(z_{t-1}) = (F(z_{t-1}) - F(0))/(1 - F(0))$. The table is based on 2500 replications. Critical values are taken from Phillips and Ouliaris (1990, Table IIa) for the ADF test and from Osterwald-Lenum (1992, Table 0) for the LR_{trace} tests.

Table 4. *Mean and standard deviation of $\hat{\beta} - \beta$, case II* [a]

				Mean (standard deviation)			
T	γ	ρ_2	$c =$	0	2	4	6
100	0.1	−0.2	OLS	0.067(0.147)	0.072(0.156)	0.066(0.187)	0.086(0.281)
			VECM	−0.161(0.430)	−0.158(0.446)	−0.110(0.413)	−0.125(0.655)
		−0.4	OLS	0.056(0.106)	0.057(0.112)	0.067(0.143)	0.085(0.223)
			VECM	−0.133(0.394)	−0.120(0.365)	−0.110(0.508)	−0.172(0.463)
		−0.6	OLS	0.052(0.090)	0.053(0.094)	0.059(0.120)	0.074(0.201)
			VECM	−0.102(0.333)	−0.101(0.345)	−0.152(0.616)	−0.163(0.456)
		−0.8	OLS	0.050(0.082)	0.051(0.086)	0.058(0.108)	0.066(0.183)
			VECM	−0.053(0.271)	−0.075(0.452)	−0.116(0.475)	−0.116(0.476)
	1.0	−0.2	OLS	0.051(0.089)	0.056(0.105)	0.066(0.319)	0.050(0.118)
			VECM	−0.094(0.330)	−0.127(0.463)	−0.156(0.537)	−0.105(0.523)
		−0.4	OLS	0.040(0.063)	0.049(0.078)	0.073(0.272)	0.037(0.062)
			VECM	−0.012(0.193)	−0.046(0.469)	−0.140(0.417)	−0.119(0.456)
		−0.6	OLS	0.033(0.051)	0.043(0.068)	0.075(0.243)	0.032(0.050)
			VECM	−0.012(0.051)	−0.001(0.260)	−0.102(0.531)	−0.096(0.447)
		−0.8	OLS	0.029(0.044)	0.040(0.061)	0.078(0.207)	0.029(0.044)
			VECM	−0.014(0.036)	−0.006(0.065)	−0.111(0.616)	−0.050(0.673)
250	0.1	−0.2	OLS	0.057(0.079)	0.058(0.082)	0.063(0.096)	0.071(0.146)
			VECM	−0.063(1.220)	−0.037(0.761)	−0.100(0.479)	−0.167(0.699)
		−0.4	OLS	0.043(0.061)	0.046(0.064)	0.054(0.076)	0.063(0.111)
			VECM	−0.006(0.060)	−0.003(0.067)	−0.025(0.204)	−0.134(0.410)
		−0.6	OLS	0.036(0.051)	0.038(0.054)	0.047(0.067)	0.064(0.100)
			VECM	−0.011(0.047)	−0.010(0.051)	0.003(0.154)	−0.107(0.401)
		−0.8	OLS	0.0.32(0.045)	0.034(0.047)	0.043(0.061)	0.063(0.093)
			VECM	−0.012(0.158)	−0.014(0.040)	−0.006(0.062)	−0.086(0.386)
	1.0	−0.2	OLS	0.037(0.052)	0.044(0.061)	0.065(0.099)	0.077(0.177)
			VECM	−0.010(0.057)	−0.007(0.088)	−0.090(0.780)	−0.071(0.458)
		−0.4	OLS	0.024(0.033)	0.031(0.043)	0.057(0.086)	0.074(0.142)
			VECM	−0.026(0.292)	−0.016(0.042)	−0.020(0.219)	−0.091(0.814)
		−0.6	OLS	0.019(0.026)	0.026(0.035)	0.053(0.076)	0.075(0.118)
			VECM	−0.021(0.018)	−0.022(0.160)	−0.010(0.523)	−0.102(0.376)
		−0.8	OLS	0.017(0.022)	0.023(0.031)	0.051(0.068)	0.078(0.109)
			VECM	−0.022(0.014)	−0.020(0.022)	−0.001(0.060)	−0.064(0.710)

[a] Mean (standard deviation) of $\hat{\beta} - \beta$ for series generated by (1)–(3) and (8) with $\beta = -1$, $\alpha = -2$, $\sigma^2 = 1$, $\theta = 0$, and $\rho_1 = 1$. The function $F(z_{t-1})$ is rescaled by applying the transformation $F^*(z_{t-1}) = (F(z_{t-1}) - F(0))/(1 - F(0))$. OLS and VECM refer to the estimates obtained from the cointegrating regression (9) and the vector error-correction model (11) respectively. The entries for the respective estimators are based on those replications for which the ADF and LR_{trace} statistic reject the null of no cointegration.

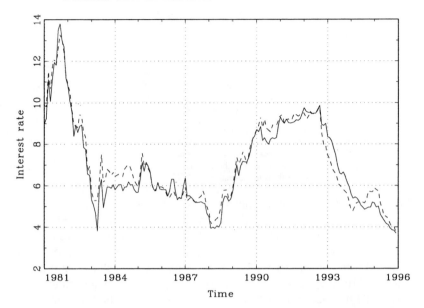

Figure 1. Monthly Dutch short- and long-term interest rates, January 1981–December 1995. Solid and dashed lines represent one-month and twelve-month interbank rates, $R_{1,t}$ and $R_{12,t}$, respectively.

equilibrium. Anderson (1997) reasons that, due to market imperfections such as transaction costs, asymmetric error correction of large and small deviations may play an important role.

To investigate the empirical usefulness of STECMs, we consider a bivariate interest-rate series for the Netherlands, consisting of one- and twelve-month interbank rates. We denote these as $R_{1,t}$ and $R_{12,t}$, respectively. Initially, we follow the practice of other authors who have used (linear) ECMs to model the relationship between interest rates of different maturity[1] and consider these series at a monthly frequency. The sample runs from January 1981 until December 1995, giving 180 observations in total. The series are graphed in Figure 1.

Table 5 shows the results from applying univariate ADF tests to the interest rates and the spread S_t, which is defined as $R_{12,t} - R_{1,t}$. The tests clearly indicate that the series are individually $I(1)$, while the spread seems to be stationary at the 5% significance level.

For a check on whether it is appropriate to use the spread as cointegrating relationship, i.e., to require the cointegrating vector to be equal to $(1, -1)$, a regression from $R_{1,t}$ on $R_{12,t}$ (including a constant as well) renders an estimate

---

[1] See Engle and Granger (1987), Campbell and Shiller (1987, 1991), Stock and Watson (1988), and Hall, Anderson, and Granger (1992), among others.

Table 5. *ADF statistics for interest rates*[a]

	$R_{1,t}$	$R_{12,t}$	S_t	5% crit. value
Level	−2.15	−2.06	−3.00	−2.88
First difference	−5.65	−8.35	—	−1.94

[a] ADF tests applied to monthly interest rates and spread. Test statistics for levels are $\hat{\tau}_\mu$ while those for the first differences are $\hat{\tau}$. Number of lagged differences in each regression was chosen such that the last lag included is significant at 5% level, using normal critical values.

of 0.999, which seems close enough to unity. The Johansen trace test is also computed, including one lagged difference of the series in the VECM (a VAR order of 2 for the levels is indicated by the Schwarz criterion), and no constants. The trace tests for testing $r = 0$ and $r = 1$ are equal to 15.30 and 1.88. When compared with the appropriate 5% critical values, we conclude that the tests point toward cointegration. The estimate of the normalized cointegrating vector is $(1, -1.029)$. Since the estimated standard error of the second element equals 0.018, we cannot reject the hypothesis that it is equal to unity. Hence, in the remainder of the analysis we assume that the spread amounts to a stationary linear combination of our two interest rates. Furthermore, the estimates of the parameters in the linear VECM (not shown here) reveal that the error-correction variable S_{t-1} is not significant in the equation for the twelve-month interest rate. For this reason, we focus on the equation for the short-term rate $R_{1,t}$ by conditioning on $R_{12,t}$.

The fitted linear conditional error-correction model (CECM) for $R_{1,t}$ is

$$\Delta R_{1,t} = \underset{(0.02)}{-0.02} + \underset{(0.04)}{0.13}\, S_{t-1} + \underset{(0.04)}{0.93}\, \Delta R_{12,t} - \underset{(0.07)}{0.16}\, \Delta R_{1,t-1}$$
$$+ \underset{(0.08)}{0.09}\, \Delta R_{12,t-1}, \tag{13}$$

$$\hat{\sigma}_\varepsilon = 0.236, \qquad DW = 2.00, \qquad SK = 0.00, \qquad EK = 4.23,$$
$$JB = 132.92(0.00), \qquad ARCH(1) = 16.53(0.00),$$
$$ARCH(4) = 19.03(0.00), \qquad LB(8) = 7.84(0.45),$$
$$LB(12) = 19.42(0.08), \qquad BIC = -2.767,$$

where standard errors are given in parentheses below the parameter estimates, $\hat{\sigma}_\varepsilon$ is the residual standard deviation, DW the Durbin–Watson statistic, SK the skewness, EK the excess kurtosis, JB the Jarque–Bera test of normality of the residuals, ARCH the LM test of no autoregressive conditional heteroskedasticity, LB the Ljung–Box test of no autocorrelation, and BIC the Schwarz criterion. The figures in parentheses following the test statistics are *p*-values.

This linear model seems quite satisfactory, with reasonable values for all coefficients. Due to the large kurtosis, normality of the residuals is strongly rejected. Closer inspection of the residuals reveals that this may be entirely caused by only three observations in the beginning of the sample, for which the residuals are very large (in absolute value). These aberrant observations may also cause the ARCH tests to reject homoskedasticity. On the other hand, it may also be that these significant test values are caused by neglected nonlinearity. In the next section, we focus on a nonlinear extension of (13).

4   Testing for smooth-transition error correction

Once the presence of an equilibrium relationship has been established, the next question is whether possible nonlinearity in the adjustment process can be detected. The objective of testing for nonlinearity is threefold. First, we want to obtain an impression of whether the error-correction process is indeed nonlinear. Second, we need to determine the appropriate transition variable, i.e., obtain an estimate of the lag d. Third, we want to obtain an idea of the most suitable form of nonlinearity in the error-correction mechanism, i.e., we want to select between the kind of asymmetry implied by (6) on the one hand and (7) or (8) on the other. In our empirical example, we confine our analysis to selecting between the logistic function (6) and the quadratic logistic function (8). The latter choice is motivated by economic arguments.

The LM-type tests developed by Luukkonen, Saikkonen, and Teräsvirta (1988) for general smooth-transition nonlinearity can be readily applied to test for smooth-transition error correction; see also Swanson (1999). The only thing to bear in mind is that in this case the transition variable is a stationary linear combination of nonstationary variables. In order to keep this chapter self-contained, we briefly discuss the derivation of the LM-type tests below.

Consider a general CECM for y_t,

$$\Delta y_t = \pi_1' w_t + F(z_{t-d}; \gamma, c)\pi_2' w_t + \eta_t, \qquad (14)$$

where $w_t = (1, \tilde{w}_t)'$, $\tilde{w}_t = (z_{t-1}, \Delta y_{t-1}, \ldots, \Delta y_{t-p+1}, \Delta x_t, \ldots, \Delta x_{t-p+1})'$, $z_t = y_t + \beta x_t$, and $\pi_i = (\pi_{i0}, \pi_{i1}, \ldots, \pi_{im})'$ for $i = 1, 2$, $m = 2p - 1$. The noise process $\{\eta_t\}$ is assumed to be normally distributed with mean zero and variance σ_η^2. Compared to (4), constants and lagged first differences of y_t and x_t have been added to allow for more general dynamic structures.

The null hypothesis of linear error correction in (14) with (6) or (8) can be formulated as $H_0 : \gamma = 0$. It is immediately seen that under the null hypothesis the model is not identified and hence the usual asymptotic theory cannot be applied to derive LM tests; see Davies (1977, 1987) for a general discussion of such identification problems. Luukkonen Saikkonen, and Teräsvirta (1988) suggest solving this by replacing the transition function $F(z_{t-d}; \gamma, c)$ in (14) by a suitable approximation around $\gamma = 0$. In the reparameterized model, the identification problem is no longer present and linearity can easily be tested.

A general test against smooth-transition error correction emerges when $F(z_{t-d})$ is replaced by a third-order Taylor approximation. Rearranging terms yields the reparameterized model,

$$\Delta y_t = \phi' w_t + \phi_1' \tilde{w}_t z_{t-d} + \phi_2' \tilde{w}_t z_{t-d}^2 + \phi_3' \tilde{w}_t z_{t-d}^3 + \eta_t. \tag{15}$$

It should be noted that when $d > p$, $\tilde{w}_t$ should be replaced by w_t because z_{t-d} is not present as a (implicit) regressor in w_t. The original null hypothesis of linearity, $H_0 : \gamma = 0$, is easily shown to be equivalent to the hypothesis that all coefficients of the auxiliary regressors $\tilde{w}_t z_{t-d}^j$, $j = 1, 2, 3$, are zero, i.e., $H_0' : \phi_1 = \phi_2 = \phi_3 = 0$. The LM-type test for this null hypothesis can be carried out in a few steps:

(1) Estimate the parameters of the model under the null hypothesis by regressing Δy_t on w_t, with z_t replaced by $\hat{z}_t = y_t + \hat{\beta} x_t$, where $\hat{\beta}$ is obtained from preliminary cointegration analysis. The value of p necessary for the construction of w_t can be taken from the linear model. Compute the sum of squared residuals $SSR_0 = \sum \hat{\eta}_t^2$, where $\hat{\eta}_t = \Delta y_t - \hat{\pi}_1' w_t$.

(2) Estimate the parameters ϕ and ϕ_j, $j = 1, 2, 3$, from the auxiliary regression

$$\hat{\eta}_t = \phi' w_t + \phi_1' \tilde{w}_t z_{t-d} + \phi_2' \tilde{w}_t z_{t-d}^2 + \phi_3' \tilde{w}_t z_{t-d}^3 + v_t, \tag{16}$$

and compute the sum of squared residuals $SSR_1 = \sum \hat{v}_t^2$.

(3) The LM-type test statistic can now be computed as

$$LM_0 = T(SSR_0 - SSR_1)/SSR_0. \tag{17}$$

The test statistic has an asymptotic χ^2 distribution with $3m$ degrees of freedom, where it is assumed that prior estimation of β does not affect the asymptotic distribution. In small samples it usually is recommended to use an F-version of the test, i.e.,

$$LM_0 = \frac{(SSR_0 - SSR_1)/(3m)}{SSR_1/(T - 4m)}, \tag{18}$$

which is approximately F-distributed with $3m$ and $T - 4m$ degrees of freedom under the null hypothesis of linearity.

To decide upon the most appropriate lag of z_t to use as transition variable, the test should be carried out for a number of different values of d, say $d = 1, \ldots, D$. If linearity is rejected for several values of d, the one with the smallest p-value is selected as the transition variable. This rule is motivated by the notion

Table 6. *Standard and outlier robust LM-type tests for smooth-transition error correction in a CECM for monthly data on the one-month interest rate*[a]

Test	Null	$d =$	p-value					
			1	2	3	4	5	6
Standard	H_0		0.002	0.039	0.968	0.880	0.485	0.721
	H_{03}		0.075	0.988	0.828	0.327	0.313	0.669
	H_{02}		0.093	0.406	0.829	0.995	0.543	0.214
	H_{01}		0.004	0.001	0.804	0.757	0.478	0.956
Robust	H_0		0.151	0.124	0.936	0.523	0.700	0.889
	H_{03}		0.390	0.976	0.555	0.188	0.452	0.578
	H_{02}		0.724	0.545	0.903	0.599	0.452	0.576
	H_{01}		0.027	0.006	0.839	0.738	0.820	0.957

[a] p-values for LM-type tests for smooth-transition error correction in one-month Dutch interest rate. The upper panel gives p-values for standard tests, the lower panel for LM-type tests which are robust to additive outliers. The null hypotheses are given in the text.

that the test might be expected to have maximum power if the true transition variable is used; see Granger and Teräsvirta (1993).

Deciding between the transition functions (6) and (8) can be done by a short sequence of tests nested within H_0. This testing sequence is motivated by the observation that if a logistic alternative is appropriate, the second-order derivative in the Taylor expansion is zero. Hence, when $\phi_2 = 0$, the model can only be a logistic model. The null hypotheses to be tested are as follows:

$$H_{03} : \phi_3 = 0,$$
$$H_{02} : \phi_2 = 0 \mid \phi_3 = 0, \tag{19}$$
$$H_{01} : \phi_1 = 0 \mid \phi_3 = \phi_2 = 0.$$

Granger and Teräsvirta (1993) suggest carrying out all three tests, independent of rejection or acceptance of the first or second test, and using the outcomes to select the appropriate transition function. The decision rule is to select the quadratic logistic function (8) only if the p-value corresponding to H_{02} is the smallest, and select the logistic function (6) in all other cases. There is, however, no guarantee that this sequence will give the right answer. For practical purposes it therefore seems useful to estimate models with both transition functions and to base a decision between the two on other criteria.

We compute the LM-type test statistics for the various null hypotheses for the one-month Dutch interest rate in the estimated CECM (13). We set d equal to 1 through 6. The first panel of Table 6 shows the p-values of the standard

LM-type tests. From the results for H_0, it is seen that linearity is rejected for both $d = 1$ and $d = 2$. Based upon the p-values, we select $d = 1$ as the appropriate transition variable. The p-values of the test sequence for testing H_{03}, H_{02}, and H_{01} indicate that the logistic function (6) is the proper choice of transition function, if we adopt the decision rule of Granger and Teräsvirta (1993). This statistical finding seems to contradict the economic reasoning of Anderson (1997), which suggests that nonlinearity in the adjustment process would arise mainly because of transaction costs. As these ought to have similar effects for positive and for negative deviations from equilibrium, the quadratic logistic function should be more suitable. Given that the p-value corresponding to the test for H_{02} is not very much larger than the p-values for the other two tests in the sequence, we choose to adopt (8) as the transition function. Notice that this illustrates the potential difficulty of comparing p-values in practice.

We estimate the parameters of our STECM by nonlinear least squares (NLS). We follow the suggestions of Teräsvirta (1994) and standardize the exponent of $F(S_{t-1})$ by dividing it by the variance of the transition variable, $\sigma_{S_{t-1}}^2 = 0.229$, so that γ is a scale-free parameter. The estimation results are

$$\Delta R_{1,t} = \underset{(0.03)}{-0.03} + \underset{(0.07)}{0.12}\, S_{t-1} + \underset{(0.04)}{0.90}\, \Delta R_{12,t} - \underset{(0.08)}{0.11}\, \Delta R_{1,t-1}$$

$$+ \underset{(0.09)}{0.11}\, \Delta R_{12,t-1} + (\underset{(0.12)}{0.32} + \underset{(0.24)}{0.42}\, S_{t-1} + \underset{(0.19)}{0.15}\, \Delta R_{12,t}$$

$$+ \underset{(0.19)}{0.25}\, \Delta R_{1,t-1} - \underset{(0.34)}{0.73}\, \Delta R_{12,t-1})\Big(1 + \exp\Big[\underset{(5.67)}{-3.74}$$

$$\times\, (S_{t-1} + \underset{(0.08)}{0.40})(S_{t-1} - \underset{(0.12)}{1.25})/\sigma_{S_{t-1}}^2\Big]\Big)^{-1} + \varepsilon_t, \qquad (20)$$

$$\hat{\sigma}_\varepsilon = 0.225, \qquad \mathrm{DW} = 1.89, \qquad \mathrm{SK} = -0.14, \qquad \mathrm{EK} = 4.88,$$

$$\mathrm{JB} = 177.25(0.00), \qquad \mathrm{ARCH}(1) = 0.01(0.92),$$

$$\mathrm{ARCH}(4) = 3.48(0.48), \qquad \mathrm{BIC} = -2.672.$$

The large standard error of the estimate of γ is due to the fact that a wide range of values of this parameter renders about the same transition function. Accurate estimation of γ then requires a large number of observations close to c_1 and c_2; see Teräsvirta (1994) for discussion. The estimate of γ is such that the transition from $F(S_{t-1}) = 0$ to $F(S_{t-1}) = 1$ is almost instantaneous at the thresholds -0.40 and 1.25. The estimates of the coefficients of the error-correction term S_{t-1} are such that adjustment is stronger if the series is in the upper or lower regime, i.e., if the spread lagged one period is larger (in absolute value). Note that for the lower regime ($S_{t-1} < -0.40$) this is counteracted considerably by the change in the intercept. In fact, S_{t-1} needs to be smaller than -0.81, approximately, for the first effect to dominate.

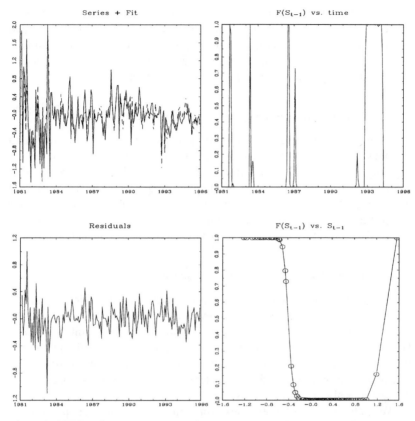

Figure 2. STECM for monthly observations on the Dutch short-term interest rate, January 1981–December 1995. The parameters of this model are given in (20).

Also notice that the ARCH test statistics have become insignificant, i.e., the previous evidence of ARCH in the linear model may have been due to neglected nonlinearity.

Figure 2 shows some graphs which serve to illustrate the estimated smooth-transition model. From the residual plot in the lower left panel it appears that the model still fails to capture some of the large interest-rate movements in the beginning of the sample. The upper right panel shows how the transition function evolves over time. It is seen that the nonlinearity mainly serves to explain the behaviour in 1993–1994, when the shape of the term structure was inverted, i.e., the short-term rate exceeded the long-term rate. Apart from this period, a few observations in the beginning of the 1980s are picked up by the nonlinear function, when the spread was more than 1.25%. From the graph in the lower right panel, it is seen that there are in fact only two observations in the regime $S_{t-1} > 1.25$. In the next section we examine whether these two observations might be regarded as outliers, or whether the monthly sampling frequency does not lead to sufficient observations in the different regimes, and hence that aggregation has resulted in "less nonlinearity".

5   Nonlinearity, outliers, and sampling frequency

In this section we investigate whether our findings in the previous section based on monthly data may be caused by only a few observations. We do this by applying tests for nonlinearity which are robust to additive outliers. We also address the potential importance of sampling frequency or aggregation level of the series. For this purpose, we investigate model (20) for weekly data. Our unreported estimates of the cointegrating vector after outlier removal and for weekly data show that $(1, -1)$ remains to be the most useful cointegrating vector.

5.1   *Testing for nonlinearity in the presence of outliers*

Using theoretical derivations and extensive Monte Carlo simulations, van Dijk, Franses, and Lucas (1999) show that evidence for nonlinearity based on the LM-type tests discussed above can be due to only a few additive outliers. For practical purposes it is important to investigate this possibility in order to prevent the empirical specification process from being governed by only a few data points. As an example, the parameter c_2 in (8) appears to be quantified on the basis of only two observations.

We apply the robust LM-type tests for nonlinearity, as they are proposed in van Dijk, Franses, and Lucas (1999), to our monthly data set, and we report the p-values of the test statistics in the second panel of Table 6. The robust test involves the same steps as the standard test outlined in the previous section. The difference is that the linear model under the null hypothesis is estimated using a robust method, which downplays the effect of additive outliers. The auxiliary regression (16) is estimated using both weighted residuals and weighted regressors, where the weights indicate the relative importance of the observations in the robust estimation procedure. The asymptotic distributions of the various tests are still χ^2 and F. The results in the second panel of Table 6 show that evidence for nonlinearity seems to vanish, as the null hypothesis of overall linearity is now rejected only at about the 12% level or more. This seems to suggest that the STECM might be abandoned entirely, and that a linear error-correction model is adequate for the monthly interest-rate series. Note however that for both $d = 1$ and 2, the test result for H_{01} rejects the null hypothesis quite convincingly, which indicates that a linear model is not completely satisfactory. One of the reasons for this finding might be that the sampling frequency of a

Table 7. *Standard and outlier-robust LM-type tests for smooth-transition error correction in a CECM for weekly data on the one-month interest rate*[a]

Test	Null	$d =$	1	2	3	4	5	6
			p-value					
Standard	H_0		0.000	0.000	0.001	0.001	0.000	0.000
	H_{03}		0.000	0.000	0.000	0.000	0.517	0.867
	H_{02}		0.000	0.000	0.001	0.006	0.018	0.006
	H_{01}		0.003	0.005	0.001	0.001	0.000	0.000
Robust	H_0		0.107	0.181	0.236	0.402	0.422	0.008
	H_{03}		0.520	0.689	0.625	0.134	0.594	0.027
	H_{02}		0.048	0.054	0.160	0.114	0.150	0.017
	H_{01}		0.240	0.324	0.212	0.402	0.607	0.454

[a] *p*-values for LM-type tests for smooth-transition error correction in weekly observations on the one-month Dutch interest rate. The upper panel gives *p*-values for standard tests, the lower panel for LM-type tests which are robust to additive outliers. The null hypotheses are given in the text.

month is too low to observe possible nonlinearity in the adjustment process very accurately.

5.2 *Sampling frequency*

So far, we have considered monthly data to fit our STECMs for the bivariate interest-rate series. Although nonlinear error correction can be motivated by arbitrage arguments, it is unclear at what speed such arbitrage would take place. When arbitrage would take, say, three weeks to become effective, and we sample our data only monthly, one can expect nonlinear adjustment to be reflected only in a single observation. Were one, however, to consider weekly data, one might obtain three data points which are informative for nonlinear modeling.

Granger and Lee (1999) investigate the effects of various types of aggregation on (tests for) nonlinearity. As it is difficult to obtain analytical results, they rely mainly on simulations. Their findings confirm the intuitive idea that systematic sampling, which is the relevant type of aggregation for our interest-rate example, decreases the evidence of nonlinearity.

To evaluate our empirical STECM in (20) in the light of sampling frequency, we collect weekly observed data for the same bivariate interest-rate series. As with the monthly series, we calculate standard and robust LM-type tests for the various hypotheses on nonlinear error correction, and we report the results in Table 7. From the first panel of this table, which contains the standard tests, we can conclude that there is substantial evidence for nonlinearity in these weekly data. For the robust tests, shown in the second panel, we observe that the p-values are generally smaller than the comparable ones in Table 6, although the overall evidence for nonlinearity is still weak. Only when d equals 6 we can reject H_0, H_{03}, and H_{02} quite convincingly, and when $d = 1$ we can reject H_{02} at the 5% level.

In order to compare the effect of sampling frequency, we decide to estimate the same model as in (20) for the weekly data. The estimation results are

$$\Delta R_{1,t} = \underset{(0.01)}{-0.01} + \underset{(0.02)}{0.05} S_{t-1} + \underset{(0.03)}{0.81} \Delta R_{12,t} - \underset{(0.04)}{0.04} \Delta R_{1,t-1}$$

$$+ \underset{(0.04)}{0.07} \Delta R_{12,t-1} + \left(\underset{(0.02)}{0.12} + \underset{(0.03)}{0.10} S_{t-1} + \underset{(0.08)}{0.41} \Delta R_{12,t} \right.$$

$$+ \underset{(0.10)}{0.21} \Delta R_{1,t-1} - \underset{(0.12)}{0.21} \Delta R_{12,t-1} \right) \left(1 + \exp\left[-\underset{(9.02)}{7.38} \right.\right.$$

$$\times (S_{t-1} + \underset{(0.04)}{0.42})(S_{t-1} - \underset{(0.04)}{1.03}) \Big/ \sigma^2_{S_{t-1}} \Big] \Big)^{-1} + \varepsilon_t, \tag{21}$$

$$\hat{\sigma}_\varepsilon = 0.131, \qquad \text{DW} = 1.97, \qquad \text{SK} = 0.32, \qquad \text{EK} = 4.80,$$

$$\text{JB} = 762.59(0.00), \qquad \text{ARCH}(1) = 63.45(0.00),$$

$$\text{ARCH}(4) = 89.10(0.00), \qquad \text{BIC} = -4.063.$$

Compared with the estimated model for the monthly data, two things are most noteworthy. First, the coefficients for the error-correction variable (as well as the intercepts) are smaller, which intuitively makes sense, and, second, the estimate for the threshold c_2 has become smaller, as well as the corresponding standard error.

In Figure 3 we present similar graphs to those in Figure 2. The most relevant difference between these two figures appears in the lower panel on the right, containing the function $F(\cdot)$ versus the transition variable S_{t-1}. As opposed to the model for the monthly data, there are now several observations in the upper regime, and hence we can have more confidence in the precision of the estimate of c_2 in the transition function (8). In other words, it pays to consider less-aggregated data for this bivariate interest-rate series. Also, our empirical findings support the simulation evidence in Granger and Lee (1999).

6   Concluding remarks

In this chapter we have analyzed the empirical specification of a smooth-transition error-correction model for a bivariate Dutch interest-rate series, where we used monthly and weekly observed data. Using simulation experiments, we substantiated the conjecture that standard linearity-based cointegration tests can

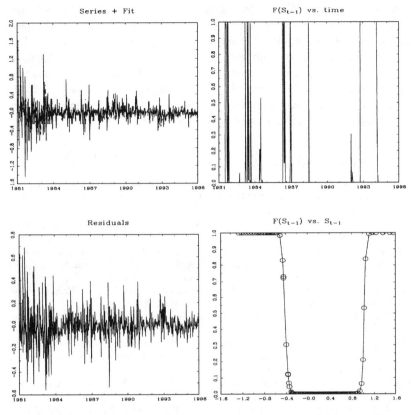

Figure 3. STECM for weekly observations on the Dutch short-term interest rate, January 1981–December 1995. The parameters of this model are given in (21).

be used to test for the presence of cointegration and to estimate the corresponding cointegrating vector. From our empirical results we must conclude that tests for nonlinearity should be used with caution when one aims to specify the nonlinear adjustment function in the STECM. First of all, our (unreported, tentative) estimation results show that key parameters like transition lag and type of transition function may not always be indicated by formal test results. We therefore recommend that the practitioner estimate various models and base model selection also on the empirical plausibility of the estimated transition function. Secondly, additive outliers can spuriously suggest nonlinearity, and may lead to the specification of complicated nonlinear functions for only one or two data points. We recommend the use of a robust test for smooth transition nonlinearity in order to avoid putting too much effort into fitting a small number of observations. In fact, it may be that a robust test suggests linearity or another form of nonlinearity. When robust tests give such deviating results, one may consider other sampling frequencies, if such data are available. In fact, the third conclusion from our empirical results is that less-aggregated data can lead to more precise estimates of nonlinear adjustment functions. In practice, the optimal level of sampling can be based on the available data at hand. Whether any theoretical arguments for some optimal level of aggregation for nonlinear modeling exist is left for further research. Finally, it seems worthwhile to consider robust tests for cointegration and nonlinearity jointly.

REFERENCES

Anderson, H. M. (1997), "Transaction costs and Nonlinear Adjustment towards Equilibrium in the US Treasury Bill Market", *Oxford Bulletin of Economics and Statistics* 59: 465–84.
Balke, N. S., and T. B. Fomby (1997), "Threshold cointegration", *International Economic Review* 38: 627–46.
Banerjee, A., J. J. Dolado, D. F. Hendry, and G. W. Smith (1986), "Exploring Equilibrium Relationships in Econometrics through Static Models: Some Monte Carlo Evidence", *Oxford Bulletin of Economics and Statistics* 48: 253–77.
Campbell, J. Y., and R. J. Shiller (1987), "Cointegration and Tests of Present Value Models", *Journal of Political Economy* 95: 1062–88.
    (1991), "Yield Spreads and Interest Rate Movements: A Bird's Eye View", *Review of Economic Studies* 58: 495–514.
Corradi, V., N. R. Swanson, and H. White (1995), "Testing for Stationarity–Ergodicity and for Comovements between Nonlinear Discrete Time Markov Processes", Working Paper, University of Pennsylvania.
Davies, R. B. (1977), "Hypothesis Testing When a Nuisance Parameter is Present Only under the Alternative", *Biometrika* 64: 247–54.
    (1987), "Hypothesis Testing When a Nuisance Parameter Is Present Only under the Alternative", *Biometrika* 74: 33–43.
Dickey, D. A., and W. A. Fuller (1979), "Distribution of the Estimators for Autoregressive Time Series with a Unit Root", *Journal of the American Statistical Association* 74: 427–31.
Dwyer, G. P., P. Locke, and W. Yu (1996), "Index Arbitrage and Nonlinear Dynamics between the S & P 500 Futures and Cash", *Review of Financial Studies* 9: 301–32.
Engle, R. F., and C. W. J. Granger (1987), "Co-integration and Error-Correction: Representation, Estimation and Testing", *Econometrica* 55: 251–76.
Escribano, A., and S. Mira (1996),"Nonlinear Cointegration and Nonlinear Error-Correction Models", Working Paper Series in Statistics and Econometrics 96-54, Universidad Carlos III de Madrid.
Granger, C. W. J. (1981), "Some Properties of Time Series Data and Their Use in Econometric Model Specification", *Journal of Econometrics* 16: 121–30.
Granger, C. W. J., and T.-H. Lee (1999), "The Effect of Aggregation on Nonlinearity", *Econometric Reviews*, to appear.
Granger, C. W. J., and T. Teräsvirta (1993), *Modelling Nonlinear Economic Relationships*. Oxford: Oxford University Press.
Gregory, A. W. (1994), "Testing for Cointegration in Linear Quadratic Models", *Journal of Business and Economic Statistics* 12: 347–60.

Hall, A. D., H. M. Anderson, and C. W. J. Granger (1992), "A Cointegration Analysis of Treasury Bill Yields", *Review of Economics and Statistics* 74: 116–26.

Hansen, G., and J.-R. Kim (1996), "Nonlinear Cointegration Analysis of German Unemployment", Working Paper 93, Institute of Statistics and Econometrics, Christian Albrechts University, Kiel.

Jansen, E. S., and T. Teräsvirta (1996), "Testing Parameter Constancy and Superexogeneity in Econometric Equations", *Oxford Bulletin of Economics and Statistics* 58: 735–68.

Johansen, S. (1988), "Statistical Analysis of Cointegration Vectors", *Journal of Economic Dynamics and Control* 12: 231–54.

 (1991), "Estimation and Hypothesis Testing of Cointegration Vectors in Gaussian Vector Autoregressive Models", *Econometrica* 59: 1551–80.

Kunst, R. M. (1992), "Dynamic Patterns in Interest Rates: Threshold Cointegration with ARCH", Discussion Paper, Institute for Advanced Studies, Vienna.

 (1995), "Determining Long-Run Equilibrium Structures in Bivariate Threshold Autoregressions: A Multiple Decision Approach", Discussion Paper, Institute for Advanced Studies, Vienna.

Luukkonen, R., P. Saikkonen, and T. Teräsvirta (1988), "Testing Linearity against Smooth Transition Autoregressive Models", *Biometrika* 75: 491–9.

Osterwald-Lenum, M. (1992), "A Note with Quantiles of the Asymptotic Distribution of the Maximum Likelihood Cointegration Rank Test Statistics", *Oxford Bulletin of Economics and Statistics* 54: 461–72.

Phillips, P. C. B., and S. Ouliaris (1990), "Asymptotic Properties of Residual Based Tests for Cointegration", *Econometrica* 58: 165–93.

Pippenger, M. K., and G. E. Goering (1993), "A Note on the Empirical Power of Unit Root Tests under Threshold Processes", *Oxford Bulletin of Economics and Statistics* 55: 473–81.

Stock, J. H., and M. W. Watson (1988), "Testing for Common Trends", *Journal of the American Statistical Association* 83: 1097–107.

Swanson, N. R. (1999), "Finite Sample Properties of a Simple LM Test for Neglected Nonlinearity in Error-Correcting Regression Equations", *Statistica Neerlandica* 53: 76–95.

Teräsvirta, T. (1994), "Specification, Estimation, and Evaluation of Smooth Transition Autoregressive Models", *Journal of the American Statistical Association* 89: 208–18.

Teräsvirta, T., and H. M. Anderson (1992), "Characterizing Nonlinearities in Business Cycles Using Smooth Transition Autoregressive Models", *Journal of Applied Econometrics* 7: S119–36.

van Dijk, D., P. H. Franses, and A. Lucas (1999), "Testing for Smooth Transition Nonlinearity in the Presence of Additive Outliers", *Journal of Business & Economic Statistics* 17, 217–35.

Yadav, P. K., P. F. Pope, and K. Paudyal (1994), "Threshold Autoregressive Modeling in Finance: The Price Difference of Equivalent Assets", *Mathematical Finance* 4: 205–21.

The Essential Reading and Language Arts Glossary I

A Student Reference Guide

Academic Vocabulary Builders

red brick®
LEARNING

Academic Vocabulary Builders are published by Red Brick Learning
7825 Telegraph Road, Bloomington, Minnesota 55438
http://www.redbricklearning.com

Library of Congress Cataloging-in-Publication Data
The essential reading and language arts glossary I: a student reference guide. — Pbk. ed.
    p. cm. — (Academic vocabulary builders)
    Includes index.
    Summary: "The Level I glossary covers essential content terms in the key subject area
of reading and language arts for elementary level students"—Provided by publisher.
    ISBN-13: 978-1-4296-2714-6 (pbk.)
    ISBN-10: 1-4296-2714-X (pbk.)
    1. Vocabulary — Study and teaching (Elementary) 2. Vocabulary — Juvenile literature.
3. Reading (Elementary) 4. Language arts (Elementary) I. Title: Essential reading and
language arts glossary 1.
LB1574.5.E87 2009
372.44 — dc22   2008021341

Cover Design
Ted Williams

Design and Illustration
Sasha Blanton

Photo Credits
Shutterstock/HomeStudio, cover, i (diary); Lukasz Kwapien, cover, i (four books);
    Phase4Photography, cover, i (pencil and yellow paper)

1 2 3 4 5 6 13 12 11 10 09 08

Table of Contents

About this book:

This book will help you learn essential words you will need to understand to do well on state tests. These essential words will also help you to do well in school.

There are 150 Reading and Language Arts words and definitions in the book. They are listed in alphabetical order under six main topics.

Here is a sample word with its features:

Easy to read definitions

Pictures to help understanding

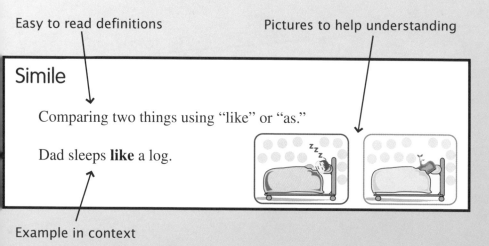

Simile

Comparing two things using "like" or "as."

Dad sleeps **like** a log.

Example in context

Word Recognition

Affix

A group of letters that attaches to other words and changes what they mean.

The words **preschool** and **helpful** each have an **affix**.

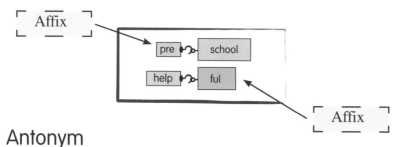

Antonym

A word that is the opposite of another word.

The **antonym** of **big** is **small**.
The **antonym** of **hot** is **cold**.

Base word

A word attached to an **affix.**

The **base word** of preschool is **school**.
The **base word** of helpful is **help**.

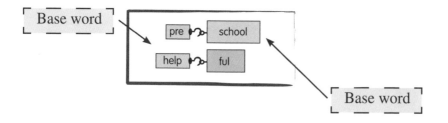

WORD
RECOGNITION

READING
STRATEGIES

LITERARY
CONCEPTS

WRITING

LANGUAGE
CONVENTIONS

RESEARCH

WORD
RECOGNITION

READING
STRATEGIES

LITERARY
CONCEPTS

WRITING

LANGUAGE
CONVENTIONS

RESEARCH

Compound word

A big word that is made of two small words. The small words have different meanings when they are not together.

 **+** **=**

sand   box   **sandbox**

Consonant

All of these letters are **consonants**.

> **b, c, d, f, g, h, j, k, l, m, n, p, q, r, s, t, v, w, x, z, (and sometimes y)**

Context clues

Words in a sentence that help you understand hard-to-read words.

In the Sonoran, the sun burned our backs. The sand made us feel like our feet were on fire.

In this sentence, the **context clues** "sun," "burned," and "sand" help you know that the Sonoran is a desert, even if you have never seen the word "Sonoran" before.

Contraction

A word that is made by combining two words, taking out some letters, and replacing them with an **apostrophe**.

Do + not = **don't**

Ending

Something you add to a base word that does not change its meaning.

If you add the **ending -ed** to the word **play**, the word you have made, **played**, still means the same thing.

Homograph

A word that looks the same as another word but may be said differently or mean something different.

Actors take a **bow** on stage. (said like *bau*)
Keith tied his shoelaces in a **bow**. (said like *bO*)

said like *bau*   said like *bO*

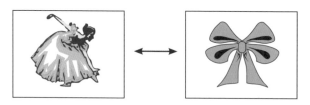

WORD
RECOGNITION

READING
STRATEGIES

LITERARY
CONCEPTS

WRITING

LANGUAGE
CONVENTIONS

RESEARCH

WORD
RECOGNITION

READING
STRATEGIES

LITERARY
CONCEPTS

WRITING

LANGUAGE
CONVENTIONS

RESEARCH

Homonym

A word that looks and sounds the same as another word but means something different.

Lie and **lie** are **homonyms**:

Don't **lie**. Always tell the truth.
Lie down if you are sleepy.

Homophone

A word that is said the same as another word but looks different **and** has a different meaning.

Juan saw a **deer** in the woods.
My grandma always calls me **dear**.

deer   **dear**

Multiple-meaning words

A word that looks and sounds the same as another word but has a different meaning. A **multiple-meaning word** is the same as a **homonym**.

I turned on the **light** so I could see.
The bag is **light** when I pick it up.

Word Recognition

WORD
RECOGNITION

READING
STRATEGIES

LITERARY
CONCEPTS

WRITING

LANGUAGE
CONVENTIONS

RESEARCH

Prefix

An **affix** that is put at the beginning of a word and changes its meaning.

The words **impossible** and **unhappy** both have a **prefix** at the beginning.

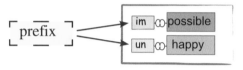

Root

A word part that means the same thing in different words.

The **root** word **use** means the same thing in all of these words:

use
user
useful

Suffix

An **affix** that is put at the end of a word and changes its meaning.

The words **breakable** and **childlike** both have a **suffix** at the end.

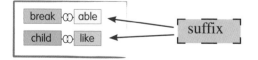

5

WORD
RECOGNITION

READING
STRATEGIES

LITERARY
CONCEPTS

WRITING

LANGUAGE
CONVENTIONS

RESEARCH

Word Recognition

Syllable

The smallest part of a word. It usually has one **vowel**
sound.

bird-watch-er   three **syllables**
dog-gie   two **syllables**

Syllables make words, but sometimes there is only
one **syllable** in a word.

dog   one **syllable**

Synonym

A word that means the same or almost the same thing
as another word.

My dad's truck is **big**!
Yeah! It's **huge**!

Big and **huge** are
synonyms.

Vowel

All of these letters are **vowels**.

> **a, e, i, o, u**, (and sometimes **y**)

Word

A group of letters that means something.

Advertisement

A message of words and pictures that makes you want to buy something.

I saw a game in a TV **advertisement,** and now I want it.

Caption

Words found under a picture. They help you understand the picture better.

I didn't understand a picture, so I read the **caption** under it. Now I understand what the picture means.

The Sphinx is an ancient statue in Egypt.

Caption

Cause

Something that makes another thing happen.

I got a good grade, so my dad took me to the park.

In this **sentence** the part that says *I got a good grade* is the **cause** of what happens next. Words like "because" and "so" often show **causes** and **effects**.

WORD RECOGNITION

READING STRATEGIES

LITERARY CONCEPTS

WRITING

LANGUAGE CONVENTIONS

RESEARCH

WORD
RECOGNITION

READING
STRATEGIES

LITERARY
CONCEPTS

WRITING

LANGUAGE
CONVENTIONS

RESEARCH

Chapter

A section of a book.

Compare

To show how people, places, or things are similar.

I can **compare** a park and a beach; both are fun places to visit.

Contrast

Showing how people, places, or things are different.

I can **contrast** a park and a beach; one is sandy while the other is soft and green.

Details

Information that tells about the **main idea**.

Details that tell about a birthday party:
Balloons
Blowing out candles
Eating cake

Draw conclusion

To use clues to help you put things together.

Effect

Something that happens because of what happened before it.

I got a good grade, so my dad took me to the park.

In this **sentence**, the part that says *my dad took me to the park* is the **effect**. Words like "because" and "so" often show **causes** and **effects**.

Essential information

The most important **facts** and **details** to know.

Fact

Something that is always true. (also see **opinion**)

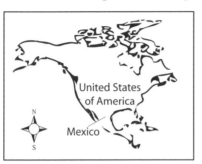

Mexico is south of the United States.

Glossary

A book that lists words along with definitions. A **glossary** usually has words from one topic, while a **dictionary** has words from all topics.

The book you are reading right now is a **glossary**.

Graphic organizer

A **chart** or graph that helps you to organize information.

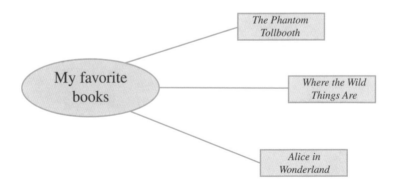

Reading Strategies

Heading

A **title** of a section of writing.

Artist Wins!

Heading

Illustration

A picture that helps show something in writing.

Some **illustrations** are drawings or **charts**. Others are photographs.

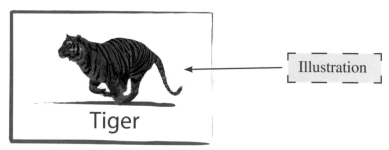

Tiger

Illustration

Index

An **alphabetical** list that helps you find things in a book.

The **index** for this book starts on page 51.

WORD RECOGNITION

READING STRATEGIES

LITERARY CONCEPTS

WRITING

LANGUAGE CONVENTIONS

RESEARCH

Infer

To figure out the main idea of a **paragraph** or piece of writing even though it is not written down exactly.

Information

Words you see or hear that tell you something. Pictures are also **information**.

Key words

The most important words or phrases that you need to find and remember.

The Amazon River is the **second largest** river in the world. It is located in **South America**. It has the **most water** of any river in the world.

Reading Strategies

Main idea

The most important idea in a piece of writing.

Nonessential information

The **facts** and **details** in a piece of writing that are not as important as the **main idea**.

Opinion

Someone's ideas. When someone writes or says something that you cannot look up in a **dictionary** or an **encyclopedia**, they are stating an **opinion**. (also see **fact**)

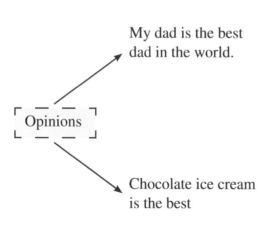

My dad is the best dad in the world.

Opinions

Chocolate ice cream is the best

Feelings, likes, and dislikes are all **opinions**.

WORD RECOGNITION

READING STRATEGIES

LITERARY CONCEPTS

WRITING

LANGUAGE CONVENTIONS

RESEARCH

Order

The steps something takes to get from beginning to end. (see **sequence**)

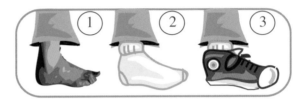

Predict

Using clues to guess what will happen in a **story**.

Jen put her bathing suit in a bag. She found her towel, a pail and shovel, and sunscreen. She and her dad put all of these things in the car. They drove away.

After reading this passage, you can **predict** that Jen is going to the beach.

Reread

To read again. Re- is a **prefix** that means *again*.

Reading Strategies

Retell

To tell a **story** you have read or heard in your own words as best you can remember it.

Sequence

The **order** in which things happen. (also see **order**)

1. I brushed my teeth.
2. I took a bath.
3. I went to bed.

Similar

Very close to the same, but not exactly.

Two pens are the same, but a pen and a pencil are **similar**.

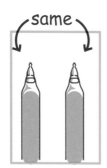
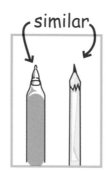

WORD RECOGNITION

READING STRATEGIES

LITERARY CONCEPTS

WRITING

LANGUAGE CONVENTIONS

RESEARCH

WORD RECOGNITION

READING STRATEGIES

LITERARY CONCEPTS

WRITING

LANGUAGE CONVENTIONS

RESEARCH

Slang

A word or words that are used in everyday speech but are not usually accepted in writing.

Yeah is **slang** for yes.

Summary

A short form of a **story**. It still has all the most important parts in it but does not have many **details**.

> In the story *Alice in Wonderland*, Alice follows a rabbit down a hole. She meets many interesting characters including a Cheshire cat, a Mad Hatter, and the Queen of Hearts. Alice learns a lot on her journey through Wonderland.

Table of contents

A list of all the **chapter** headings in a book along with their page numbers.

In this book, the **table of contents** is on page iii.

Time line

A graphic organizer that shows the **sequence** of events on a line.

Brushed my teeth   Took a bath   Went to bed

Title

What a book, **story**, or other piece of writing is named.

Title — Three Little Pigs / Plants ← Title

Topic

What a piece of writing is about.

The **topic** of the paper is "Plants."

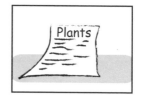

Plants

Word origin

The place and time where a word originally comes from.

Long ago, people used to write with feathers called **"penna"** in Latin. That is where we get the English word **"pen."**

WORD RECOGNITION

READING STRATEGIES

LITERARY CONCEPTS

WRITING

LANGUAGE CONVENTIONS

RESEARCH

Literary Concepts

WORD RECOGNITION

READING STRATEGIES

LITERARY CONCEPTS

WRITING

LANGUAGE CONVENTIONS

RESEARCH

Act

The main parts of a **play**. An **act** is often broken into **scenes**.

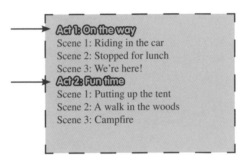

Act 1: On the way
Scene 1: Riding in the car
Scene 2: Stopped for lunch
Scene 3: We're here!
Act 2: Fun time
Scene 1: Putting up the tent
Scene 2: A walk in the woods
Scene 3: Campfire

Audience

The people who come to watch a **play** or movie.

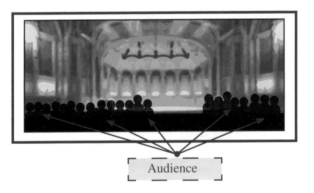

Audience

Character

A person in a **story**. In a **play**, **characters** are played by actors.

Characters

Conflict

The problem that **characters** must solve during the **story** or **play**.

Dialogue

The words that **characters** say to each other.

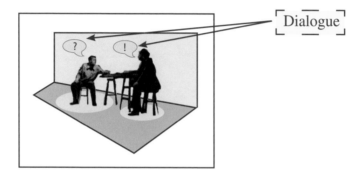

Drama (or Play)

A **story** that is written to be acted in a theater.

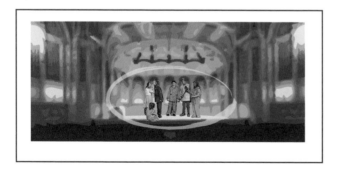

This is a picture of a **drama**.

Literary Concepts

Exaggeration

To describe something that could not be true in real life.

I'm so hungry, I could eat a horse!

Fable

A **story** that teaches a lesson. In a **fable**, you will usually read about talking animals and supernatural powers.

In the **fable** *The Tortoise and the Hare*, the tortoise knows that just going fast is not as important as finishing the race.

Fairy tale

A **story** with amazing **characters** like fairies, wizards, and goblins.

Sidebar navigation:

WORD RECOGNITION

READING STRATEGIES

LITERARY CONCEPTS

WRITING

LANGUAGE CONVENTIONS

RESEARCH

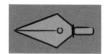
Fantasy

A **story** that could not have happened in real life. It may contain magic, talking animals, or make-believe worlds.

Fairy tales, **fables**, and **tall tales** are all different types of **fantasy**.

Fiction

A **story** made up from the writer's imagination.

Figurative language

Language that does not mean exactly what it says. This includes **metaphors**, **symbols**, and **similes**.

For example, a poem might say the sky is a shining blue ocean. It's not really an ocean, but it looks like it.

WORD RECOGNITION

READING STRATEGIES

LITERARY CONCEPTS

WRITING

LANGUAGE CONVENTIONS

RESEARCH

Literary Concepts

Folk tale

A **story** that is told from person to person for many years.

Legend

A **story** that many people believe to be true about a **character** from the past.

In Scotland, there is a **legend** about a giant animal who lives in Loch Ness.

The Loch Ness Monster

Literature

A collection of all the **stories** and **poetry** that writers in a language have written down.

WORD
RECOGNITION

READING
STRATEGIES

LITERARY
CONCEPTS

WRITING

LANGUAGE
CONVENTIONS

RESEARCH

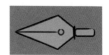

Metaphor

Figurative language that compares one thing to another. **Similes** are a kind of **metaphor**, but most **metaphors** do not use "like" or "as."

The sand was a **boiling sea** in the hot sun.

Sand does not boil when it is hot. **The sand was a boiling sea** is a **metaphor**.

Myth

A traditional **story** that tries to explain natural events, like floods or hurricanes. Often in **myths**, these natural events are caused by powerful animals or superhuman beings.

There is a Native American **myth** about a bird that makes thunder when it flaps its wings.

WORD RECOGNITION

READING STRATEGIES

LITERARY CONCEPTS

WRITING

LANGUAGE CONVENTIONS

RESEARCH

Narrative

A **story** or a book.

Nonfiction

A piece of writing that is based on **facts**.

Newspapers are **nonfiction**.

Onomatopoeia

A word that is written and said just like a sound.

splash
buzz
howl

Personification

Figurative language that describes a thing as if it were a person.

The sun **smiled** down on us.

Plot

The events that happen in a **story**.

Poetry

Writing that usually has a **rhythm**, uses **figurative language**, sometimes **rhymes**, and is usually arranged in lines.

Look into the garden,
Where the grass was green;
Covered by the snowflakes,
Not a blade is seen.

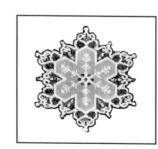

WORD RECOGNITION

READING STRATEGIES

LITERARY CONCEPTS

WRITING

LANGUAGE CONVENTIONS

RESEARCH

Point of view

A **story** told the way one **character** sees it.

Prose

Any writing that is not **poetry**.

Realistic fiction

A **fiction story** that has details that make it seem real.

> Kendra dropped her book bag next to the front door. She was hungry after a long school day. She went into the kitchen and looked for a snack. Just then, her father came down the steps and into the kitchen. "Hi, Dad," she said.

In this example, Kendra is not a real person. This **story** is **fiction**, but it sounds very much like real life. It is **realistic fiction**.

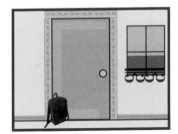

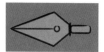

Rhyme

Words with endings that sound the same.

cat ·······►

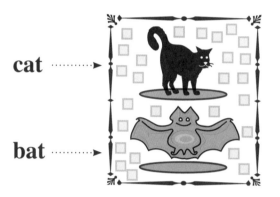

Cat and *bat* **rhyme**.

bat ·······►

Rhythm

A pattern of similar sounds or words. Also the sound a poem or **story** makes when you read it aloud.

One and **two** and **three** and **four**,
Who comes **knock** ing **at** the **door**?

Scene

A part of an **act**.

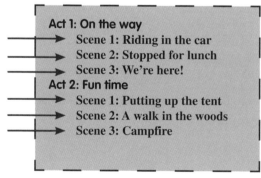

Act 1: On the way
    Scene 1: Riding in the car
    Scene 2: Stopped for lunch
    Scene 3: We're here!
Act 2: Fun time
    Scene 1: Putting up the tent
    Scene 2: A walk in the woods
    Scene 3: Campfire

WORD RECOGNITION

READING STRATEGIES

LITERARY CONCEPTS

WRITING

LANGUAGE CONVENTIONS

RESEARCH

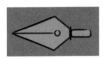

Literary Concepts

Setting

The time and place in which part of a **play** or **story** takes place.

The **setting** shown in this picture is **in the woods at night**.

Simile

Comparing two things using "like" or "as."

Dad sleeps **like** a log.

Story

A **narrative** that is shorter than a book. It usually has a beginning, a middle, and an end.

WORD RECOGNITION

READING STRATEGIES

LITERARY CONCEPTS

WRITING

LANGUAGE CONVENTIONS

RESEARCH

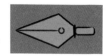

Symbol

Something in a **story** that represents a thing or an idea.

Hearts and roses are **symbols** of love.

Tall tale

A **story** that couldn't happen in real life because the characters are **exaggerated** or the things that happen are **exaggerated**. **Tall tales** are often funny.

Paul Bunyan stories are **tall tales**.

Theme

The message of the **story**. A **story** might have more than one **theme**.

In *The Tortoise and the Hare*, the **theme** is that it is good to go slow and finish the race.

WORD RECOGNITION

READING STRATEGIES

LITERARY CONCEPTS

WRITING

LANGUAGE CONVENTIONS

RESEARCH

Body

The middle paragraphs of an **essay**, where you write about your **main idea**.

> **Our Flag**
>
> The flag of the United States has a long history. The flag has looked very different over the years. But no matter how it looked, two things were always the same—the stars and stripes.
>
> The first U.S. flag had 13 white stars on a blue background and 13 red and white stripes. The number 13 was important back then because there were 13 colonies that became the very first states of the United States. The United States was much smaller back then. As the United States got more land and more states, the flag changed. Each time, the number of stars showed how many states were part of the United States.
>
> Today, our flag has 50 stars and 13 stripes. Each of these parts has a meaning. The 50 stars stand for the 50 states the United States has now. The 13 stripes stand for the first 13 colonies.

Brainstorming

Writing down everything you know about a topic so that you can put your ideas in order.

Chart

A picture that shows **information**, like a graph.

This is a **chart**.

Chart

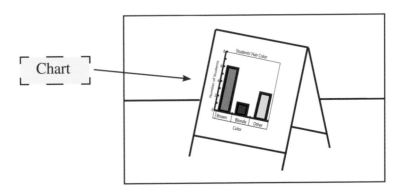

Conclusion

The end of a **story** or an **essay**. In an essay, the **conclusion** is the last **paragraph**.

> **Our Flag**
> The flag of the United States has a long history. The flag has looked very different over the years. But no matter how it looked, two things were always the same—the stars and stripes.
> The first U.S. flag had 13 white stars on a blue background and 13 red and white stripes. The number 13 was important back then because there were 13 colonies that became the very first states of the United States. The United States was much smaller back then. As the United States got more land and more states, the flag changed. Each time, the number of stars showed how many states were part of the United States.
> Today, our flag has 50 stars and 13 stripes. Each of these parts has a meaning. The 50 stars stand for the 50 states the United States has now. The 13 stripes stand for the first 13 colonies.

Describe

To explain the **details** of a thing using your five senses.

Details of a tree:
Tall, green, hard, bumpy

Diagram

A picture that helps you understand how things are connected.

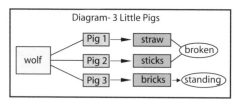

Draft

A version of a **story** or **essay**. Every time you write or change your writing, you create a draft.

Writing

Editing

Going back over your writing to make sure you have made the right word choices and that your writing makes sense when you read it.

Essay

A short piece of nonfiction writing about one **topic**.

> **Our Flag**
> The flag of the United States has a long history. The flag has looked very different over the years. But no matter how it looked, two things were always the same—the stars and stripes.
> The first U.S. flag had 13 white stars on a blue background and 13 red and white stripes. The number 13 was important back then because there were 13 colonies that became the very first states of the United States. The United States was much smaller back then. As the United States got more land and more states, the flag changed. Each time, the number of stars showed how many states were part of the United States.
> Today, our flag has 50 stars and 13 stripes. Each of these parts has a meaning. The 50 stars stand for the 50 states the United States has now. The 13 stripes stand for the first 13 colonies.

Introduction

The start of a book or **essay**.

> **Our Flag**
> The flag of the United States has a long history. The flag has looked very different over the years. But no matter how it looked, two things were always the same—the stars and stripes.
> The first U.S. flag had 13 white stars on a blue background and 13 red and white stripes. The number 13 was important back then because there were 13 colonies that became the very first states of the United States. The United States was much smaller back then. As the United States got more land and more states, the flag changed. Each time, the number of stars showed how many states were part of the United States.
> Today, our flag has 50 stars and 13 stripes. Each of these parts has a meaning. The 50 stars stand for the 50 states the United States has now. The 13 stripes stand for the first 13 colonies.

WORD RECOGNITION

READING STRATEGIES

LITERARY CONCEPTS

WRITING

LANGUAGE CONVENTIONS

RESEARCH

Journal

A book that you write in every day. You can write anything you want in a **journal**. **Journals** are not usually edited.

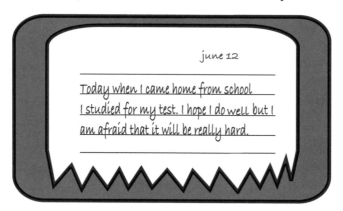

june 12

Today when I came home from school
I studied for my test. I hope I do well but I
am afraid that it will be really hard.

Learning log

A **journal** in which you write what you have learned every day.

Letter

A piece of writing addressed to another person. Most **letters** have a special form.

Dear Beth,
　　What have you been up to? I
have started playing soccer. I like to
be outside in the sun. My dad comes
to see me play and he cheers me on.

　　　　　Yours truly,
　　　　　Sue

Writing

Narrating

Telling a **story**.

Narrator

The person or **character** who tells a **story**.

Organize

To put your ideas in an **order** that makes it easier to understand.

You can **organize** your ideas with a **table** or an **outline**.

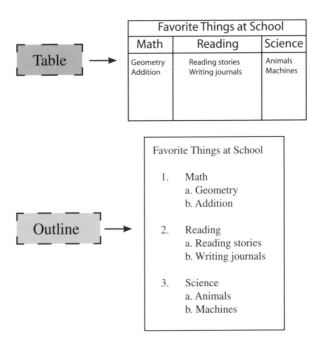

Table

Favorite Things at School		
Math	Reading	Science
Geometry Addition	Reading stories Writing journals	Animals Machines

Outline

Favorite Things at School

1. Math
    a. Geometry
    b. Addition

2. Reading
    a. Reading stories
    b. Writing journals

3. Science
    a. Animals
    b. Machines

WORD RECOGNITION

READING STRATEGIES

LITERARY CONCEPTS

WRITING

LANGUAGE CONVENTIONS

RESEARCH

Outline

A written list of ideas in an **order** that makes sense.

Favorite Things at School

1. Math
    a. Geometry
    b. Addition

2. Reading
    a. Reading stories
    b. Writing journals

3. Science
    a. Animals
    b. Machines

Persuade

To try to make someone do what you want them to do.

Presentation

A talk about someone or something. Sometimes, people show pictures or objects during a **presentation**.

The teacher gives a **presentation** to the class.

WORD RECOGNITION

READING STRATEGIES

LITERARY CONCEPTS

WRITING

LANGUAGE CONVENTIONS

RESEARCH

Proofread

To check your writing to make sure there are no spelling or punctuation mistakes.

Our ——— ~~The U.S.~~ Flag two

    The flag of the United States has a long history. The flag has looked very different over the years. But no matter how it looked, ~~three~~ things were always the same—the stars and stripes. White stars on a blue background
    The first U.S. flag had 13 stars and 13 red and white stripes. The number 13 was important back then because there were 13 colonies that became the very first states of the United States. The United States was much smaller back then. As the United States got more land and more states, the flag changed. Each time, the number of stars showed how many states were part of the United States.
    Today, our flag has 50 stars and 13 stripes. Each of these parts has a meaning. The 50 stars stand for the 50 states the United States has now. The 13 stripes stand for the first 13 colonies.

Publish

To print your writing somewhere so that anyone can read it.

Everything pictured has been **published**.

Rearrange

To put your ideas in a different **order**.

Relate

To connect one idea to another.

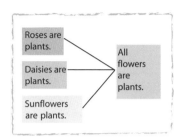

Roses are plants.

Daisies are plants.

Sunflowers are plants.

All flowers are plants.

Report

To tell others about something you have learned.

Revise

To make changes to your writing. When you **revise**, you write another **draft**.

Our ——— ~~The U.S.~~ Flag   two

The flag of the United States has a long history. ⎮The flag has looked very different over the years. But no matter how it looked, ~~three~~ things were always the same—the stars and stripes.   White stars on a blue background

The first U.S. flag had 13 stars and 13 red and white stripes. The number 13 was important back then because there were 13 colonies that became the very first states of the United States. The United States was much smaller back then. As the United States got more land and more states, the flag changed. Each time, the number of stars showed how many states were part of the United States.

Today, our flag has 50 stars and 13 stripes. Each of these parts has a meaning. The 50 stars stand for the 50 states the United States has now. The 13 stripes stand for the first 13 colonies.

Our Flag

The flag of the United States has a long history. The flag has looked very different over the years. But no matter how it looked, two things were always the same—the stars and stripes.

The first U.S. flag had 13 white stars on a blue background and 13 red and white stripes. The number 13 was important back then because there were 13 colonies that became the very first states of the United States. The United States was much smaller back then. As the United States got more land and more states, the flag changed. Each time, the number of stars showed how many states were part of the United States.

Today, our flag has 50 stars and 13 stripes. Each of these parts has a meaning. The 50 stars stand for the 50 states the United States has now. The 13 stripes stand for the first 13 colonies.

Table

A picture that helps you organize your ideas.

Favorite Things at School		
Math	Reading	Science
Geometry Addition	Reading stories Writing journals	Animals Machines

Transition

A word that tells a reader that you will move from one idea to another in writing.

Some **transition** words include "**first**" and "**next**."

Language Conventions

Adjective

A word that describes a **noun** or **pronoun**.

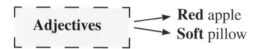

Adjectives → **Red** apple
→ **Soft** pillow

Adverb

A word that describes a **verb** or **adjective**.

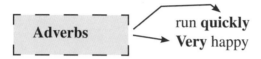

Adverbs → run **quickly**
→ **Very** happy

Adverbs usually end in –ly.

Alphabetical (ABC) order

In order by first letter in the order of the alphabet.

Apples

Bananas

Cheese

WORD
RECOGNITION

READING
STRATEGIES

LITERARY
CONCEPTS

WRITING

LANGUAGE
CONVENTIONS

RESEARCH

Apostrophe

A **punctuation mark** used to show that a letter has been dropped from a word.

Don't
Can't

n't

Apostrophes also show that something or somebody belongs to another thing or person.

Sue's desk
Pepe's kitten

e's

Comma

A **punctuation mark** used to separate word groups in a **sentence** or separate items in a list.

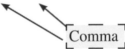

Comma

Amy went to school, and then she went home.
Mom got milk, eggs, and cheese from the store.

Comma

Command

A **sentence** that tells you to do something.

Clean your room.
Do your homework.

Conjunction

A word that connects two **sentences**.
Common conjunctions are *and*, *but*,
and *or*.

Dad left his hat, **and** he left his keys.
Mom left her bag, **but** she did not leave her keys.
Dad left his hat, **or** Mom left her hat.

Exclamation

A **sentence** that shows a strong feeling. You can use an
exclamation point instead of a **period** at the end of this
kind of sentence.

I want to play!
Don't touch the fire!

Interjection

A word or group of words that gets people's attention
without using a complete **sentence**.

Hi!
Look out!

Noun

A part of speech that tells the name of any person, place, or thing.

cow
bedroom
Japan
grandma
bottle

Paragraph

A group of **sentences** that tell about the same **main idea**.

> The first U.S. flag had 13 white stars on a blue background and 13 red and white stripes. The number 13 was important back then because there were 13 colonies that became the very first states of the United States. The United States was much smaller back then. As the United States got more land and more states, the flag changed. Each time, the number of stars showed how many states were part of the United States.
>
> Today, our flag has 50 stars and 13 stripes. Each of these parts has a meaning. The 50 stars stand for the 50 states the United States has now. The 13 stripes stand for the first 13 colonies.

Parentheses

Punctuation marks that close off some words because they are not part of the **sentence**.

I waved at my little brother
(because he was far away and could not see me) to come closer.

Parentheses

Language Conventions

Parts of speech

The name for the categories of how words work.

The **parts of speech** are:

Noun	Pronoun
Verb	Conjunction
Adjective	Interjection
Adverb	Preposition

Period

A **punctuation mark** that tells you where the end of a sentence is. Most sentences end in **periods**.

Sara asked for more juice. ←
Gino is in the house. ←   Periods

Plural

More than one of a **noun**. Most **plurals** end in -s. They are called regular **plurals**.

dog   **dogs**

Some **plurals** do not end in -s. They are called irregular **plurals**.

mouse   **mice**

Predicate

The part of a **sentence** that tells what the **subject** is doing.

Marco **picked up the phone**.

Preposition

A **part of speech** that connects a **noun** or **pronoun** to another word in a **sentence**.

Frank put a book **inside** his bag.
He took his bag **to** school.
He put the bag **under** his desk.

Pronoun

A **part of speech** that takes the place of a **noun**.

<u>Rob</u> loves to write <u>stories</u>.

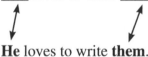

<u>He</u> loves to write **them**.

He and **them** are **pronouns**.

Proper noun

A **noun** that names a particular person, place, or thing. **Proper nouns** begin with a capital letter.

Paris
José
Mr. Kim's store
Time Magazine
Texas

Punctuation mark

A **symbol** that helps you understand how words go together.

- . period
- ! exclamation point
- ? question mark
- , comma
- ' apostrophe
- " " quotation marks
- ( ) parentheses

Question

A **sentence** that asks something. When you write a **question**, you end the **sentence** with a **question mark**.

Where is my jacket?
Do all birds have wings?

Language Conventions

WORD RECOGNITION

READING STRATEGIES

LITERARY CONCEPTS

WRITING

LANGUAGE CONVENTIONS

RESEARCH

Quotation marks

Punctuation marks that let you know when a **character** in a story is talking.

"Hello," said the woman.

 Quotation marks

Sentence

A set of words that tells a complete idea. All **sentences** have both a **subject** and a **predicate**.

My book is on the floor.
Did you drop it?

Singular

One thing.

dog   mouse

46

Statement

A **sentence** that gives **information**.

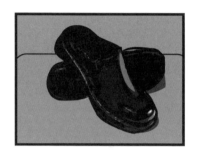

I have eight pairs of black shoes.
Janna danced with her brother.

Subject

The part of a **sentence** that tells who or what is doing the action.

Janna danced with her brother.

Verb

An action word.

Sam **kicks** the ball.

WORD RECOGNITION

READING STRATEGIES

LITERARY CONCEPTS

WRITING

LANGUAGE CONVENTIONS

RESEARCH

Dictionary

A book that lists every word in a language in **alphabetical** order. **Dictionaries** also tell you what those words mean, how to say them, and where the word comes from in history.

Encyclopedia

Writing that gives information about all different types of people, places, and things. **Encyclopedias** are in **alphabetical order** and are usually a group of books.

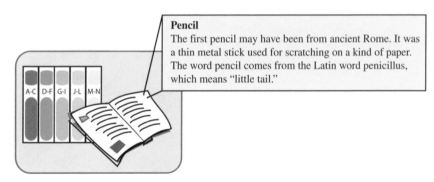

Pencil
The first pencil may have been from ancient Rome. It was a thin metal stick used for scratching on a kind of paper. The word pencil comes from the Latin word penicillus, which means "little tail."

A-C D-F G-I J-L M-N

Evidence

Facts that you can show others to prove something.

Plants need sunlight to grow.
Plants use sunlight to make food.

Media

Different ways of giving **information**. TV, radio, the Internet, newspapers, and magazines are all **media**.

Reference materials

Anything you can go to for information.

Thesaurus

A book that can help you find **synonyms** and **antonyms** of other words.

Index

Index

Index

Notes

Notes

Notes

Notes